New Techniques and Equipment in Large Offshore Aquaculture Platform

New Techniques and Equipment in Large Offshore Aquaculture Platform

Guest Editors

Zhenhua Ma

Jianguang Qin

MDPI

Basel • Beijing • Wuhan • Barcelona • Belgrade • Novi Sad • Cluj • Manchester

Guest Editors
Zhenhua Ma
Chinese Academy of Fishery Sciences
Guangzhou
China

Jianguang Qin
Flinders University
Adelaide
Australia

Editorial Office
MDPI AG
Grosspeteranlage 5
4052 Basel, Switzerland

This is a reprint of the Special Issue, published open access by the journal *Journal of Marine Science and Engineering* (ISSN 2077-1312), freely accessible at: https://www.mdpi.com/journal/jmse/special_issues/G2X4D2Q8YH.

For citation purposes, cite each article independently as indicated on the article page online and as indicated below:

Lastname, A.A.; Lastname, B.B. Article Title. *Journal Name* **Year**, *Volume Number*, Page Range.

ISBN 978-3-7258-2749-7 (Hbk)
ISBN 978-3-7258-2750-3 (PDF)
https://doi.org/10.3390/books978-3-7258-2750-3

Cover image courtesy of Zhenhua Ma

Contents

About the Editors

Zhenhua Ma

Zhenhua Ma graduated from Flinders University in Australia with a Ph.D. in Aquaculture. He is currently a Level-3 Professor and Ph.D. Supervisor. He holds several key positions, including Deputy Director of the South China Sea Fisheries Research Institute at the Chinese Academy of Fishery Sciences (CAFS), Director of the Sanya Tropical Fisheries Research Institute, Director of the Tropical Fisheries Research and Development Center at the South China Sea Fisheries Research Institute of CAFS, and Chief Expert of the CAFS Innovation Team for Deep-Sea Aquaculture Technology and Species Development. Additionally, he serves as a member of the Hainan Provincial Committee of the Chinese People's Political Consultative Conference (CPPCC), a specially appointed expert for the CAST "Haizhi" Program, Director of the Center for Fishery Environment and Aquatic Product Quality Supervision and Testing under the Ministry of Agriculture and Rural Affairs (Guangzhou), a member of the Third Sub-Committee on Ornamental Fish of the National Technical Committee on Aquaculture Standardization, and the Executive Deputy Director of the Hainan Key Laboratory of Efficient Utilization and Processing of Deep-Sea Fishery Resources. His primary research focuses on aquatic animal breeding, deep-sea aquaculture technology, and strategic studies. Professor Ma has led over 30 major scientific projects, including those funded by the National Natural Science Foundation of China and significant science and technology programs in Guangxi and Hainan provinces. He has received more than ten awards, including First and Second Prizes of the Hainan Science and Technology Progress Award. He has published 206 academic papers, 127 of which are indexed by SCI, and holds 18 authorized patents, including seven national invention patents. He has also supervised 28 graduate students.

Jianguang Qin

Professor Jianguang Qin obtained his bachelor's degree at Dalian Fisheries College, China, in 1982, his master's degree at the University of Oklahoma in 1988, and his Ph.D. at Ohio State University in 1994. After a 3-year postdoc at the University of Hawaii, he joined Flinders University in 1997. Jian is a pioneer of the aquaculture program at Flinders and has led the aquaculture research at Flinders to become world-class in the area of fishery sciences. He has supervised 23 Ph.D. students and 20+ honors students to completion and has published 4 books,19 book chapters, and 400+ journal papers. His research areas cover the food, feeding, and nutrition of aquatic animals in an attempt to discover trophic relationships in food webs and apply new findings to applied science. His research disciplines include ecology, physiology, behavior, biotechnology, and toxicology. Research organisms of interest include bacteria, algae, zooplankton, fish, oysters, abalone, and freshwater crayfish.

Preface

The global demand for seafood has reached unprecedented levels, driven by population growth, increasing consumer awareness of the health benefits associated with fish consumption, and the quest for sustainable protein sources. However, traditional nearshore aquaculture operations are increasingly constrained by space limitations, environmental degradation, and regulatory pressures. In this context, offshore aquaculture, which involves cultivating fish in deeper, open waters far from coastlines, presents a promising frontier. This approach offers a sustainable solution to the challenges faced by nearshore aquaculture, paving the way for large-scale production in open seas, where conditions are often ideal for fish growth and where environmental impacts can be managed more effectively.

The primary goal of *New Techniques and Equipment in Large Offshore Aquaculture Platform* is to provide a comprehensive overview of the latest advancements in offshore aquaculture technology and equipment, highlighting key innovations that are shaping the industry. The contributions presented in this Special Issue stem from a collaborative effort by researchers, engineers, and industry experts who are pioneering solutions to some of the most pressing challenges in the field. This preface serves to contextualize the importance of these innovations, outline the structure of the Special Issue, and provide readers with a glimpse into the transformative potential of far-reaching marine aquaculture.

The Need for Offshore Aquaculture

The shift from nearshore to offshore aquaculture is a response to the limitations imposed by traditional coastal aquaculture. Nearshore operations are heavily impacted by pollution, habitat destruction, and overuse of limited space, which leads to conflicts with other marine users. As global aquaculture production continues to rise, these issues have become more acute, prompting the industry to explore sustainable alternatives. Offshore aquaculture enables production to scale without encroaching on coastal ecosystems, allowing fish farms to be situated in cleaner, less disturbed waters. The deeper waters of offshore locations also provide more stable conditions, with fewer fluctuations in temperature, oxygen levels, and salinity compared to coastal environments. These benefits, coupled with the opportunity to reduce the environmental footprint of aquaculture, underscore the need for expanding into offshore territories.

Technological Innovations and Their Impact

The transition to offshore aquaculture, however, comes with significant challenges. Operating in open seas requires technologies and equipment that can withstand harsh marine conditions, such as high winds, strong currents, and large waves. This necessitates the development of robust aquaculture platforms and mooring systems, advanced environmental control systems, and innovative feeding and monitoring technologies. The contributions in this Special Issue address these challenges head-on, offering solutions that are both practical and innovative.

The technological advancements covered in this Special Issue span several critical areas. From the breeding of resilient fish species and the construction of large-scale platforms to the implementation of precision feeding systems and AI-driven monitoring tools, each chapter delves into a specific aspect of offshore aquaculture. The aim is to present a holistic view of how these innovations work together to enable sustainable, productive, and economically viable aquaculture operations.

Species Selection and Breeding Technologies

One of the foundational elements of successful offshore aquaculture is the selection and breeding of fish species that can thrive in open-sea conditions. Species cultivated in offshore environments must be resilient to challenges such as fluctuating water temperatures, strong currents, and high salinity levels. This Special Issue explores various breeding programs and genetic research efforts aimed at developing species with enhanced growth rates, disease resistance, and environmental tolerance.

These advancements in species selection are crucial, as they allow aquaculture operations to expand their range and adapt to the specific conditions of different offshore locations. Additionally, breeding programs that focus on reducing the ecological impact of aquaculture species help to align industry practices with broader sustainability goals, ensuring that the benefits of offshore aquaculture extend beyond mere production efficiency.

Environmental Control Systems

Environmental control is a cornerstone of effective offshore aquaculture. Unlike nearshore farms, where environmental conditions are more easily managed, offshore platforms face unpredictable and sometimes extreme conditions. Maintaining optimal water quality, temperature, and oxygen levels is essential to the health and growth of farmed fish. This Special Issue introduces cutting-edge environmental control technologies, including sensor networks, automated feeding systems, and real-time monitoring solutions.

These technologies allow operators to respond swiftly to changes in the marine environment, adjusting conditions as needed to promote fish health and reduce waste. For instance, sensors that continuously monitor parameters such as dissolved oxygen and nutrient levels enable precise adjustments, ensuring that fish are kept in a stable environment. By leveraging data analytics and machine learning, these systems provide a level of control that was previously unattainable, marking a significant step forward in aquaculture management.

Platform Design and Structural Innovation

The structural design of offshore aquaculture platforms is another area of focus in this Special Issue. Building platforms that can withstand the physical stresses of open-sea conditions is essential for the long-term viability of offshore aquaculture. Engineers must consider factors such as wave force, currents, and storm resilience when designing these structures. The platforms discussed in this Special Issue include innovations such as rotatable net cages, which reduce biofouling and improve water flow, and self-cleaning systems that enhance water quality while minimizing labor requirements.

These advancements in platform design contribute to the sustainability and efficiency of offshore aquaculture operations. By ensuring that platforms are durable and require minimal maintenance, these designs reduce the overall cost and environmental impact of offshore aquaculture. This Special Issue provides detailed insights into the engineering principles and testing methods used to create these resilient structures, offering valuable guidance for future developments in platform design.

Feeding Technologies and Fish Health Management

Effective feeding strategies are critical to the success of offshore aquaculture, as feed costs represent a substantial portion of operational expenses. Precision feeding systems, which use real-time data from environmental sensors, have been developed to regulate feed distribution more accurately. These systems ensure that fish receive the optimal amount of nutrients, minimizing waste and reducing the environmental impact of aquaculture.

In addition to feeding technologies, this book examines advancements in fish health management, including the use of diagnostic tools and monitoring systems to detect early signs of disease and stress. By integrating data from various sources, such as water quality sensors and fish behavior monitors, operators can make informed decisions that promote the health and welfare of farmed fish. These innovations contribute to higher survival rates, faster growth, and improved product quality, which are essential for the economic viability of offshore aquaculture.

Integration of Artificial Intelligence and Robotics

The role of artificial intelligence (AI) and robotics in offshore aquaculture is a transformative development that is reshaping the industry. AI-driven systems enable real-time monitoring and decision-making, allowing operators to optimize feeding schedules, monitor fish health, and predict environmental changes. Robotics, meanwhile, is used to perform tasks such as cleaning cages and inspecting equipment, reducing the need for manual labor in challenging offshore environments.

This Special Issue explores the integration of AI and robotics in aquaculture, providing examples of how these technologies enhance operational efficiency and sustainability. For instance, AI-based image analysis can be used to assess fish populations within cages, enabling precise feed allocation and harvest planning. Bio-inspired robotic systems are also being developed to interact with fish in ways that minimize stress, supporting animal welfare in aquaculture. These technologies represent the future of offshore aquaculture, offering new possibilities for sustainable, large-scale production.

Case Studies and Practical Applications

In addition to theoretical insights and technological discussions, this Special Issue includes practical case studies that demonstrate the application of new techniques and equipment in real-world scenarios. Each case study provides an in-depth look at a specific aspect of offshore aquaculture, from the design of a vacuum fish pump that minimizes fish damage to the development of mooring systems that ensure platform stability. These case studies serve as valuable resources for researchers, engineers, and industry practitioners, offering lessons learned and best practices that can be applied to future projects.

The case studies also highlight the collaborative nature of innovation in aquaculture, showcasing how researchers and industry professionals are working together to address common challenges. This spirit of collaboration is essential to the continued advancement of offshore aquaculture, as it enables the sharing of knowledge and the development of solutions that benefit the entire industry.

The Future of Offshore Aquaculture

As the demand for sustainable seafood continues to grow, the innovations presented in this Special Issue will play an increasingly important role in meeting that demand. Offshore aquaculture has the potential to become a cornerstone of global food security, providing a reliable source of high-quality protein while minimizing environmental impact. The technologies and techniques discussed in this Special Issue represent a significant step toward realizing this potential, offering scalable solutions that can be adapted to diverse marine environments around the world.

However, the path forward is not without challenges. Issues such as regulatory hurdles, environmental concerns, and the need for continued research into species selection and disease management must be addressed to ensure the long-term success of offshore aquaculture. This Special Issue aims to contribute to this ongoing effort by providing a comprehensive resource on the latest advancements in the field. By fostering a deeper understanding of the technologies and practices that underpin offshore aquaculture, it is our hope that this Special Issue will inspire further innovation and support the sustainable growth of the industry.

Acknowledgments

The editors and contributors of *New Techniques and Equipment in Large Offshore Aquaculture Platform* would like to extend their gratitude to the researchers, engineers, and industry experts whose dedication and expertise have made this Special Issue possible. We are also grateful to the organizations and institutions that have supported this work, including universities, research institutes, and government agencies around the world. Their commitment to advancing the field of aquaculture is a testament to the importance of collaboration and innovation in addressing global food security challenges.

Zhenhua Ma and Jianguang Qin
Guest Editors

Journal of
Marine Science and Engineering

MDPI

Editorial

New Techniques and Equipment in Large Offshore Aquaculture Platform

Zhenhua Ma [1,2,*] and Jianguang Qin [3,*]

1 Sanya Tropical Fisheries Research Institute, Sanya 512426, China
2 South China Sea Fisheries Research Institute, Chinese Academy of Fishery Sciences, Guangzhou 510300, China
3 College of Science and Engineering, Flinders University, Bedford Park, SA 5001, Australia
* Correspondence: zhenhua.ma@hotmail.com (Z.M.); jian.qin@flinders.edu.au (J.Q.)

Citation: Ma, Z.; Qin, J. New Techniques and Equipment in Large Offshore Aquaculture Platform. *J. Mar. Sci. Eng.* **2024**, *12*, 2119. https://doi.org/10.3390/jmse12122119

Received: 11 November 2024
Accepted: 13 November 2024
Published: 21 November 2024

The expansion of far-reaching marine aquaculture represents a critical frontier for the sustainable growth of global aquaculture [1–3]. This sector is becoming increasingly important in the face of the space constraints, environmental degradation, and regulatory restrictions that limit nearshore aquaculture. The development of advanced technologies and innovative equipment has transformed the aquaculture landscape, enabling operations to be conducted in more challenging offshore environments [4,5]. This Special Issue aims to showcase the latest advancements in far-reaching marine aquaculture technologies and equipment, focusing on key areas such as aquaculture species, breeding technologies, environmental control systems, large-scale platform construction, and cutting-edge aquacultural innovations. The collective results of these studies highlight the potential for future research and expansion of these techniques to other regions and species, driving the aquaculture industry towards more sustainable and productive practices.

A central theme in this Special Issue is the development of new species suitable for far-reaching marine aquaculture and the associated farming techniques required to optimize their growth and health in offshore environments. As aquaculture expands into deeper waters, selecting and developing species that thrive under these conditions is essential. This process requires targeted breeding programs, enhanced disease management protocols, and optimized feeding strategies that account for the specific needs of species in dynamic marine environments. The research presented here highlights the progress made in cultivating these species, including improvements in breeding methodologies and stress management, which are aimed at maximizing productivity while minimizing the ecological impact. The selection of species that exhibit resilience to the unique challenges of offshore aquaculture, such as fluctuating water temperatures, strong currents, and variable salinity, is of particular importance.

Environmental control technologies form another cornerstone of far-reaching marine aquaculture [6]. In offshore settings, maintaining optimal environmental conditions is challenging due to the unpredictable nature of the marine environment. Parameters such as water temperature, salinity, dissolved oxygen levels, and nutrient availability must be carefully monitored and controlled to ensure the health and growth of farmed species. Innovations in sensor technology, automated feeding systems, and real-time environmental monitoring have significantly enhanced the stability of conditions in far-reaching aquaculture environments. These technologies allow precise adjustments to be made in response to environmental fluctuations, ensuring that aquaculture systems remain efficient and productive. This section of the Special Issue delves into the integration of these advanced systems, exploring their role in improving overall aquaculture management and efficiency.

The structural design and construction of offshore aquaculture platforms have undergone significant advancements to ensure the safety and sustainability of operations in open-sea environments. These platforms must be designed to withstand harsh marine conditions, including strong currents, waves, and storms. Recent research has focused on the dynamic motion characteristics of these platforms, particularly the interaction between breeding tanks and wave forces. Optimizing platform stability, durability, and structural integrity is crucial to the long-term success of offshore aquaculture. Innovative designs, such as rotatable net cages and self-cleaning systems, have been developed to enhance water quality and reduce the accumulation of biofouling, which can negatively affect both fish health and operational efficiency. This section explores the latest developments in platform design and their impact on aquaculture productivity and sustainability.

Feeding technologies and fish health management are also crucial to the success of far-reaching marine aquaculture [7]. Precision feeding systems, which utilize real-time data from environmental sensors, have been developed to facilitate more accurate feed distribution, ensuring that fish receive the optimal amount of nutrients while minimizing feed wastage and environmental impact. Additionally, significant progress has been made in understanding the nutritional requirements of species farmed in offshore environments, leading to the development of specialized feed formulations that enhance growth rates and improve fish health. This Special Issue also covers studies on the physiology of fish and their responses to environmental stressors, providing valuable insights into the development of more resilient aquaculture systems capable of withstanding the challenges of offshore farming.

The integration of emerging technologies, such as artificial intelligence (AI), machine learning, and robotics, has revolutionized aquaculture management. AI-driven systems are being used to monitor fish behavior, track health metrics, and optimize feeding regimens in real time, improving both efficiency and sustainability. For instance, image sonar and deep-learning models have been employed to estimate fish populations within cages, allowing for more precise feed allocation and harvest planning. Bio-inspired robotic systems are also being developed to interact with farmed fish, enabling more accurate assessments of their welfare and behavior. These technological innovations enhance operational efficiency and contribute to the sustainable management of offshore aquaculture ecosystems.

The advancements in far-reaching marine aquaculture technologies and equipment presented in this Special Issue represent a major step forward in the development of sustainable, large-scale offshore aquaculture. The research featured here spans a wide range of topics, from species selection and environmental control to platform design and the integration of cutting-edge technologies. Together, these studies provide a comprehensive overview of the current state of far-reaching marine aquaculture and offer valuable insights for future research and application. As the global demand for seafood continues to rise, these innovations will play an increasingly important role in ensuring that aquaculture can meet this demand in an environmentally responsible and economically viable manner. The findings and technologies presented in this Special Issue are poised to shape the future of aquaculture, promoting its expansion into new marine environments and species and driving the industry towards a more sustainable and productive future.

In the first contribution to this Special Issue, Hong et al. present an innovative approach to improving fish pump technology, which plays a crucial role in aquaculture, particularly for deep-sea vessels and cage culture systems. Their research focuses on enhancing the hydraulic performance of a newly designed vacuum fish pump while minimizing damage to the fish during operation. Key aspects of this study include an in-depth analysis of flow dynamics, fluid simulations of the pump body, and the flow channel structure. The authors investigate the effects of variables such as inlet flow rate, pipeline negative pressure, and the impact forces on the tank's inner walls. Utilizing 167 calculation models based on the Latin hypercube sampling method, they conduct multi-objective optimization using the NSGA-II algorithm to identify optimal structural parameters for the pump. The results indicate that, under the best conditions, the direction of incident water flow is positioned near the upper end of the tank, which reduces the speed of the water–fish mixture as it enters the tank, significantly reducing fish collision damage. The optimal flow velocity at the inlet was found to be approximately 2.5 m/s, with a consistent negative pressure gradient between the tank and inlet pipeline. This design ensures effective fish suction and lifting while also maintaining high survival rates and minimizing the physical harm sustained by the fish. Hong et al.'s work significantly contributes to aquaculture equipment design, providing insights into how hydraulic performance can be optimized for safer and more efficient fish handling.

In their paper, Shen et al. address the persistent challenge of biofouling in aquaculture net cages, which can negatively impact both structural integrity and fish growth. To mitigate this, they propose a novel fixed aquaculture platform equipped with a rotatable horizontal cylindrical cage to facilitate easier cleaning and maintenance. Using ANSYS software, the team develop a numerical model to simulate the platform's structural response under various conditions, such as waves and currents at three typical attack angles. They calculate the strain, acceleration, and displacement of the structure to evaluate its hydrodynamic performance. The study reveals that as the wave height increases, so do the strain, acceleration, and displacement of the cage, while the wave period has minimal impact. The most unfavorable loading occurs when forces are applied perpendicular to the long axis of the cage. Interestingly, while acceleration increases with water depth, the strain response decreases. Furthermore, when the rotational constraint of the cylindrical cage is released, the cage experiences higher acceleration compared to when it is fixed, though the effect on structural strain and load is minimal. Shen et al.'s research offers valuable insights into aquaculture platform design, highlighting how cage rotation and water depth influence hydrodynamic responses and ultimately aiding in the development of safer, more efficient aquaculture systems.

In their work, Liu et al. address the challenge of fish re-identification (re-ID) under varying environmental and camera conditions, focusing on large yellow croaker. Traditional re-ID methods often struggle with domain distribution differences between fish images captured in different culture settings, limiting the effectiveness of existing training data. The authors propose a novel approach that combines CycleGAN (Cycle Generative Adversarial Network) with transfer learning to overcome these limitations. Their method involves constructing two datasets: one from controllable environments and another from actual farming conditions. The CycleGAN framework is employed to transform images from the source domain to the target domain, enabling data amplification. They further improve identity loss judgment using IDF (Identity Foreground Loss) and narrow the distribution gap between domains with MMD (Maximum Mean Discrepancy). Transfer learning is then applied to the expanded dataset to accurately identify large yellow croaker across different conditions. The experimental results are highly promising, demonstrating recognition accuracies of 96.9% in controlled environments and 94% in real farming conditions. This approach enhances the reliability of fish identification across various settings and lays the groundwork for future advancements in fish behavior tracking and phenotype measurement, providing essential technical support for the development of intelligent aquaculture systems.

In their manuscript, Tao et al. examine the critical impact of sloshing responses on the design and operational efficiency of aquaculture vessels. Sloshing, the movement of liquid inside a partially filled tank, significantly affects both the safety of culture equipment and the overall efficiency of aquaculture operations. To analyze these dynamics, the authors conducted experiments using a 1/50 scale model of a novel aquaculture vessel in a controlled wave basin. They explored sloshing behavior under two wave directions—beam and head waves—and two fill levels of 81.5% and 47.4%. The study investigated both time-domain and frequency-domain characteristics of sloshing under regular and extreme sea conditions. Their findings revealed that the sloshing response is primarily driven by the wave frequency, with more pronounced effects under beam wave conditions than head waves. The response was also stronger for a half load than a full load. The complexity of the sloshing mechanism stems from the interaction between external waves, vessel motion, and internal liquid movement. The results of irregular wave tests supported the conclusions of regular wave tests, but they also highlighted the presence of stronger nonlinearity and more prominent higher natural modes, with greater amplitude. Tao et al.'s research offers crucial insights into optimizing aquaculture vessel design by understanding sloshing behavior providing valuable data for enhancing operational safety and efficiency in dynamic marine environments.

In their contribution, Wang et al. investigate the influence of different photoperiods on the growth performance and physiological responses of juvenile scalloped spiny lobsters (*Panulirus homarus*), a species that is of high economic value in aquaculture. Over 56 days the researchers exposed 90 lobsters to varying light–dark cycles (0L:24D, 6L:18D, 12L:12D 18L:6D, 24L:0D) and carefully measured growth rates, digestive and immune enzyme activities, as well as antioxidant enzyme responses. The study found no significant differences in survival rate, molting frequency, or meat yield among the different photoperiod treatments However, the 12L:12D photoperiod resulted in the highest weight gain rate (WGR) and specific growth rate (SGR), indicating that this balanced light–dark cycle is optimal for lobster growth. Under continuous darkness (0L:24D), pepsin activity remained elevated in gastric tissues, while trypsin and chymotrypsin activities were highest in the hepatopancreas. Other key enzymes, such as α-amylase, lipase, and acid phosphatase, showed peak activities under different light conditions, with optimal lipase activity recorded at 12L:12D and the highest α-amylase activity at 18L:6D. In terms of antioxidant capacity, the study showed that the total antioxidant capacity (T-AOC), catalase (CAT), and superoxide dismutase (SOD) activities were highest under the 12L:12D photoperiod. Additionally, the highest levels of glutathione peroxidase (GSH-Px) were observed under 18L:6D. Malondialdehyde (MDA), an indicator of oxidative stress, peaked at 12L:12D. These results suggest that the 12L:12D light–dark cycle is essential for achieving optimal growth and maintaining physiological balance in juvenile scalloped spiny lobsters. Wang et al.'s findings offer valuable guidance for optimizing lighting conditions in aquaculture settings to enhance the growth and health of this species.

In their work, Pino et al. delve into the use of zebrafish (*Danio rerio*) as a model organism for neurobehavioral research, particularly in the context of stress, fear, and anxiety responses. Their research explores the effects of bio-inspired mini robotic fish, designed with various components, on zebrafish behavior. The primary goal of the study is to optimize robotic biomimicry to minimize stress and improve fish welfare, which could have broader applications in aquaculture. While previous studies predominantly focused on externally controlled fish models, Pino et al. introduce novel prototypes of freely actuated swimming robots. These robots allow for more natural interactions between the bio-inspired robots and the zebrafish. By analyzing the zebrafish's behavioral responses, the researchers identify specific robotic components that may induce anxiety in fish. These findings not only enhance our understanding of zebrafish stress responses but also offer valuable insights for designing future bio-inspired robotic systems in aquaculture, aiming to minimize stress and ensure the welfare of aquatic species.

In their study, Wang et al. investigate the effects of acute seawater acidification on the antioxidant defenses, metabolic performance, and liver histology of juvenile yellowfin tuna (*Thunnus albacares*). The experiment exposes the tuna to a pH gradient ranging from 8.1 to 6.6 over 48 h, simulating acidification stress. The results reveal that a pH of 7.1 significantly disrupts the antioxidant and metabolic systems of the fish compared to the control group. Specifically, at this pH level, notable increases are observed in glutathione reductase (GR), total antioxidant capacity (T-AOC), lactate dehydrogenase (LDH), hexokinase (HK), pyruvate kinase (PK), sodium–potassium ATPase (Na^+K^+-ATP), and calcium-magnesium ATPase ($Ca^{2+}Mg^{2+}$-ATP). Despite these changes, levels of low-density lipoprotein cholesterol (LDL-C), high-density lipoprotein cholesterol (HDL-C), and triglycerides (TGs) remained relatively stable across all treatment groups. However, the study also detected elevated transaminase levels at pH 7.1, indicating potential liver damage corroborated by signs of liver tissue degeneration and hepatocyte vacuolation. These findings suggest that acute acidification leads to a reduction in antioxidant capacity and a suppression of metabolic activity in juvenile yellowfin tuna, ultimately causing oxidative damage. Wang et al.'s research provides critical insights into the physiological responses of yellowfin tuna to seawater acidification, offering a foundational understanding of the mechanisms behind acidification stress. The study also underscores the broader implications for sustainable tuna farming in the face of changing ocean conditions.

Sun et al.'s work examines the impact of acute ammonia nitrogen (NH3-N) exposure on the kidney's antioxidant capacity, phosphatase activity, and related gene expression in juvenile yellowfin tuna (*Thunnus albacares*). A total of 180 juvenile tuna are exposed to varying NH_3-N concentrations (5 and 10 mg/L) for 6, 24, and 36 h, with natural seawater (0 mg/L) serving as the control. The study measures the levels of malondialdehyde (MDA), a marker of lipid peroxidation, and antioxidant enzymes such as superoxide dismutase (SOD), catalase (CAT), glutathione peroxidase (GSH-PX), alkaline phosphatase (AKP), and acid phosphatase (ACP) to evaluate changes in antioxidant activity and immune-related phosphatase function in the trunk kidney. At 36 h, MDA, SOD, CAT, and GSH-PX levels significantly increase in the 5 mg/L group compared to the control, with even higher elevations in the 10 mg/L group after 24 and 36 h. Phosphatases, which play a key role in immune responses, also show marked activity increases—AKP surges at 6 h, and ACP rises significantly at 36 h in the 5 mg/L group. Using real-time fluorescence quantitative PCR, the authors observe significant upregulation of antioxidant genes SOD2 and GPX1b in the 5 mg/L group at 6 and 36 h, while similar upregulation is seen in the 10 mg/L group at 36 h. Additionally, immune cytokine gene expression shows an increase in Interleukin 10 (IL-10) in the 5 mg/L group at 6 h, whereas Interleukin 6 receptor (IL-6r) expression decreases. Both IL-10 and IL-6r levels were suppressed in the 10 mg/L group. The findings suggest that low concentrations of NH3-N (around 5 mg/L) enhance antioxidant defenses, phosphatase activity, and gene expression, while higher concentrations (10 mg/L) may have a suppressive effect. In juvenile yellowfin tuna farming, managing NH3-N levels is critical for health; at concentrations between 5 and 10 mg/L, stress exposure should be limited to 24 h, whereas concentrations below 5 mg/L allow for extended exposure up to 36 h without adverse effects.

In their contribution to the Special Issue, Yang et al. investigate the effects of licorice (*Glycyrrhiza uralensis*) as a feed additive on the growth performance and gene expression of Asian seabass (*Lates calcarifer*), a species that holds significant economic importance in China's aquaculture industry. While licorice has been known for its immunological benefits in aquaculture, its impact on growth has been less explored. This study aims to fill that gap by evaluating the influence of various concentrations of licorice (0%, 1%, 3%, and 5%) in artificial feed on the expression of growth-related genes over 56 days. The key growth-related genes analyzed include growth hormone–releasing hormone (GHRH), growth hormone (GH), growth hormone receptor (GHR), insulin-like growth factor 1 (IGF1), IGF2, IGF2 receptor (IGF2R), and myostatins (MSTN1 and MSTN2). The results demonstrate that diets supplemented with 3% and 5% licorice significantly improve survival rates and weight gain in comparison to the control group. Notably, licorice supplementation enhances the expression of GHR and IGF1 in the liver, while a 5% licorice diet downregulates IGF2 expression. As the licorice concentration increases, IGF2R and MSTN1 expression initially decreases and then increases, while MSTN2 expression is inhibited. Moreover, licorice supplementation significantly downregulates the expression of GH and GHRH. Yang et al.'s findings suggest that the inclusion of licorice in the diet of Asian seabass, particularly at optimal levels, can boost growth-related gene expression, thereby enhancing both the weight gain rate and specific growth rate of the fish. This research offers valuable insights into the potential of licorice as a feed additive to improve the survival and growth of Asian seabass in aquaculture settings.

In their work, Zhu et al. propose an innovative method for accurately estimating fish quantities in aquaculture cages using image sonar technology. To meet the challenges of real-time fish counting in dynamic environments, the authors utilize forward-looking image sonar combined with the YOLO target detection model, enhanced by an attention mechanism and a backpropagation (BP) neural network. This integrated approach enables continuous, automated estimation of fish numbers within cages. The research involves conducting a quantitative experiment in the South China Sea to create a comprehensive dataset for training both the YOLO model and the neural network. The improved YOLOv8 model achieves an average detection accuracy (mAP50) that is 3.81% higher than that of the original algorithm. Additionally, the neural network's accuracy in estimating fish quantities reaches 84.63%, outperforming cubic polynomial fitting by 0.72%. Zhu et al.'s work significantly advances aquaculture management, providing a precise method for assessing fish quantities. This allows for more informed decision-making with regard to feeding strategies and harvest planning, contributing to the overall efficiency and sustainability of aquaculture operations.

In response to increasing limitations in coastal aquaculture environments and the need for larger-scale operations, Zhang et al. explore the potential of deep-sea aquaculture using large-scale aquaculture vessels. These vessels offer new opportunities for expanding aquaculture into deeper waters, but their success requires the optimization of tank design to ensure ideal fish growth conditions and effective removal of particulate matter. This balance is essential for maximizing productivity and profitability in aquaculture vessels. Zhang et al.'s study specifically investigates how the structural ratio of aquaculture tanks impacts flow field characteristics and particulate removal efficiency. Using FLOW-3D software (solver version: 11.2.0.16), they conduct numerical simulations on the flow field of an 8000-ton aquaculture vessel at anchor. Their simulations evaluate the impact of the corner ratio on key parameters such as flow velocity, turbulence intensity, tank utilization, and particulate removal efficiency. The results indicate that tanks with corner structures exhibit better overall flow dynamics, including higher flow velocities, stronger turbulence, and improved particulate discharge efficiency. However, the study also finds that further increases do not significantly improve these flow characteristics once the corner length exceeds one-third of the tank length. The findings provide a valuable reference for the structural design and optimization of aquaculture tanks, contributing to the efficiency and effectiveness of large-scale deep-sea aquaculture vessels.

In their paper, Hu et al. introduce a novel approach for accurately estimating the distribution of *Trachinotus ovatus* in marine cages, leveraging omnidirectional scanning sonar and deep-learning techniques. Their method provides precise fish location data by segmenting the water into layers and applying clustering algorithms to analyze the data from each layer. The process begins with omnidirectional scanning sonar, which performs a spiral detection within the cages to capture fish image data. These images are then labeled to construct a training dataset for an enhanced CS-YOLOv8s model. Once trained, the CS-YOLOv8s model is employed to identify and locate fish in the images. Next, the cage environment is divided into water layers, with depth intervals of 40 cm. The DBSCAN clustering method is applied to the identification coordinates for each layer, generating precise location data for fish at various depths. Finally, the data from all layers are combined to form a comprehensive map of fish distribution within the cage. The method is tested through extensive experimentation and accurately estimates the distribution of *Trachinotus ovatus*, with results closely matching manual observations. Hu et al.'s approach represents a significant advancement in using sonar and deep-learning technology for aquaculture, providing a more efficient and automated method for managing fish populations in marine cages.

In their work, Liu et al. address the challenges of deep-sea aquaculture, which offers significant potential to alleviate the spatial and environmental pressures associated with near-shore operations while producing higher-quality aquatic products. Despite China's relatively flat coastline, where aquaculture typically occurs in waters 30–50 m deep, the frequent typhoons and adverse sea conditions make designing effective mooring systems for deep-sea cages challenging. Liu et al.'s study investigates multiple fish cage configurations, focusing on a 1×4 layout and a 2×2 layout, and proposes three distinct mooring system design schemes. The researchers compare mooring line tension under different self-storage conditions, examining factors such as line accumulation and contact with the steel structure on the leeward side. Furthermore, flexible net models are compared to rigid net models to assess how net deformation influences cage movement and mooring line tension. Liu et al. also analyze the dynamic response of the optimal mooring design under irregular wave conditions, offering valuable insights into the safety and economical design of mooring systems for multiple fish cages. Their findings provide essential guidance for improving the reliability and resilience of deep-sea aquaculture systems, particularly in regions prone to challenging sea conditions.

In the final contribution, Martzikos et al. explore the integration of renewable energy sources with aquaculture on floating multi-use platforms, aiming to create sustainable offshore infrastructure. From March 2021 to January 2022, a 1:15 scale prototype was tested in Reggio Calabria, Italy, providing insights into the structures' behavior under various wave conditions. The study utilizes Artificial Neural Networks (ANNs) to predict changes in mooring loads at critical points, analyzing metocean data to evaluate different ANN models and optimization techniques. The best predictive model achieves a Normalized Root Mean Square Error (NRMSE) of 1.7–4.7%, demonstrating the effectiveness of ANNs in forecasting offshore platform dynamics. This research underscores the potential of machine learning for sustainable ocean system management, paving the way for advancements in data-driven marine resource management.

Funding: This work was supported by the Hainan Province Natural Science Foundation Enterprise Talent Project (324QY579); the Research on Breeding Technology of Candidate Species for Guangdong Modern Marine Ranching (2024-MRB-00-001); the Science and Technology special fund of Hainan Province (ZDYF2022XDNY349, ZDYF2024XDNY247); the Central Public-interest Scientific Institution Basal Research Fund, CAFS (2024XT04, 2023TD58); Central Public-interest Scientific Institution Basal Research Fund, South China Sea Fisheries Research Institute, CAFS (2024RC15); the Project of Sanya Yazhou Bay Science and Technology City (SKJC-2022-PTDX-015, SCKJ-JYRC-2023-42); the earmarked fund for HNARS (HNARS-03-Z02); the National Key Research and Development Program of China (2022YFD2400501); Hainan Provincial Natural Science Foundation of China (320QN360); Central Guidance Funds for Regional Science and Technology Development (GUIKE ZY22096005); Guangxi Provincial Science and Technology Base and Talent Program (GUIKR AD21238026); and National Natural Science Foundation of China (32460927).

Acknowledgments: As Guest Editors of the Special Issue "New Techniques and Equipment in Large Offshore Aquaculture Platform", we wish to extend our sincere gratitude to all the authors whose valuable contributions made the publication of this issue possible. Their work has significantly enriched and enhanced the coverage of this publication.

Conflicts of Interest: The authors declare no conflicts of interest.

List of Contributions:

1. Hong, Y.; Zhu, Y.; Zhang, C.; Yang, M.; Jiang, T. Hydrodynamic Characteristic Analysis and NSGA-II Optimization of a Vacuum Fish Pump. *J. Mar. Sci. Eng.* **2023**, *11*, 1361. https://doi.org/10.3390/jmse11071361
2. Shen, K.; Bi, C.; Jiang, Z.; Guo, S.; Wang, B. Hydrodynamic Response Analysis of a Fixed Aquaculture Platform with a Horizontal Cylindrical Cage in Combined Waves and Currents. *J. Mar. Sci. Eng.* **2023**, *11*, 1413. https://doi.org/10.3390/jmse11071413
3. Liu, S.; Qian, C.; Tu, X.; Zheng, H.; Zhu, L.; Liu, H.; Chen, J. Identification of Large Yellow Croaker under Variable Conditions Based on the Cycle Generative Adversarial Network and Transfer Learning. *J. Mar. Sci. Eng.* **2023**, *11*, 1461. https://doi.org/10.3390/jmse11071461
4. Tao, Y.; Zhu, R.; Gu, J.; Wei, Q.; Hu, F.; Xu, X.; Zhang, Z.; Li, Z. Sloshing Response of an Aquaculture Vessel: An Experimental Study. *J. Mar. Sci. Eng.* **2023**, *11*, 2122. https://doi.org/10.3390/jmse11112122
5. Wang, Y.; Yang, R.; Fu, Z.; Ma, Z.; Bai, Z. The Photoperiod Significantly Influences the Growth Rate, Digestive Efficiency, Immune Response, and Antioxidant Activities in the Juvenile Scalloped Spiny Lobster (*Panulirus homarus*). *J. Mar. Sci. Eng.* **2024**, *12*, 389. https://doi.org/10.3390/jmse12030389

6. Pino, A.; Vidal, R.; Tormos, E.; Cerdà-Reverter, J.M.; Marín Prades, R.; Sanz, P.J. Towards Fish Welfare in the Presence of Robots: Zebrafish Case. *J. Mar. Sci. Eng.* **2024**, *12*, 932. https://doi.org/10.3390/jmse12060932
7. Wang, X.; Yang, R.; Fu, Z.; Zhao, L.; Ma, Z. Antioxidant and Metabolic Response to Acute Acidification Stress of Juvenile Yellowfin Tuna (*Thunnus albacares*). *J. Mar. Sci. Eng.* **2024**, *12*, 970. https://doi.org/10.3390/jmse12060970
8. Sun, Y.; Fu, Z.; Ma, Z. The Effects of Acute Ammonia Nitrogen Stress on Antioxidant Ability, Phosphatases, and Related Gene Expression in the Kidney of Juvenile Yellowfin Tuna (*Thunnus albacares*). *J. Mar. Sci. Eng.* **2024**, *12*, 1009. https://doi.org/10.3390/jmse12061009
9. Yang, R.; Zhao, W.; Wang, Y.; Fu, Z.; Hu, J.; Zhou, S.; Li, M.; Ma, Z. Effect of Licorice on Gene Expression Related to the Growth of Asian Seabass *Lates calcarifer*. *J. Mar. Sci. Eng.* **2024**, *12*, 1036. https://doi.org/10.3390/jmse12071036
10. Zhu, G.; Li, M.; Hu, J.; Xu, L.; Sun, J.; Li, D.; Dong, C.; Huang, X.; Hu, Y. An Experimental Study on Estimating the Quantity of Fish in Cages Based on Image Sonar. *J. Mar. Sci. Eng.* **2024**, *12*, 1047. https://doi.org/10.3390/jmse12071047
11. Zhang, F.; Cui, M.; Liu, H.; Zhang, C. The Effect of Corner Structure on the Optimisation of Fishable Flow Field in Aquaculture Tanks. *J. Mar. Sci. Eng.* **2024**, *12*, 1185. https://doi.org/10.3390/jmse12071185
12. Hu, Y.; Hu, J.; Sun, P.; Zhu, G.; Sun, J.; Tao, Q.; Yuan, T.; Li, G.; Pang, G.; Huang, X. A Method for Estimating the Distribution of *Trachinotus ovatus* in Marine Cages Based on Omnidirectional Scanning Sonar. *J. Mar. Sci. Eng.* **2024**, *12*, 1571. https://doi.org/10.3390/jmse12091571
13. Liu, F.; Jiang, Z.; Cheng, T.; Xu, Y.; Zhu, H.; Wang, G.; Sun, G.; Zhang, Y. Study on the Dynamic Response of Mooring System of Multiple Fish Cages under the Combined Effects of Waves and Currents. *J. Mar. Sci. Eng.* **2024**, *12*, 1648. https://doi.org/10.3390/jmse12091648
14. Martzikos, N.; Ruzzo, C.; Malara, G.; Fiamma, V.; Arena, F. Applying Neural Networks to Predict Offshore Platform Dynamics. *J. Mar. Sci. Eng.* **2024**, *12*, 2001. https://doi.org/10.3390/jmse12112001

References

1. Gentry, R.R.; Froehlich, H.E.; Grimm, D.; Kareiva, P.; Parke, M.; Rust, M.; Gaines, S.D.; Halpern, B.S. Mapping the global potential for marine aquaculture. *Nat. Ecol. Evol.* **2017**, *1*, 1317–1324. [CrossRef] [PubMed]
2. Boyd, C.E.; D'Abramo, L.R.; Glencross, B.D.; Huyben, D.C.; Juarez, L.M.; Lockwood, G.S.; McNevin, A.A.; Tacon, A.G.J.; Teletchea, F.; Tomasso, J.R., Jr.; et al. Achieving sustainable aquaculture: Historical and current perspectives and future needs and challenges. *J. World Aquac. Soc.* **2020**, *51*, 578–633. [CrossRef]
3. Gentry, R.R.; Rassweiler, A.; Ruff, E.O.; Lester, S.E. Global pathways of innovation and spread of marine aquaculture species. *One Earth* **2023**, *6*, 20–30. [CrossRef]
4. Yue, K.; Shen, Y. An overview of disruptive technologies for aquaculture. *Aquac. Fish.* **2022**, *7*, 111–120. [CrossRef]
5. Rastegari, H.; Nadi, F.; Lam, S.S.; Ikhwanuddin, M.; Kasan, N.A.; Rahmat, R.F.; Mahari, W.A.W. Internet of things in aquaculture: A review of the challenges and potential solutions based on current and future trends. *Smart Agric. Technol.* **2023**, *4*, 100187. [CrossRef]
6. Fujita, R.; Brittingham, P.; Cao, L.; Froehlich, H.; Thompson, M.; Voorhees, T. Toward an environmentally responsible offshore aquaculture industry in the United States: Ecological risks, remedies, and knowledge gaps. *Mar. Policy* **2023**, *147*, 105351. [CrossRef]
7. Rombenso, A.; Araujo, B.; Li, E. Recent advances in fish nutrition: Insights on the nutritional implications of modern formulations. *Animals* **2022**, *12*, 1705. [CrossRef] [PubMed]

Journal of
Marine Science and Engineering

MDPI

Article

Hydrodynamic Characteristic Analysis and NSGA-II Optimization of a Vacuum Fish Pump

Yang Hong [1], Ye Zhu [1], Chenglin Zhang [1,2], Meng Yang [1] and Tao Jiang [1,*]

1 Fishery Machinery and Instrument Research Institute, Chinese Academy of Fishery Sciences, Shanghai 200092, China; hongyang@fmiri.ac.cn (Y.H.); zhuye@fmiri.ac.cn (Y.Z.); zhangchenglin@fmiri.ac.cn (C.Z.); yangmeng@fmiri.ac.cn (M.Y.)
2 College of Engineering Science and Technology, Shanghai Ocean University, Shanghai 201306, China
* Correspondence: jiangtao@fmiri.ac.cn

Abstract: The fish pump is an important piece of power equipment for aquaculture, especially for deep-sea engineering vessels and cage culture. Fish pump research focuses on reducing fish body damage, improving survival rates, and increasing fish lifting efficiency. The research object in this paper is a new type of vacuum fish pump, with the aim of improving the hydraulic performance of the vacuum fish pump and reducing the damage to the fish body. The dependent variables include the dynamic change process of the flow state and flow field under diachronic conditions, the fluid simulation analysis of the vacuum pump body and the flow channel structure, the inlet flow rate of the fish pump, the negative pressure of the pipeline, and the impact force of the water flow on the inner wall of the tank. The independent variables include the operating conditions of the pump body and the fish pump. The Latin hypercube sampling method is used to extract 167 sets of calculation models for the independent variables, and multi-objective optimization is performed based on the NSGA-II algorithm for the hydrodynamic performance of the fish pump. On the basis of ensuring the fish body damage rate, the structural parameters of the vacuum fish pump with the optimal hydrodynamic performance under 167 sets of parameter values were obtained. The optimized parameters were then entered into the solver again, and the results showed that, in the optimal structural parameters under certain conditions, the direction of the incident water flow in the vacuum fish pump tank is close to the upper end of the tank body, which will reduce the speed of the fish-water mixed flow when entering the tank, thereby reducing the collision damage to the fish body. Currently, the water flow velocity at the water inlet is about 2.5 m/s, and the negative pressure value distribution gradient between the tank body and the water inlet pipeline is quite consistent, which can achieve good fish suction and fish lifting effects.

Keywords: equipment aquaculture; fish lifting; vacuum fish pump; hydraulic characteristics; particle swarm algorithm

Citation: Hong, Y.; Zhu, Y.; Zhang, C.; Yang, M.; Jiang, T. Hydrodynamic Characteristic Analysis and NSGA-II Optimization of a Vacuum Fish Pump. *J. Mar. Sci. Eng.* **2023**, *11*, 1361. https://doi.org/10.3390/jmse11071361

Academic Editor: María Isabel Lamas Galdo

Received: 25 May 2023
Revised: 28 June 2023
Accepted: 28 June 2023
Published: 4 July 2023

1. Introduction

It is difficult for the traditional methods and catch-and-catch methods to meet the actual needs of field operations with the development of the aquaculture industry, particularly with the emergence of large-scale deep-water cages, deep-sea farming boats, and other fish farming models [1]. Therefore, achieving the mechanization and automation of fish captures as well as guaranteeing the catch's survival and non-damage rate is an important topic in the research field of fishery catch equipment [2]. The fish pump is becoming a research hot spot for many domestic scholars as a key piece of equipment to improve the mechanization level of aquaculture. There are various forms of fish pumps, including vacuum pumps, centrifugal impeller pumps, jet pumps, etc. In-depth research has been conducted, and the design and research process of the fish pump has been focused on protecting the fish body from harm, increasing the fish body's survival rate, and increasing the fishing efficiency [3].

The design and development of the fish pump needs to focus on solving the problems of lossless fish body transmission, enhancing adaptability, improving intelligence levels, and establishing continuous fishing systems for pelagic fisheries. Tian Changfeng [4], based on the structural characteristics of single-channel pumps, summarized the design methods of single-channel pumps through a large number of experimental studies and design practices and analyzed the structure of single-channel pump impellers and the reasons for their excellent non-clogging performance. They then proposed a new method for improving the hydraulic design of single-channel pumps. Summerfelt [5] used 3D design software to make a solid model of the fish pump and presented the local details, which laid the foundation for the subsequent design and manufacture of the fish pump. Chu Shupo [6] explored the effect of the volute structure on the performance of the fish pump, studied the characteristics of the pump's internal flow field, and analyzed the effect of the control law of the vacuum outlet form on its performance. Ding Ziyang [7] and Zhang Liang [8] analyzed the pressure and velocity distribution inside the fish as well as the trajectory of the trace in the pump. Numerous scholars have also conducted extensive research on the analysis of the hydrodynamic characteristics of structures. Dutta [9] used the open-source computational fluid dynamics (CFD) modeling tool REEF3D to simulate the oscillatory flow, and the CFD model solves the Reynolds–Averaged Navier–Stokes (RANS) equations in all three dimensions. Further analysis was conducted using CFD to study the effect of blockage ratio on the hydrodynamic characteristics of different oscillatory flow regimes. Dutta [10] used three-dimensional simulations to investigate scour in combined wave-current flows around rectangular piles with various aspect ratios. The simulation model solved the RANS equations using the k–ω turbulence model and included the Exner equation to compute bed elevation changes. The model also used the level-set approach to accurately capture the free surface and included a hydrodynamic module with a morphological module to simulate the scour process.

The currently used fish pump has problems such as a high fish body damage rate, low efficiency, a bulky body, and high energy consumption. This paper aims at improving the hydrodynamic performance of the vacuum fish pump. CFD analysis entails performing fluid simulation analysis on the vacuum pump body and flow channel structure as well as conducting research on the working conditions of fish suction and analyzing the flow field, pressure, speed, and numerical simulation parameters in the flow channel. The dependent variables include the hydrodynamic performance of the fish pump and the influence of fish body damage, the inlet flow velocity of the fish pump, the negative pressure of the pipeline, and the impact force of the water flow on the inner wall of the tank. The independent variables include the working environment of the pump body and the fish pump. The structural form of the fish pump with the optimum performance is obtained using the multi-objective optimization of the fish pump based on the particle swarm optimization algorithm.

2. Overview of Design Scheme

Taking a vacuum fish pump as the research object, Figure 1 depicts the specific structure of this type of vacuum fish pump. The structure of the vacuum fish pump is mainly composed of a power unit, a vacuum pump body, and connecting pipes. The vacuum pump body is the basic unit of the fish-sucking pump structure, and it causes less harm to the fish body when sucking fish [11]. The fluid simulation analysis of the vacuum pump body and the flow channel structure is conducted using the standard turbulent flow model and the Euler multiphase flow method. Additionally, the fish suction working condition is studied, and the flow field, pressure, velocity, and numerical simulation parameters in the flow channel are analyzed. Taking the impact on the hydrodynamic performance of the vacuum fish pump, the inlet flow rate of the fish pump, the negative pressure of the pipeline, and the impact force of the water flow on the inner wall of the tank as the dependent variables, the tank volume, the diameter of the water inlet, and the exhaust speed are selected. The diameter of the mouth, the height, and the angle of the pump body are taken as independent variables, and the multi-objective optimization of the fish pump based on the particle swarm algorithm is used to analyze and obtain the optimal structure of the fish pump.

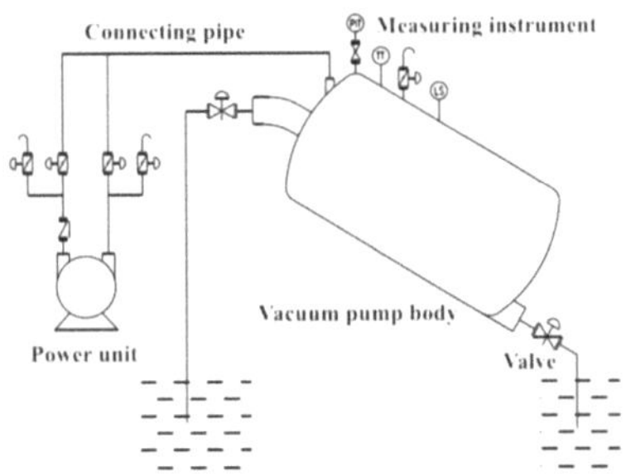

Figure 1. Design scheme of a vacuum fish pump.

3. Basis of Theoretical Analysis

In this paper, we use the commercial computational fluid dynamics software ANSYS fluent for finite element (FEM) analysis and calculation. The Eulerian multiphase model in ANSYS Fluent allows for the modeling of multiple separate yet interconnected phases. The phases can be liquids, gases, or solids in nearly any combination. In contrast to the discrete phase model, each phase receives a Eulerian treatment. The number of secondary phases in the Eulerian multiphase model is exclusively limited by memory needs and convergence behavior. As long as sufficient memory is available, any number of secondary phases can be modeled.

3.1. Volume Fraction Equation

The description of multiphase flow as interpenetrating continua incorporates the concept of phasic volume fractions, denoted here by α_q. Volume fractions represent the space occupied by each phase, and the laws of conservation of mass and momentum are satisfied by each phase individually. The derivation of the conservation equations can be executed by ensemble averaging the local instantaneous balance for each of the phases or by using the mixture theory approach. The volume of phase q, V_q is defined by:

$$V_q = \int_V \alpha_q dV \tag{1}$$

where:

$$\sum_{q=1}^{n} \alpha_q = 1 \tag{2}$$

The effective density of the phase q is:

$$\hat{\rho} = \alpha_q \rho_q \tag{3}$$

where ρ_q is the physical density of phase q, α_q is the volume fraction of phase q.

The volume fraction equation may be solved either through implicit or explicit time discretization.

3.2. Conservation Equations

This section presents the general conservation equations from which the equations are derived, followed by the solved equations.

3.2.1. Conservation of Mass

The continuity equation for phase q is:

$$\frac{\partial}{\partial t}(\alpha_q \rho_q) + \nabla(\alpha_q \rho_q \vec{v}_q) = \sum_{p=1}^{n} (\dot{m}_{pq} - \dot{m}_{qp}) + S_q \quad (4)$$

where $\vec{v}_q$ is the velocity of phase q and $\dot{m}_{pq}$ characterizes the mass transfer from the pth to q^{th} phase, and $\dot{m}_{pq}$ characterizes the mass transfer from phase q to phase p, and you are able to specify these mechanisms separately.

By default, the source term S_q on the right-hand side of Equation (4) is zero, but it can specify a constant or user-defined mass source for each phase. A similar term appears in the momentum and enthalpy equations.

3.2.2. Conservation of Momentum

The momentum balance for phase q yields:

$$\begin{aligned} \tfrac{\partial}{\partial t}(\alpha_q \rho_q \vec{v}_q) + \nabla(\alpha_q \rho_q \vec{v}_q \vec{v}_q) = & -\alpha_q \nabla p + \nabla \bullet \overline{\overline{\tau}}_q + \alpha_q \rho_q \vec{g} \\ & + \sum_{p=1}^{n} (\vec{R}_{pq} + \dot{m}_{pq} \vec{v}_{pq} - \dot{m}_{qp} \vec{v}_{qp}) \\ & + (\vec{F}_q + \vec{F}_{\text{lift},q} + \vec{F}_{\text{wl},q} + \vec{F}_{\text{vm},q} + \vec{F}_{\text{td},q}) \end{aligned} \quad (5)$$

where $\overline{\overline{\tau}}_q$ is the qth phase stress-strain tensor, $\vec{v}_{pq}$ is the interphase velocity, $\vec{g}$ is the acceleration due to gravity, $\vec{F}_q$ is an external body force, $\vec{F}_{\text{lift},q}$ is a lift force, $\vec{F}_{\text{wl},q}$ is a wall lubrication force, $\vec{F}_{\text{vm},q}$ is a virtual mass force, and $\vec{F}_{\text{td},q}$ is a turbulent dispersion force. $\vec{R}_{pq}$ is an interaction force between phases, and p is the pressure shared by all phases.

3.3. *Equations Solved by ANSYS Fluent*

The equations for fluid-fluid and granular multiphase flows, as solved by ANSYS fluent, are presented here for the general case of an n-phase flow.

3.3.1. Continuity Equation

The volume fraction of each phase is calculated from a continuity equation:

$$\frac{1}{\rho_{\text{rq}}}\left(\frac{\partial}{\partial t}(\alpha_q \rho_q) + \nabla(\alpha_q \rho_q \vec{v}_q) = \sum_{p=1}^{n} (\dot{m}_{pq} - \dot{m}_{qp})\right) \quad (6)$$

where ρ_{rq} is the phase reference density, or the volume averaged density of the qth phase in the solution domains. The solution of this equation for each secondary phase, along with the condition that the volume fractions sum to one allows for the calculation of the primary-phase volume fraction.

3.3.2. Fluid-Fluid Momentum Equations

The conservation of momentum for a fluid phase q is:

$$\begin{aligned} \tfrac{\partial}{\partial t}(\alpha_q \rho_q \vec{v}_q) + \nabla(\alpha_q \rho_q \vec{v}_q \vec{v}_q) = & -\alpha_q \nabla p + \nabla \bullet \overline{\overline{\tau}}_q + \alpha_q \rho_q \vec{g} \\ & + \sum_{p=1}^{n} (K_{pq}(\vec{v}_p - \vec{v}_q) + \dot{m}_{pq} \vec{v}_{pq} - \dot{m}_{qp} \vec{v}_{qp}) \\ & + (\vec{F}_q + \vec{F}_{\text{lift},q} + \vec{F}_{\text{wl},q} + \vec{F}_{\text{vm},q} + \vec{F}_{\text{td},q}) \end{aligned} \quad (7)$$

Here $\vec{g}$ is the acceleration due to gravity and $\overline{\overline{\tau}}_q$ is the qth phase stress-strain tensor, $\vec{F}_q$ is an external body force, $\vec{F}_{\text{lift},q}$ is a lift force, $\vec{F}_{\text{wl},q}$ is a wall lubrication force, $\vec{F}_{\text{vm},q}$ is a virtual mass force, and $\vec{F}_{td,q}$ is a turbulent dispersion force. $\vec{R}_{pq}$ is an interaction force between phases, and p is the pressure shared by all phases, K_{pq} is the momentum exchange coefficient between phase p and phase q.

3.4. Approximate Model Optimization Design Method

The Kriging model, second-order response surface methodology (RSM) model, and radial basis Function (RBF) model are used to develop the surrogate model of the structural parameters of the vacuum fish pump and the impact force on the bottom and two sides of the tank body. The RBF model requires plenty of sample points since the prediction accuracy and robustness of the second order RSM model are very poor for highly nonlinear problems [12]. Furthermore, the optimization problem of a vacuum fish pump is often highly nonlinear, and the number of sample points is very limited; thus, it is necessary or even required to use a surrogate model that satisfies all these requirements [13]. Although the Kriging model lacks transparency, its prediction accuracy and robustness are not affected by changes in sample scale; hence, it was chosen to develop the surrogate model.

3.4.1. Kriging Model

The Kriging model is an unbiased estimation model with the smallest estimated variance. It can be based on the dynamic structure of known data samples, fully consider the relevant characteristics of variables within the value range and analyze the trends and dynamics of known data samples. A good fit for nonlinear problems between the response variable and the design variable [14,15]. The Kriging model includes both regression and a nonparametric part.

$$y\,(X) = F(\hat{\beta},\, X) + Z(X) \tag{8}$$

Among them: X is the training sample given by the approximate model; $F(\hat{\beta},\, X)$ is the regression model determined by the known function group about X, which can be expressed as:

$$\begin{aligned} F(\hat{\beta},\, X) &= \hat{\beta}_1 f_1\,(x_1) + \hat{\beta}_1 f_1\,(x_1) \\ &+ \hat{\beta}_n f_n\,(x_n) = \hat{\beta} f^T\,(X) \end{aligned} \tag{9}$$

$\hat{\beta}$ is the regression coefficient; $f_i\,(x_i)$ is the basis function determined in advance; n is the number of sample points of the training sample. $Z(X)$ is a random process with a mean of 0 and a variance of σ^2, and the covariance between two interpolation points is:

$$\mathrm{Cov}\left[Z(x_i), Z(x_j)\right] = \sigma^2 R\left[R(x_i,\, x_j)\right] \tag{10}$$

where: σ^2 is the variance of the random process; R is a symmetric positive definite diagonal matrix of order n x n; $R(x_i,\, x_j)$ is the spatial correlation function of any two sampling points x_i and, x_j among the k sample points.

3.4.2. Second-Order RSM Model

The response surface method is based on the design of experiments, and it uses a specific display function to establish the relationship between the response parameter and the variable. The polynomial model can be used to simulate the real functional relationship in a relatively small area, thus simplifying the complex model [16]. In the actual application process, because there are one or more inflection points in the polynomial response surface approximation model of degree 3 or above that will interfere with the prediction results, the second-order polynomial response surface model is often used in engineering applications, and its function expression is [17]:

$$\hat{y}\,(x) = \beta_0 + \sum_{i=1}^{10} \beta_i x_i + \sum_{i=11}^{10} \beta_i x_{i-10}^2 \tag{11}$$

Among them, the coefficients are calculated by the least square method:

$$\beta = \left(x^T, x\right)^{-1} x^T y = [\beta_1, \beta_2, \ldots \beta_{20}] \tag{12}$$

3.4.3. RBF Model

The radial basis function surrogate model is formed from a series of functions developed by the same method through linear weighted superposition [18], which is characterized by good flexibility, simple structure, and less calculation. The mathematical expression of the radial basis function model is:

$$y = \sum_{i}^{n} w_i \varphi\left(r^i\right) = w^T \varphi \tag{13}$$

Among them, $\varphi = \varphi(r^i) = \varphi(\|x - x_n\|)^T$ is the basis function, and the prediction accuracy obtained by different basis functions is different; $w_i = (w_i, w_2, \ldots w_n)^T$ is the weight coefficient.

3.5. NSGA-II Model

The vacuum fish pump optimization is a multi-objective optimization and multi-attribute decision-making problem. In terms of vacuum fish pump design, the requirements for change in total pressure inside the tank body and change in flow rate at the inlet conflict with each other. The inherent parallel mechanism and global optimization characteristics of genetic algorithms have attracted the interest of researchers in the field of multi-objective optimization. In 1993, Srinivas and Deb proposed a non-dominated sorting genetic algorithm, which has since been widely used in solving numerous problems. However, NSGA has many shortcomings, which make it difficult to obtain satisfactory results when dealing with high-dimensional, multimodal, and other problems. In 2000, Deb made improvements to NSGA and obtained the NSGA-II Algorithm, which further improved the computational speed and robustness of the algorithm. Therefore, the NSGA-II algorithm is used to optimize the structure of a vacuum fish pump.

The basic flow of the NSGA-II algorithm is as follows [19]:

(1) Set the current evolutionary generation $t = 0$, randomly initialize the *t-th* generation population *Pt*, sort all individuals according to the non-domination relationship, and calculate the individual crowding distance.

(2) Select *0.5 N* from *Pt* using the two-way league method to perform crossover and mutation operations on individuals to generate a preserved population.

(3) Merge populations *Pt* and *Qt* to obtain merged population *Rt*, and perform non-dominated sorting on all individuals in population *Rt*, and calculate the individual crowding distance.

(4) Select N individuals from *Rt* according to the sorting results to generate a new population $R(t+1)$, $t = t + 1$.

(5) Judge the relationship between *t* and Gen_{max}. If $t > Gen_{max}$_max, then output R_{t+1}, if $t \leq Gen_{max}$, the algorithm returns to step (2) for cyclic execution.

4. Prototype Experiment

In order to validate the actual working performance of the design model of the vacuum fish pump, a solid prototype of the fish pump was made at a ratio of 1:1, and several field tests of the prototype were conducted in an aquaculture fish tank. The test plan and the test instruments used are shown in Figure 2 below. An electromagnetic flow meter and a pressure gauge are installed near the upper part of the water inlet of the vacuum fish pump to measure the actual flow and pressure of the water inlet. The lifting platform installed at the bottom and the crane changed the suction height and angle of the fish pump for multiple tests.

Figure 2. Experimental setup and scheme.

The main design parameters of the vacuum fish pump are shown in Table 1.

Table 1. Fish pump prototype test model parameters.

Structural Parameters	Value	Structural Parameters	Value
Fish outlet diameter/mm	95	Tank exhaust speed (m^3/min)	0.4–1.8
Air suction port diameter/mm	88	Pump body material	316L Stainless steel
Vacuum pump inlet diameter/mm	75	gross weight/kg	80

The data obtained from the test are shown in Table 2 below. According to the test results, the water suction height and angle of the vacuum fish pump will directly affect the negative pressure and water absorption performance of the fish pump, and according to experience, the main body of the fish pump. The structure will also have a great impact on its working performance [20]. Therefore, a number of parameters that have a greater impact on the fish pump (water absorption height, angle, volume of the pump body, diameter of the water inlet, and diameter of the exhaust port) were selected as self-contained parameters below. As independent variables, consider the flow velocity at the water inlet, the negative pressure value, the impact force of the inner wall of the tank, etc. Additionally, the Latin hypercube sampling method is used to select 167 sets of calculation models, wherein 117 sets of calculation models are selected to develop a Surrogate model, and 50 sets of calculation models are selected to verify the effectiveness of the Kriging model. Finally, multi-objective parameter optimization based on the particle swarm optimization algorithm is used to obtain the optimal structure of the vacuum fish pump.

Table 2. Test data of fish pump prototype.

Test Number	Tank Placement Height (mm)	Angle (°)	Negative Pressure Change Value (Kpa)	Time Interval
1	780	0	−0.8–(−23.9)	12:56:45–12:59:05
2	870	5	−0.8–(−22.9)	13:00:34–13:02:29
3	970	10	−1.1–(−24.1)	13:08:23–13:10:24
4	1760	10	−0.9–(−29.6)	13:39:23–13:41:16
5	1700	5	−0.8–(−29.9)	13:43:03–13:45:13

5. Hydrodynamic Characteristics Analysis and Performance Optimization of Vacuum Fish Pump

5.1. Calculation Instance

5.1.1. The Finite Element Model of the Device

In this paper, we use the commercial computational fluid dynamics software ANSYS FEM analysis and calculation and compute 12 core parallel calculations on a single server. The geometric modeling of the vacuum fish pump is executed using 3D modeling software. Structures such as brackets and flanges are simplified in order to facilitate calculation and simulation. When performing CFD simulation analysis, the SST implicit turbulence model and Euler multiphase flow model are used to set the water inlet of the pump as a pressure inlet and the initial gauge pressure. Set the exhaust port as the mass flow outlet, and set the inner mesh surface of the pump body as the wall boundary. Table 3 summarizes the fluid domain's boundary conditions, and Figure 3 shows the finite element model derived after processing.

Table 3. Fluid domain's boundary conditions.

Physical Boundaries of the Domain	Boundary Conditions
Fluid domain	The fluid domain is a mixed region of air and water, and the volume fraction of the water domain is defined as "0" during initialization
Inlet	Pressure inlet(Set the volume fraction of the water inlet to 0)
Outlet	Mass flow outlet (air-0.021 kg/s (20 °C, 101.3 Kpa))
Wall boundary	Tank wall surface–Main wall Tank bottom wall surface–Bottom wall Other wall surfaces–Wall

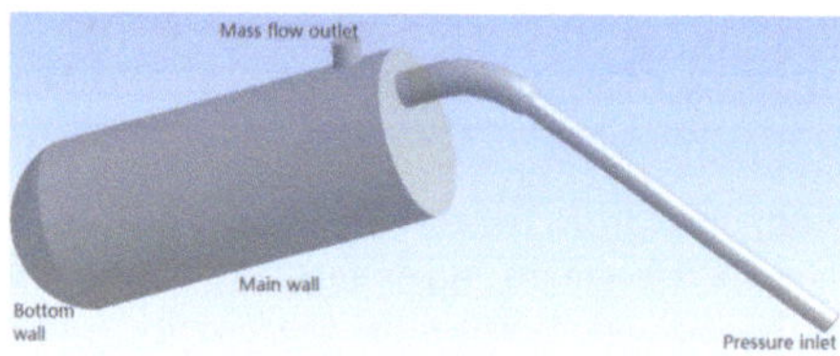

Figure 3. Flow field analysis finite element model.

The grid division of the computational domain is shown in Figure 4. The grid is divided into the form of a structured grid. The grid independence test is performed to determine the specific grid size. Three grid sizes of 0.05 m, 0.03 m, and 0.01 m were selected for numerical simulation. By comparing the calculation results of the three grid sizes, it can be seen that the average calculation error of the 0.03 m grid size relative to the 0.01 m grid size is 4.18%. The average calculation error of the 0.05 m grid size relative to the 0.01 mm grid size is 2.64%. Considering the calculation accuracy and calculation time cost, the main grid size is proposed to be 0.01 m, with a 0.002 m grid size selected for local grid refinement at the stress-concentrated parts. The number of grid cells is 85,285.

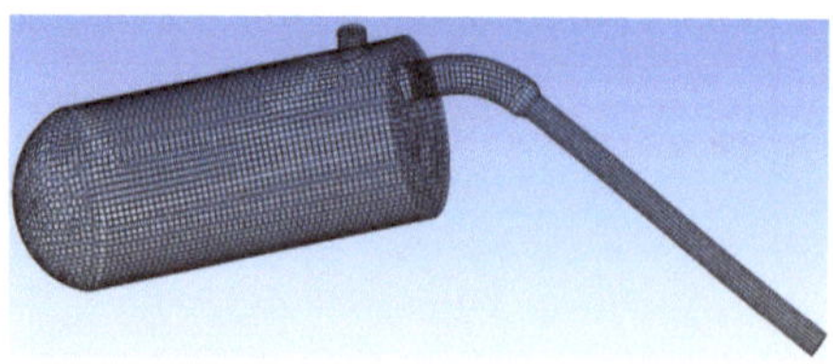

Figure 4. Model mesh division.

5.1.2. Calculation Model Reliability Analysis

The negative pressure value of the water inlet of the vacuum pump in the experimental results above is compared with the value obtained by the calculation simulation, as shown in Figure 5, in order to verify the accuracy and reliability of the CFD calculation model. It can be seen from the figure that the error between the CFD simulation results and the experimental results is 8–12%. The error might be caused by a difference in the layout of the water inlet pipe between the actual experiment and the installation of the measuring instrument. The error is within the acceptable range [21]. The study findings reveal that the accuracy and reliability of the numerical calculation model have been verified.

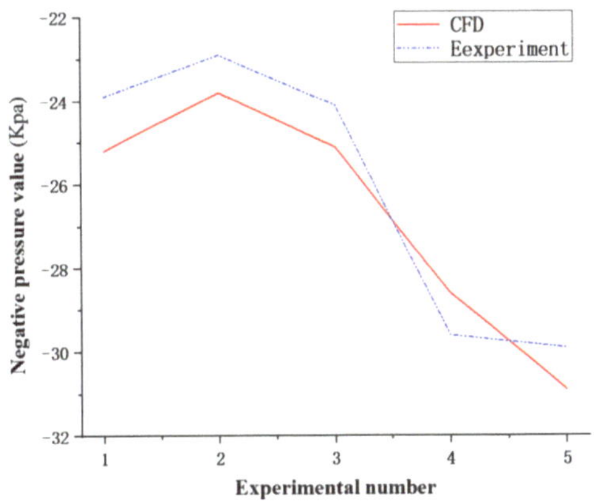

Figure 5. Comparison of experimental values and CFD simulation values.

5.2. Analysis of Hydrodynamic Characteristics of Fish Pump

The internal flow field characteristics have varied performances at different time points due to the mutual influence between the change of the vacuum degree in the tank of the vacuum fish pump and the negative pressure in the tank, the flow velocity, and the impact force of the tank wall. Therefore, it is necessary to analyze the dynamic change process of the flow state and flow field of the vacuum fish pump under different duration conditions through calculation and simulation [22]. The simulation model with the same structural parameters, operating height, and angle as the test prototype model is selected for calculation and analysis. Figures 6–8 below show the internal flow state, flow field velocity, and pressure distribution cloud diagrams of the vacuum fish pump simulation model in three time periods (1 s, 3 s, and 5 s).

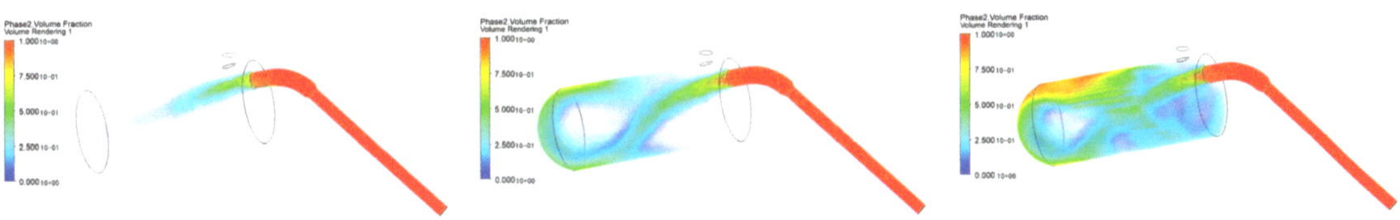

(**a**) 1 s (**b**) 3 s (**c**) 5 s

Figure 6. Instantaneous flow state distribution inside the fish pump. (**a**) 1 s (**b**) 3 s (**c**) 5 s.

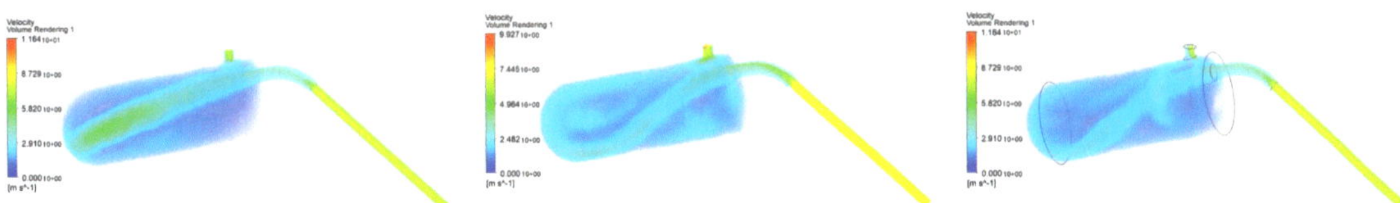

(**a**) 1 s (**b**) 3 s (**c**) 5 s

Figure 7. Instantaneous flow velocity distribution inside the fish pump. (**a**) 1 s (**b**) 3 s (**c**) 5 s.

(**a**) 1 s (**b**) 3 s (**c**) 5 s

Figure 8. Instantaneous pressure distribution inside the fish pump. (**a**) 1 s (**b**) 3 s (**c**) 5 s.

Figures 6–8 shows that when the vacuum fish pump operates for 1 second, a huge negative pressure is generated at the moment when the tank is exhausted. At this time, the negative pressure value of the suction port is about −24 Kpa in the vacuum tank, while the pressure in the body is about −45 Kpa higher than that of the water inlet. As a result, the water flow at the water inlet enters the tank quickly. The direction of the flow velocity entering the tank is approximately parallel to the wall of the tank body, and the incident water flow has a velocity of about 5.2 m/s. When the vacuum fish pump is set to 4 s, the negative pressure value of the tank body tends to be stable, the water flow velocity at the water inlet gradually decreases to about 3.6 m/s, and the direction of the flow velocity entering the tank body gradually approaches the tank wall surface. When the vacuum fish pump is set to 5 s, the negative pressure value of the tank gradually decreases, as does the uniformity of the pressure value distribution in the water intake pipe, resulting in a drop in water flow velocity at the water inlet to about 2.9 m/s. At this time, the flow state of the water entering the tank is relatively dispersed, and the tail flow of the inlet water produces a secondary backflow during the contact and process with the bottom of the tank.

5.3. Performance Optimization of Fish Pump Based on NSGA-II

From the previous discussion, it can be seen that, there is a mutual influence between the negative pressure value inside the vacuum fish pump, the water inlet flow rate, and the impact force of the tank wall under different time durations. The water inlet flow rate and negative pressure value will be beneficial to the enhancement of the fish lifting ability of the fish pump. However, a high-water inlet flow rate and negative pressure value will cause the direction of the jet flow in the tank to be close to the bottom surface of the tank, which will increase the fish pump's lifting ability. The impact force of the fish-water mixed flow, when it enters the vacuum fish pump, will cause greater damage to the fish body, and the excessively high-water inlet flow rate will also increase the collision probability of the fish body when it enters the pipeline, also causing damage to the fish body. This results in an increased damage rate. Therefore, it is necessary to optimize the structural parameters of the vacuum fish pump under the premise of ensuring a low fish body damage rate.

An NSGA-II multi-objective parameter optimization was performed using the impact force value of fish body collision, negative pressure value, water inlet flow rate, the impact force of pool bottom and side wall as dependent variables, and exhaust velocity, pump placement angle, the height of exhaust port diameter, and water inlet diameter as independent variables.

First, the Latin hypercube sampling method is used to sample the independent variables. Table 4 below shows the range of values. Figure 9 shows part of the data obtained by sampling, and a total of 167 sets of calculation parameters were extracted (Chen Xiaolong, 2020).

Table 5 below shows a subset of the data obtained after sampling the independent variables using the Latin hypercube sampling method. A total of 167 sets of calculation parameters were extracted. The data was then entered into Ansys-Fluent for parameter batch modeling and ultimately into the vacuum fish pump calculation model for solution iterations to obtain the dependent variable's solution results.

Table 4. Latin hypercube sampling value.

	Design Argument	Range Value
1	Inlet radius	25–100 mm
2	Outlet radius	15–80 mm
3	Inlet height	300–2000 mm
4	Tank volume	0.12–1.26 m^3
5	Initial negative pressure	Standard atmosphere (101.3 Kpa)
6	Tank exhaust velocity	0.4–1.8 m^3/min (20 °C, 101.3 Kpa)
7	Tank placement Angle	(−15°–15°)

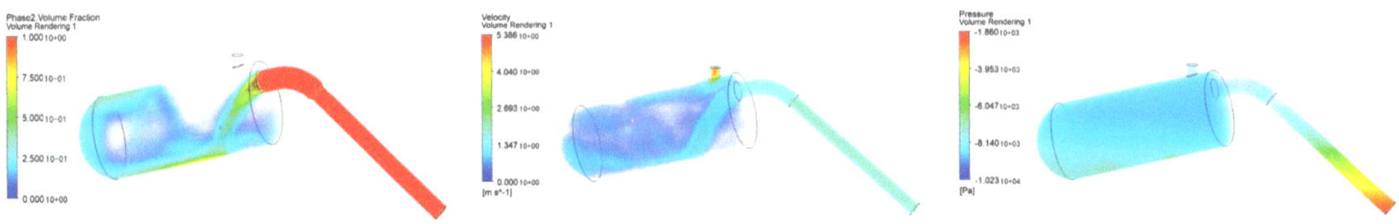

(**a**) Fluid distribution (**b**) Velocity distribution (**c**) pressure distribution

Figure 9. The instantaneous flow state and flow field distribution inside the optimized fish pump. (**a**) Fluid distribution (**b**) Velocity distribution (**c**) pressure distribution.

Table 5. A subset of the data from the Latin hypercube sampling.

P1–Inlet Diameter (mm)	P2–Outlet Diameter (mm)	P3–Tank Height (mm)	P4–Tank Degree (°)	P5–Mass Flow of Inlet (kg s^{-1})
54	44	0	0	0.039
59.6	69.16666667	−218.5185185	11.11111111	0.03612963
51.4	78.7962963	1103.703704	1.111111111	0.021203704
70.83333333	59.53703704	−155.5555556	0	0.038425926
37.5	71.57407407	474.0740741	5.555555556	0.032685185
98.61111111	76.38888889	1040.740741	14.44444444	0.012018519
56.94444444	33.05555556	788.8888889	13.33333333	0.026944444
84.72222222	47.5	348.1481481	7.777777778	0.015462963
76.38888889	54.72222222	1292.592593	−1.111111111	0.013166667
90.27777778	21.01851852	222.2222222	−2.222222222	0.029240741
95.83333333	61.94444444	662.962963	−3.333333333	0.030388889
79.16666667	28.24074074	537.037037	4.444444444	0.017759259
48.61111111	42.68518519	1355.555556	−7.777777778	0.018907407
29.16666667	18.61111111	600	−4.444444444	0.028092593
93.05555556	35.46296296	1166.666667	−5.555555556	0.034981481
68.05555556	40.27777778	977.7777778	3.333333333	0.0235
62.5	23.42592593	−29.62962963	10	0.008574074
87.5	25.83333333	1229.62963	−8.888888889	0.037277778
34.72222222	57.12962963	96.2962963	−6.666666667	0.022351852
65.27777778	49.90740741	914.8148148	−14.44444444	0.024648148
81.94444444	64.35185185	−92.59259259	−10	0.014314815
54.16666667	16.2037037	285.1851852	12.22222222	0.033833333
45.83333333	30.64814815	725.9259259	−13.33333333	0.01087037
73.61111111	52.31481481	159.2592593	−12.22222222	0.016611111
43.05555556	45.09259259	411.1111111	2.222222222	0.025796296
26.38888889	66.75925926	1418.518519	6.666666667	0.031537037
31.94444444	37.87037037	851.8518519	8.888888889	0.009722222

This paper uses RSM, Kriging, and RBF to construct the surrogate model by entering the dependent variable solution results from 167 data sets into the surrogate model. The Kriging algorithm with the highest convergence accuracy is selected by comparing the iterative results of various surrogate models. Table 6 below shows the comparisons of the root mean square for different surrogate models.

The multi-objective parameter optimization based on NSGA-II is then performed after developing the surrogate model. Table 7 below shows the final parameter optimization results.

Table 6. The comparisons of root mean square of different surrogate models.

Model	Item				
	Impact Force on the Bottom and Two Sides of the Tank Body f [N]	**Pressure on the Bottom Surface of the Tank Body pb [Kpa]**	**Pressure on the Walls on Both Sides of the Tank Body ps [Kpa]**	**Change in Total Pressure inside the Tank Body tp [Kpa/s]**	**Change in Flow Rate at the Inlet v [m/s^2]**
Kriging	0.180	0.178	0.178	0.178	0.177
RSM	0.218	0.733	0.697	0.786	0.220
RBF	0.076	8.787	8.772	8.807	0.230

Table 7. Parameter optimization results of particle circle algorithm.

Structure Parameters of Vacuum Fish Pump	Optimal Solution of the Argument
Inlet radius/mm	68.347
Outlet radius/mm	40.025
Inlet height/mm	966.05
Tank placement Angle/°	3.474
Tank exhaust velocity m^3/min	1.093 (20 °C, 101.3 KPa)

Bring the optimal structural parameters of the vacuum fish pump obtained above into the fluent solver once again, then use post-processing to obtain the internal flow field and flow state distribution diagram of the fish pump with a time duration of 3 s, as shown in Figure 9 below. The volume fraction distribution, internal velocity distribution, and pressure distribution of the vacuum fish pump optimization model under the condition of 3 s duration are shown in the figure from left to right. It can be clearly seen from the figure that the direction of the incident water flow in the vacuum fish pump tank is close to the upper end of the tank body, which will reduce the speed of the fish-water mixed flow when entering the tank, thereby reducing the collision damage to the fish body. At this time, the water flow velocity at the water inlet is about 2.5 m/s, and the negative pressure value distribution gradient between the tank body and the water inlet pipeline is relatively uniform, which can achieve good fish suction and fish lifting effects.

6. Conclusions

Taking a vacuum fish pump as the research object, the fluid simulation analysis of the tank and channel construction of the vacuum fish pump was carried out using the combination of CFD and NSGA-II, and the effect of vacuum suction was analyzed. The hydrodynamic performance of the fish pump and multiple structural parameter variables of fish damage were optimized, and the following main conclusions were obtained:

(1) Several prototype experiments were carried out to assess the actual working performance of the design model of the vacuum fish pump. The experimental values were compared with the CFD calculation model, which proved the accuracy and reliability of the calculation model.

(2) The dynamic change process of the flow state and flow field of the vacuum fish pump was analyzed under different chronological conditions through calculation and simulation. There is a mutual influence relationship between the impact forces on the tank wall and the internal flow field characteristics, which have different performances at different time points.

(3) The structural parameters of the vacuum fish pump, including exhaust velocity, tank placement angle, height, exhaust port diameter, and water inlet diameter, were selected. The impact force is the dependent variable. Latin hypercube sampling is utilized

to sample the independent variable in conjunction with the impact force value of the fish body collision damage, and a multi-objective parameter optimization based on the NSGA-II algorithm is performed. The structural parameters of the vacuum fish pump are optimized under the premise of ensuring the fish body damage rate and the structural parameters of the vacuum fish pump with the optimal hydrodynamic performance under 167 sets of parameter values are obtained, and the optimized parameters are substituted into the solver again. The results show that, under the condition of optimal structural parameters, the direction of the incident water flow in the vacuum fish pump tank is close to the upper end of the tank body, which reduces the speed of the fish-water mixed flow when entering the tank, thereby reducing the fish body collision damage. When the water flow velocity at the water inlet is about 2.5 m/s, the negative pressure value distribution gradient between the tank body and the water inlet pipeline is relatively uniform, allowing for good fish suction and a fish lifting effect.

Author Contributions: Y.H.: Experimental design, Data analysis, Writing and Editing; Y.Z.: Data curation, Participate in experiments; C.Z.: Experimental analysis, Guidance; M.Y.: Analysis of results, Visualization; T.J.: Experimental design guidance, Supervision. All authors have read and agreed to the published version of the manuscript.

Funding: This research was funded by National Key R & D Program of China (NO.: 2022YFD2001702).

Data Availability Statement: The author is unable to publish a link to the archived dataset due to data privacy and workplace regulations, and hereby declares.

Conflicts of Interest: The authors declare no conflict of interest.

References

1. Chen, X.; Tian, C.; Liu, X.; Huang, Y.; Che, X.; Yang, J.; Hong, Y. Research progress and development suggestions of fish pump. *Fish. Mod.* **2020**, *47*, 7–11.
2. Guo, X.; Tan, X. Analysis of main structure type and application prospect of fish pump. *Fish. Inf. Strategy* **2013**, *28*, 214–218.
3. Barrut, B.; Blancheton, J.P.; Champagne, J.Y.; Grasmick, A. Water delivery capacity of a vacuum airlift-Application to water recycling in aquaculture systems. *Aquac. Eng.* **2012**, *48*, 31–39. [CrossRef]
4. Tian, C.; Chen, X.; Che, X.; Liu, X. Design and experiment of single tank vacuum suction pump in aquaculture pond. *Fish. Mod.* **2020**, *47*, 39–44.
5. Summerfelt, S.T.; Davidson, J.; Wilson, G.; Waldrop, T. Advances in fish harvest technologies for circular tanks. *Aquac. Eng.* **2018**, *40*, 62–71. [CrossRef]
6. Chu, S.; Xu, Z.; Tang, T.; Wang, Z.; Zhan, Z. Vacuum suction pump for far-reaching Marine culture platform. *Ship Eng.* **2020**, *42*, 68–71.
7. Ding, Z.; Xu, L.; Gao, M.; Wang, G. Design and simulation analysis of vacuum fish pump. *J. Phys. Conf. Ser.* **2021**, *2113*, 012009. [CrossRef]
8. Zhang, L.; Wang, Q.; Li, X. Waveform Prediction of Blade Tip-Timing Sensor Based on Kriging Model and Static Calibration Data. *Math. Probl. Eng.* **2023**, *2023*, 9632212. [CrossRef]
9. Dutta, D.; Afzal, M.S.; Alhaddad, S. 3D CFD Study of Scour in Combined Wave–Current Flows around Rectangular Piles with Varying Aspect Ratios. *Water* **2023**, *15*, 1541. [CrossRef]
10. Dutta, D.; Bihs, H.; Afzal, M.S. Computational Fluid Dynamics modelling of hydrodynamic characteristics of oscillatory flow past a square cylinder using the level set method. *Ocean Eng.* **2022**, *253*, 111211. [CrossRef]
11. Zhang, Y.; Zhou, F.; Li, J.; Kang, J.; Liu, C.; Li, N.; Pan, S. Novel efficient energy saving approach for liquid ring vacuum pump in coal mine gas drainage. *Process Saf. Environ. Prot.* **2023**, *171*, 926–937. [CrossRef]
12. Peng, X.; Xu, L.; Gao, J.; Liu, C.; Zhou, J. Reliability analysis of city gas pipelines subjected to surface loads based on the Kriging model. Proceedings of the Institution of Mechanical Engineers, Part O. *J. Risk Reliab.* **2023**, *237*, 69–79.
13. Liu, A.; Li, Z.; Wang, N.; Zhang, Y.; Krankowski, A.; Yuan, H. SHAKING: Adjusted spherical harmonics adding KrigING method for near real-time ionospheric modeling with multi-GNSS observations. *Adv. Space Res.* **2023**, *71*, 67–79. [CrossRef]
14. Oyewola, O.M.; Petinrin, M.O.; Labiran, M.J.; Bello-Ochende, T. Thermodynamic optimisation of solar thermal Brayton cycle models and heat exchangers using particle swarm algorithm. *Ain Shams Eng. J.* **2023**, *14*, 101951. [CrossRef]
15. Safarik, J.; Snasel, V. Acceleration of Particle Swarm Optimization with AVX Instructions. *Appl. Sci.* **2023**, *13*, 734. [CrossRef]
16. Mohammadi, S.; Hejazi, S.R. Using particle swarm optimization and genetic algorithms for optimal control of non-linear fractional-order chaotic system of cancer cells. *Math. Comput. Simul.* **2023**, *206*, 538–560. [CrossRef]
17. Laskin, A.A.; Raykov, A.A.; Burmistrov, A.V.; Salikeev, S.I. Conductance of Channels of a Dry Screw Vacuum Pump in the Molecular Gas Flow Mode. *J. Mach. Manuf. Reliab.* **2022**, *51*, 520–524. [CrossRef]

18. Kumar, P.; Rao, B.; Burman, A.; Kumar, S.; Samui, P. Spatial variation of permeability and consolidation behaviors of soil using ordinary kriging method. *Groundw. Sustain. Dev.* **2023**, *20*, 100856. [CrossRef]
19. Zhang, D.; Liu, Y.; Kang, J.; Zhang, Y.; Meng, F. The effect of discharge areas on the operational performance of a liquid-ring vacuum pump: Numerical simulation and experimental verification. *Vacuum* **2022**, *206*, 111425. [CrossRef]
20. Wang, J.; Zhao, L.; Zhao, X.; Pan, S.; Wang, Z.; Cui, D.; Geng, M. Optimal design and development of two-segment variable-pitch screw rotors for twin-screw vacuum pumps. *Vacuum* **2022**, *203*, 111254. [CrossRef]
21. Burmistrov, A.; Raykov, A.; Isaev, A.; Salikeev, S.; Kapustin, E.; Fomina, M. Efficiency improvement of Roots vacuum pump working process:Computational fluid dynamics methods modeling. *Vak. Forsch. Prax.* **2022**, *34*, 32–37. [CrossRef]
22. Guo, G.; Zhang, R. Experimental study on pressure fluctuation characteristics of gas–liquid flow in liquid ring vacuum pump. *J. Braz. Soc. Mech. Sci. Eng.* **2022**, *44*, 261. [CrossRef]

Journal of
Marine Science and Engineering

MDPI

Article

Hydrodynamic Response Analysis of a Fixed Aquaculture Platform with a Horizontal Cylindrical Cage in Combined Waves and Currents

Kanmin Shen [1,2], Chunwei Bi [3,*], Zhenqiang Jiang [1,2,4], Shouan Guo [5] and Bin Wang [1,2]

1 Key Laboratory of Far-Shore Wind Power Technology of Zhejiang Province, Hangzhou 311122, China
2 Power China Huadong Engineering Corporation Limited, Hangzhou 311122, China
3 Key Laboratory of Mariculture, Ministry of Education, Ocean University of China, Qingdao 266003, China
4 College of Engineering, Ocean University of China, Qingdao 266100, China
5 State Key Laboratory of Coastal and Offshore Engineering, Dalian University of Technology, Dalian 116024, China
* Correspondence: bichunwei@ouc.edu.cn

Abstract: Biofouling on net cages adversely affects structural safety and the growth of aquacultural fish. Therefore, a novel fixed aquaculture platform with a rotatable horizontal cylindrical cage is proposed in this study, which is convenient for the cleaning of biofouling. Based on ANSYS, the numerical model of the fixed aquaculture platform was established. The response results of the strain, acceleration, and displacement of the structure under the combined action of waves and currents at three typical attack angles were calculated. The effects of water depth and cage rotation on the hydrodynamic response of the structure are discussed. The results show that the strain, acceleration, and displacement of the cage increase with the increase in wave height; however, the change with the wave period is not obvious. The direction perpendicular to the long axis of the cage is the most unfavorable load direction. The acceleration of each position increases with the increase in water depth; however, the strain response has the opposite trend. When the rotation constraint of the horizontal cylindrical cage is released, the acceleration of the cage is larger than that when the cage is fixed. The rotation of the cage has a tiny effect on the structural strain and load acting on the structure.

Keywords: fixed aquaculture platform; hydrodynamic response; attack angle; rotating cage; finite element method

Citation: Shen, K.; Bi, C.; Jiang, Z.; Guo, S.; Wang, B. Hydrodynamic Response Analysis of a Fixed Aquaculture Platform with a Horizontal Cylindrical Cage in Combined Waves and Currents. *J. Mar. Sci. Eng.* **2023**, *11*, 1413. https://doi.org/10.3390/jmse11071413

Academic Editor: Abdellatif Ouahsine

Received: 13 June 2023
Revised: 12 July 2023
Accepted: 12 July 2023
Published: 14 July 2023

1. Introduction

The development of offshore mariculture cannot be separated from the support of large-scale aquaculture facilities and equipment. The seabed of China's Bohai Sea, Yellow Sea, and East China Sea is basically located on the continental shelf, and the water depth of most of the sea areas that can be developed and utilized is within 50 m. Considering the particularity of the sea environment, the applicable water depth for offshore aquaculture facilities in China should not be too deep. Therefore, a fixed aquaculture platform is very suitable for China's industrial needs. Fixed aquaculture platforms have strong resistance to wind and waves, large aquaculture water volume, and are easy to integrate with automatic and intelligent equipment. These platforms have been rapidly developed and applied in China in recent years. In addition, biofouling is a major industrial problem facing aquaculture [1]. When fouling organisms attach to the net, this leads to a low exchange rate for water inside aquaculture facilities and a sharp increase in external environmental loads [2–5], which will directly affect the quality of cultured fish and the structural safety of facilities. However, the main structure of the aquaculture platform is located underwater, and the biofouling attached to the structure is difficult to remove. Biofouling, such as the accumulation of shellfish, hydroids, and marine mollusks, leads to unwanted organisms attaching themselves to aquaculture nets and can multiply quickly, causing a universal problem for aquaculture nets. Biofouling can impair water exchange and increase the risk of disease in farmed fish. In addition, biofouling can significantly increase the hydrodynamic load on the net. Anti-biofouling technology mainly adopts ultrasonic cleaning and high-pressure water cleaning methods. Ultrasonic cleaning, which has low labor costs and high efficiency, requires the use of professional equipment. However, the professional equipment is generally expensive, and some technical problems have not been solved. High-pressure water jet cleaning is a mainly manual operation, and the cost is relatively low. However it is inefficient and has limited performance when operating underwater. If shore-based cleaning operations are desired, the biofouled net needs to be replaced and moved onshore which will increase the cost of the cleaning process and incur additional labor costs. In view of this, this study proposes a fixed aquaculture platform with a cylindrical cage that can rotate around the horizontal axis. The structure can rotate the nets above the water surface and then clean with high-pressure water jets, which greatly improves cleaning efficiency and reduces costs (Figure 1).

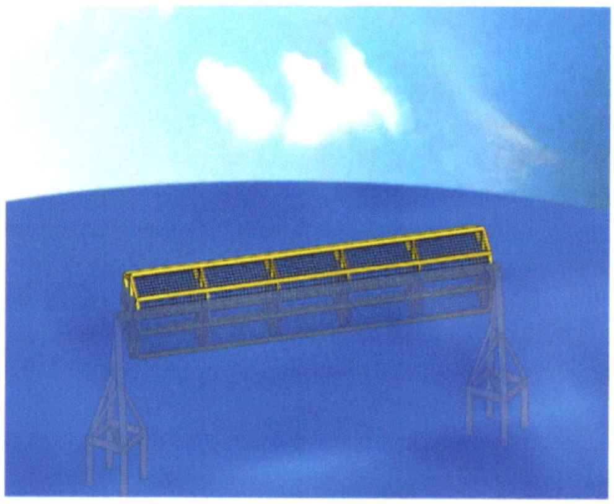

Figure 1. Sketch of the fixed aquaculture platform with a horizontal cylindrical cage.

There are many studies about the dynamic characteristics of different forms of fixed offshore engineering structures under marine environmental loads [6–10]. Different from the traditional marine engineering structure, the emerging aquaculture cage platform structure is more complex. In particular, the existence of the net system further aggravates the interaction between the wave current and the cage structure. At present, many scholars have used various means to research net cages from the aspect of flow fields [11–15], wave field distribution [16–18], and dynamic response [19–26]. However, most of the above studies are aimed at the hydrodynamic characteristics of floating cages. In terms of recent studies, Kim et al. [22] studied the hydrodynamic response of a submersible abalone aquaculture cage under environmental load. Zhao et al. [27] conducted an experimental study of the hydrodynamic characteristics of a column-stabilized fish cage combined with the finite element method and the least squares method. Zhao et al. [28] then numerically investigated nonlinear wave loads on a trestle-netting enclosure aquaculture facility using a finite element model based on stream function wave theory.

In summary, the existing research on fixed aquaculture facilities is still limited. Therefore, in this study, the hydrodynamic characteristics of the proposed fixed aquaculture platform were investigated. The effects of parameters such as attack angle and draft state on the hydrodynamic response were explored. On this basis, the influence of cage rotation was also calculated. The main purpose of the numerical simulations conducted for cages without rotation constraints was to find out the influence of cage rotation on the hydrodynamic response of the structure. The simulation results can provide technical support for the selection of materials and the design of structures. It is expected to provide a reference for engineering applications.

2. Numerical Method

The main components of the fixed aquaculture platform are the frame and the nets. They can be treated as cylinders of different diameters. The ratio of the diameter of the cylinder (D) to wave length (L) satisfies $D/L < 0.2$. Therefore, the load on the structure can be calculated using a Morison equation in combined waves and currents [29].

2.1. Governing Equation

The governing equation is as follows [30]:

$$[M]\{\ddot{u}\}+[C]\{\ddot{u}\}+[K]\{\ddot{u}\}=\{f\}, \tag{1}$$

where $[M]$ represents the mass matrix; $\{\ddot{u}\}$ represents the acceleration vector; $[C]$ represents the damping matrix; $\{\dot{u}\}$ represents the velocity vector; $[K]$ represents the stiffness matrix; $\{u\}$ represents the displacement vector; and $\{f\}$ represents the load vector.

$\{f\}$ is given by [31]:

$$\{f\}=\rho A\{\ddot{v}\}+\rho C_a A\{\ddot{v}-\ddot{u}\}+\frac{1}{2}\rho C_D D_e|\dot{v}-\dot{u}|\{\dot{v}-\dot{u}\}, \tag{2}$$

where ρ is the density of water, A represents the cross-sectional area of the component, C_a represents the added mass coefficient, $\ddot{v}$ represents the acceleration of water particles, C_D represents the drag coefficient, $\dot{v}$ represents the velocity of water particles, and D_e is the diameter of the component. In ANSYS, when waves and currents act together, the parameter KCRC can be used to determine the calculation method of a water particle's velocity.

The inertia coefficient is $C_M = C_a + 1$; thus, Equation (2) is equivalent to the following equation:

$$\{f\}=\rho C_a A\{-\ddot{u}\}+\rho C_M A\{\ddot{v}\}+\frac{1}{2}\rho C_D D_e|\dot{v}-\dot{u}|\{\dot{v}-\dot{u}\}. \tag{3}$$

Thus, the governing equation is as follows:

$$[M]\{\ddot{u}\}+[C]\{\dot{u}\}+[K]\{u\}=\rho C_a A\{-\ddot{u}\}+\rho C_M A\{\ddot{v}\}+\frac{1}{2}\rho C_D D_e|\dot{v}-\dot{u}|\{\dot{v}-\dot{u}\}. \tag{4}$$

Moving the first term from the right to the left in the above equation, the final expression is

$$[M+M_a]\{\ddot{u}\}+[C]\{\dot{u}\}+[K]\{u\}=\rho C_M A\{\ddot{v}\}+\frac{1}{2}\rho C_D D_e|\dot{v}-\dot{u}|\{\dot{v}-\dot{u}\}, \tag{5}$$

where $M_a = \rho C_a A$.

The fifth-order Stokes wave theory [32] is adopted in the numerical simulation. The η, $\dot{v}_X$, and $\dot{v}_Z$ values are given by

$$\eta=\frac{1}{k}\sum_{n=1}^{5}\lambda_n\cos[n(kx-\omega t)], \tag{6}$$

$$\dot{v}_X=c\sum_{n=1}^{5}n\lambda_n\cosh[nk(z+d)]\cos[n(kx-\omega t)], \tag{7}$$

$$\dot{v}_Z=c\sum_{n=1}^{5}n\lambda_n\sinh[nk(z+d)]\sin[n(kx-\omega t)], \tag{8}$$

where η represents the free surface, $\dot{v}_X$ represents the velocity component along the X direction, and $\dot{v}_Z$ represents the velocity component along the Z direction.

According to the above equations, the force of the structure can be calculated, and the displacement, acceleration, and strain can thus be solved.

2.2. Finite Element Model

The numerical model of the structure is simulated by pipe element members with PIPE20 and PIPE59. The PIPE59 element can be used to simulate members with wave and current loads. The element can consider axial torsion, bending, tension, and compression deformation, which is suitable for nonlinear large strain problems. Therefore, the structure above the mud surface is simulated by PIPE59. The PIPE20 element has the functions of tension, compression, torsion, and bending and can be used to simulate the pile–soil interaction [33]. Thus, PIPE20 is used to simulate the structure below the mud surface. The component dimensions of the fixed aquaculture platform model are shown in Table 1. The three views of the fixed aquaculture platform are shown in Figure 2. Pipe_1 is the pile foundation, Pipe_2 is the central rotating shaft, Pipe_3 is the transverse support of the net cage, Pipe_4 is the longitudinal support of the net cage, Pipe_5 is the radial support of the net cage, Pipe_6 is the oblique brace of the tower, Pipe_7 is the supporting rod of the tower, and Pipe_8 is the bearing column.

Table 1. Size of the model structure.

Component	Length (m)	Outside Diameter (m)	Inner Diameter (m)	Quantity
Pipe_1	12	2	1.9	8
Pipe_2	108	2.5	2.38	1
Pipe_3	100	1	0.94	8
Pipe_4	7.67	1	0.94	48
Pipe_5	10	1	0.94	48
Pipe_6	30	2	1.9	4
Pipe_7	16.89	1	0.94	8
Pipe_8	11	2	1.9	8

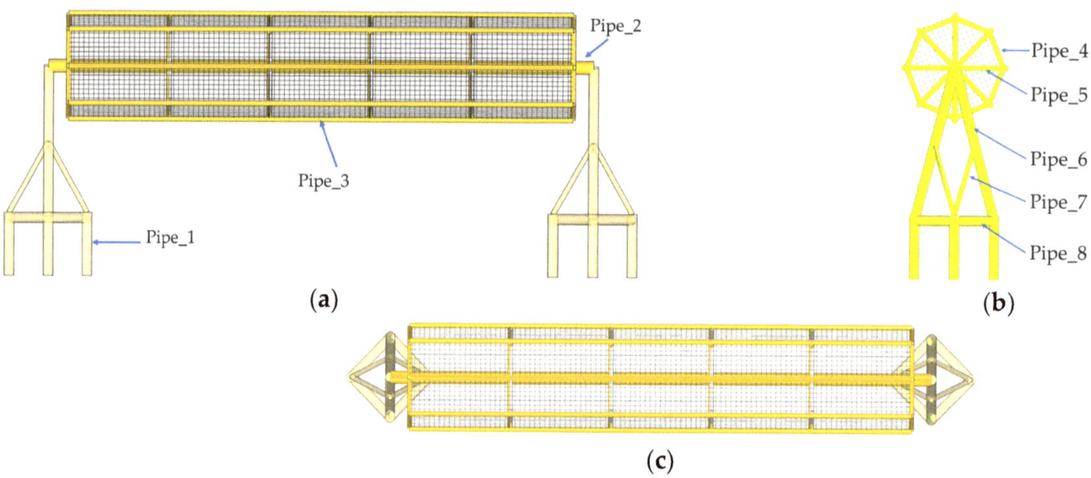

Figure 2. Three-view drawing of the fixed aquaculture platform. (**a**). Main view. (**b**). Lateral view. (**c**). Top view.

The material of the net was polyamide, and PIPE59 was used for simulations. Polyamide has the advantage of high wear resistance. Compared with steel nets, polyamide nets have the advantages of low cost and light weight. In the simulation, the gravity similarity criterion and the model calculation method of variable scale were used. Additionally, the mesh grouping method was adopted to improve calculation speed. The twine diameter was 2 mm, the length of mesh bars was 20 mm, and the net solidity ratio was $S_n = 0.2$.

The numerical model of the fixed aquaculture platform was established using ANSYS. Grid convergence was studied to improve calculation efficiency and ensure calculation accuracy. The grid of the structure is given in Figure 3. The origin of the coordinate system is on the water surface. The direction of the X axis is perpendicular to the axis of the cylindrical cage on the horizontal plane. The direction of the Y axis is parallel to the axis of the cylindrical cage on the horizontal plane. The Z axis is perpendicular to the water surface.

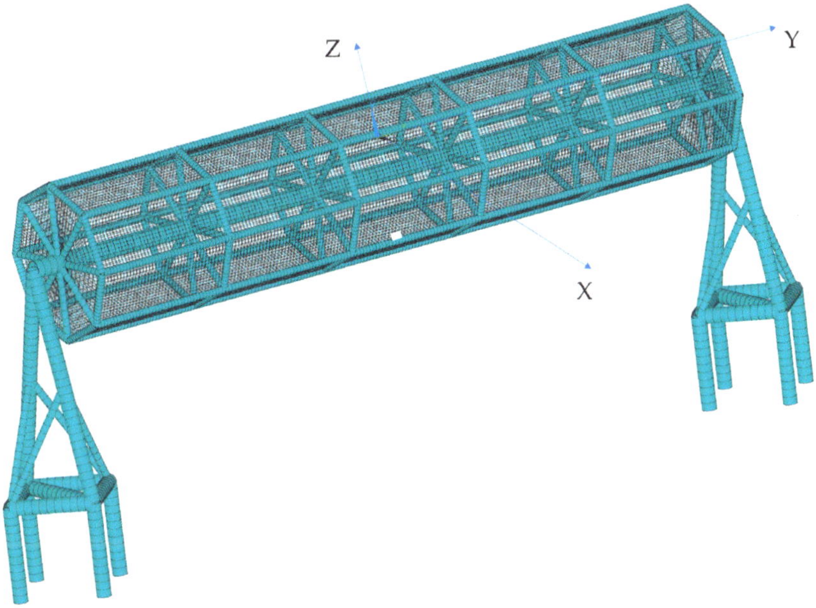

Figure 3. Finite element model of the fixed aquaculture platform with a horizontal cylindrical cage.

2.3. Load Conditions

The water depth of the fixed aquaculture platform was 33 m, and the load was set to the combined action of waves and currents. The prevailing part of the fixed aquaculture platform was located below the water surface. Thus, the wind load had a tiny effect on the hydrodynamic response of the structure and was not considered. The upper velocity was set to 1.93 m/s, the middle velocity was 1.47 m/s, and the bottom velocity was 0.96 m/s. Wave height ranged from 3 to 9 m, the wave period ranged from 9.3 to 12.4 s, and the wave was defined as a Stokes fifth-order wave (see Table 2). The hydrodynamic responses of the fixed aquaculture platform at attack angles of 0°, 45°, and 90° were studied. As shown in Figure 3, the attack angle refers to the angle with the X axis.

Table 2. Marine environmental load conditions.

Case No.	Wave Height H (m)	Wave Period T (s)	Velocity (m/s)	Attack Angle β (°)
A1	3	9.3		0
A2	3	10.8		0
A3	3	12.4	Upper velocity 1.93 m/s	0
A4	6	12.4	Middle velocity 1.47 m/s	0
A5	9	12.4	Bottom velocity 0.96 m/s	0
A6	3	12.4		45
A7	3	12.4		90

3. Experimental Verification

The accuracy of the numerical method was verified through an experiment. The experiment was carried out in the flume of the State Key Laboratory of Coastal and Offshore Engineering at Dalian University of Technology. The flume is 22 m long, 0.8 m wide, and 0.8 m high. The water depth was 0.4 m. The experimental diagram is shown in Figure 4a. The net cage was 0.4 m in diameter and 0.3 m in height, and the frame member was 0.01 m in diameter. The top of the cage was 0.07 m above the water surface. The length of the mesh bar was 12.5 mm, the twine diameter was 2 mm, and net solidity was 0.32. The diameter of the monopile was 0.05 m and the height was 0.64 m, with both ends connected to load cells. The force acting on the structure under wave conditions or current conditions was measured. The force acting on the structure is the sum of results measured by load cells.

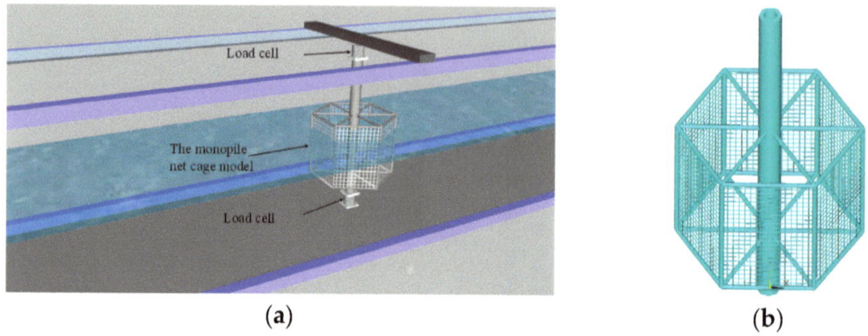

Figure 4. Experimental diagram of the monopile net cage and the numerical model. (**a**). The experimental diagram. (**b**). The numerical model.

3.1. Verification of Current Load

Figure 5 compares the force acting on the model in terms of values obtained from experimental and numerical methods under current load, and the velocity ranges from 0.1 m/s to 0.5 m/s. With the increase in flow velocity, the force on the structure increases significantly. The force on the structure at 0.5 m/s can reach about 15 times that at 0.1 m/s. The results of the experiment and numerical simulation are basically consistent. The results show that the accuracy of the numerical method meets the requirements.

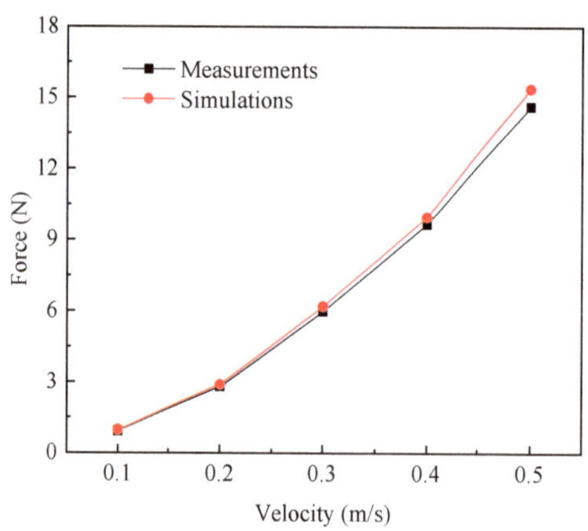

Figure 5. The force acting on the model under current load.

3.2. Verification of Wave Load

Figure 6 compares the experimental and numerical results under wave load. The range of wave height was 0.04 m~0.12 m and the period was 1.3 s. The force acting on the structure increased nonlinearly with the increase in wave height. The trend of the numerical simulation results is consistent with the experimental results, and the maximum error is 5%. It shows that the numerical method used in this paper has high reliability and accuracy and can thus accurately simulate the net cage.

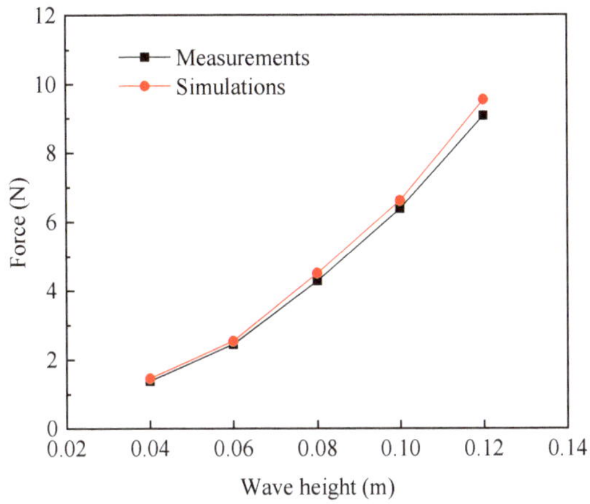

Figure 6. The force acting on the model under wave load.

4. Numerical Results

This section provides the simulation results of the fixed aquaculture platform. The measuring points of strain, acceleration, and displacement are given in Figure 7.

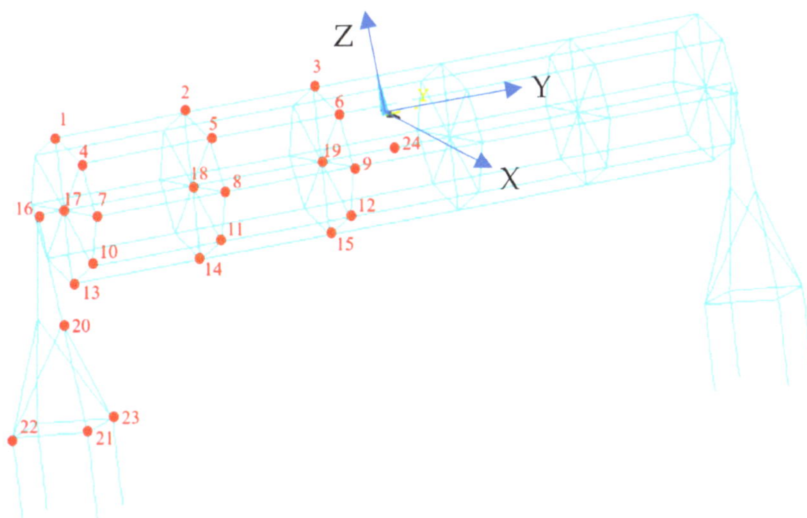

Figure 7. The measuring points of strain, acceleration, and displacement.

4.1. Strain Responses

When the attack angle was 0°, the strain results of the four representative positions (S1, S3, S13, S15) were as shown in Figures 8 and 9.

When T = 12.4 s, the strain peaks of the three representative positions (S1, S2, S3) increase significantly with the increase in wave height (see Figure 10a). When H = 3 m and the period increases from 9.3 s to 12.4 s, the strain peaks at the three representative positions are almost unchanged (Figure 10b).

The phase difference of strain at the top of the net cage is obvious (Figure 11a). However, the phase difference of strain at the bottom of the net cage is not significant (Figure 11b). As the measuring point approaches the middle of the cage, the strain response peak decreases gradually.

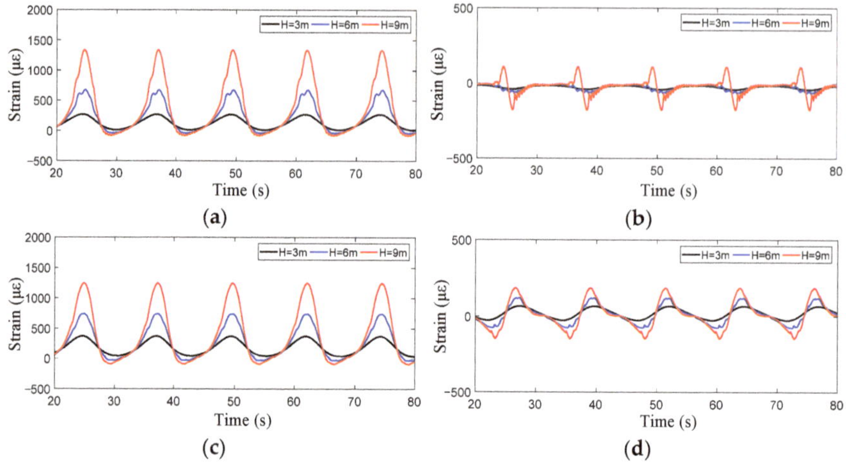

Figure 8. Strain curves with different wave heights (T = 12.4 s). (**a**). Strain curve of S1. (**b**). Strain curve of S3. (**c**). Strain curve of S13. (**d**). Strain curve of S15.

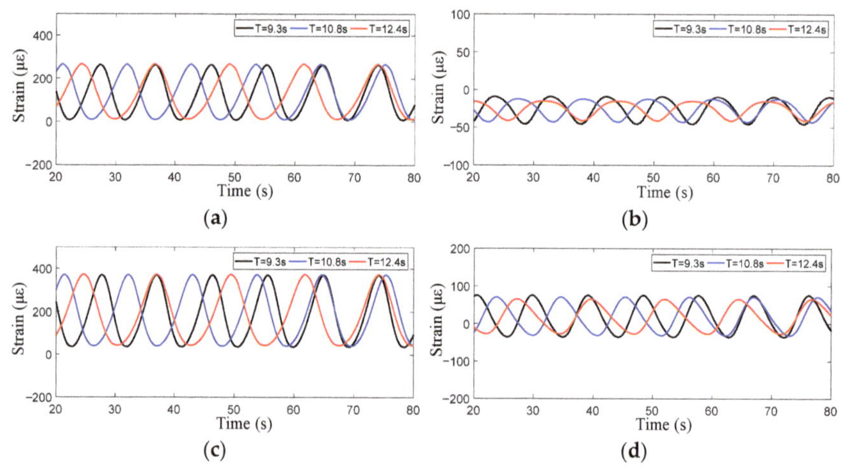

Figure 9. Strain curves with different wave periods (H = 3m). (**a**). Strain curve of S1. (**b**). Strain curve of S3. (**c**). Strain curve of S13. (**d**). Strain curve of S15.

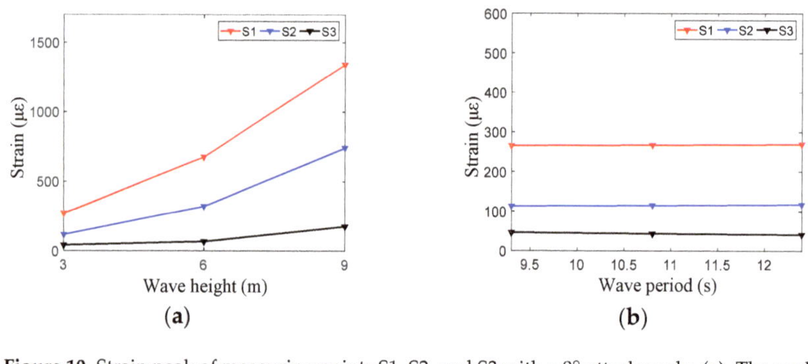

Figure 10. Strain peak of measuring points S1, S2, and S3 with a 0° attack angle. (**a**). The peak value varies with wave height. (**b**). The peak value varies with wave period.

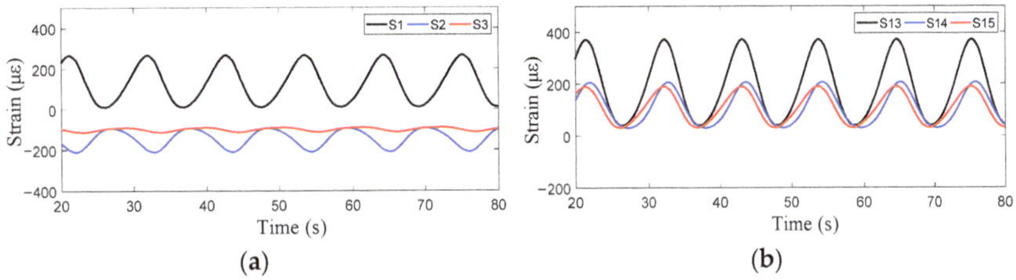

Figure 11. Time history curves of strain for the measuring points (H = 3 m, T = 12.4 s). (**a**). S1, S2, and S3. (**b**). S13, S14, and S15.

The strain curves of measuring points S1, S2, and S3 at the top of the cage and measuring points S13, S14, and S15 at the bottom with different attack angles are shown in Figure 12. When the angle of attack is 45°, the peak strain values of measuring points S1 and S2 at the top of the cage and the corresponding measuring points S13 and S14 at the bottom differ little, while the peak strain values of measuring point S3 at the top of the cage are smaller than the corresponding measuring point S15 at the bottom. When the angle of attack is 90°, the strain peak values of S1 and S2 at the top of the cage are significantly larger than those of S13 and S14 at the bottom, while the strain peak values of S3 at the top of the cage are similar to those of S15 at the bottom.

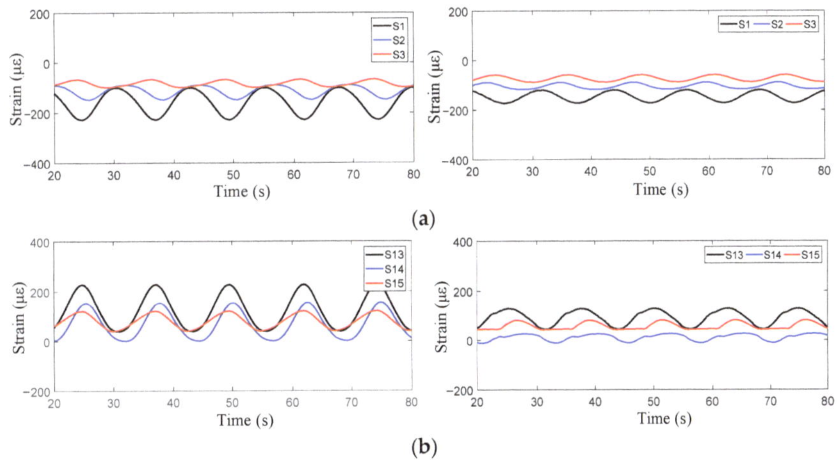

Figure 12. Comparison of strain at different attack angles (H = 3 m, T = 12.4 s). (**a**). Comparison of strain curves at S1, S2, and S3 at attack angles of 45° (**left**) and 90° (**right**). (**b**). Comparison of strain curves at S13, S14, and S15 at attack angles of 45° (**left**) and 90° (**right**).

4.2. Acceleration Responses

Acceleration monitoring of the fixed aquaculture platform is very important for structural safety evaluation. The acceleration curves of measuring points J1, J4, J7, J10, and J13 distributed from the highest position of the cage to the lowest position of the cage were compared (see Figure 13). Under this condition, the acceleration response of the cage initially decreases and then increases from top to bottom.

Representative measuring points J1, J2, and J3 at the upper part of the cage were selected to draw the curves for peak values of acceleration response (see Figure 14). When the period is constant, the wave height has an obvious effect on the peak acceleration. However, the peak acceleration of the measuring points has no obvious change with increases in the period.

The frequency spectrum curves at measuring points J1, J3, J13, and J15 were also compared (see Figure 15). The spectral peak frequency of structural acceleration response is about 0.7 Hz. The frequency domain curves of J1 and J13 are basically consistent.

The acceleration curves of J1 and J3 at other attack angles were then compared (see Figure 16). When the attack angle is 45°, the peak acceleration of J1 is less than J3. When the attack angle is 90°, there is little difference between the peak values of the two measuring points.

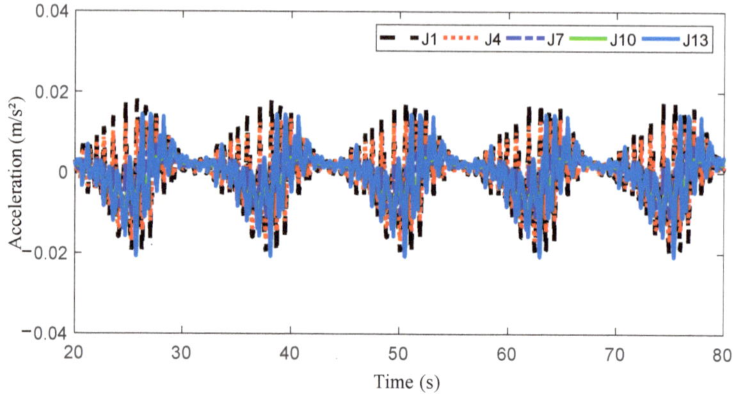

Figure 13. Time history curves of acceleration at J1, J4, J7, J10, and J13 (H = 3 m, T = 12.4 s).

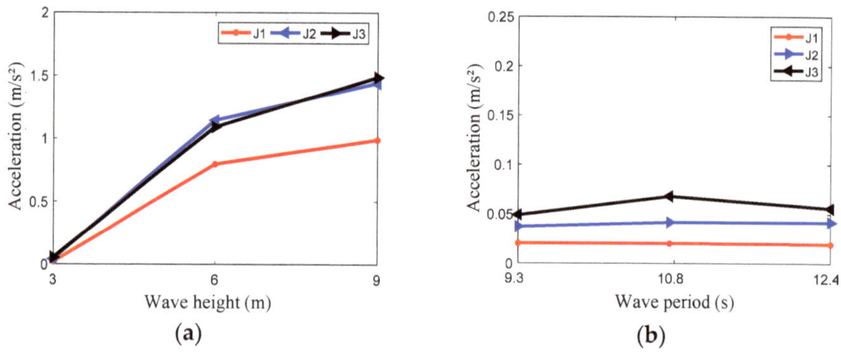

Figure 14. Comparison of peak acceleration at measuring points J1, J2, and J3. (**a**). Peak acceleration when T = 12.4 s. (**b**). Peak acceleration when H = 3 m.

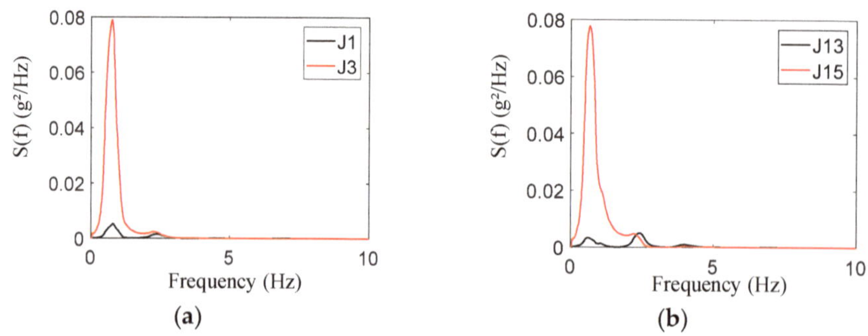

Figure 15. Comparison of acceleration frequency domain responses at J1, J3, J13, and J15. (**a**). Measuring points J1 and J3. (**b**). Measuring points J13 and J15.

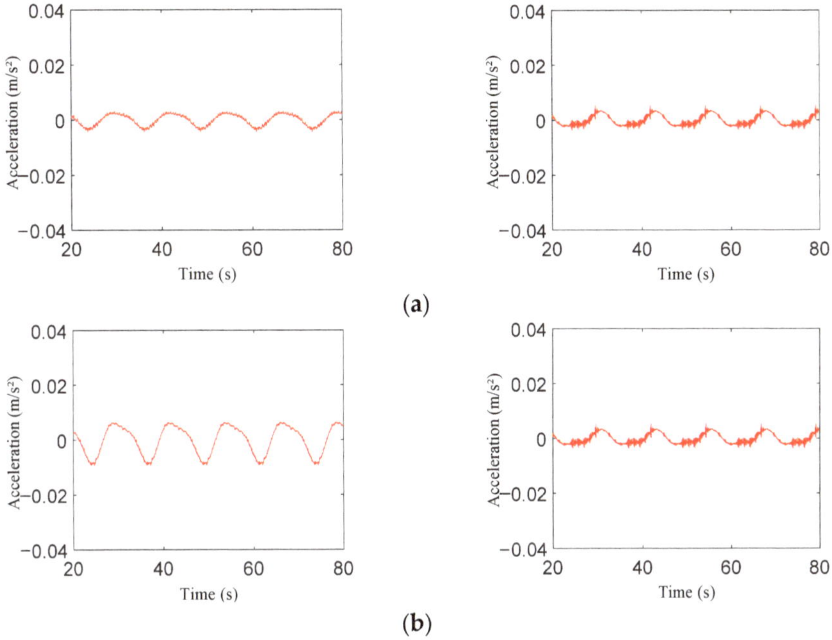

Figure 16. Time history curves of acceleration for J1 and J3 at different attack angles (H = 3 m, T = 12.4 s). (**a**). Comparison of time history curves for J1 at attack angles of 45° (**left**) and 90° (**right**). (**b**). Comparison of time history curves for J3 at attack angles of 45° (**left**) and 90° (**right**).

4.3. Displacement Responses

Taking H = 3 m, T = 12.4 s as an example, the change in displacement response along the height was compared, and measuring points U1, U13, and U17 on the cage and measuring points U16, U20, and U21 on the oblique brace were selected. On the net cage, displacement decreases with the increase in height (Figure 17). However, the peak displacement of the measuring point located in the middle of the oblique brace is the largest.

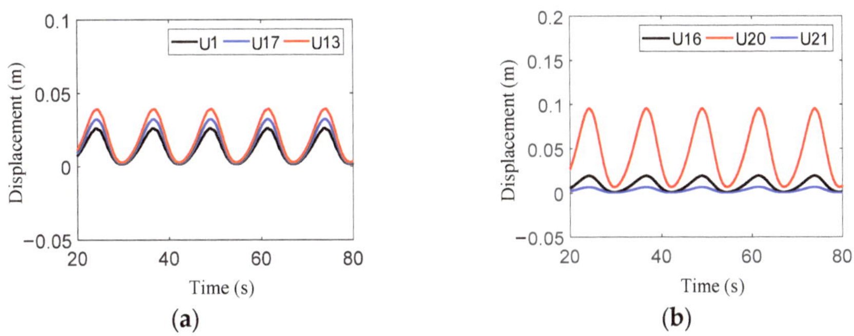

Figure 17. Comparison of displacement time history curves. (**a**). Measuring points U1, U13, and U17. (**b**). Measuring points U16, U20, and U21.

The peak displacement response is less affected by wave period (see Figure 18a). When the period is 12.4 s, it increases linearly with the change in wave height (see Figure 18b). With the change in wave height, the displacement of U1 and U13 increases slightly, while that of measuring points U3 and U15 increases significantly.

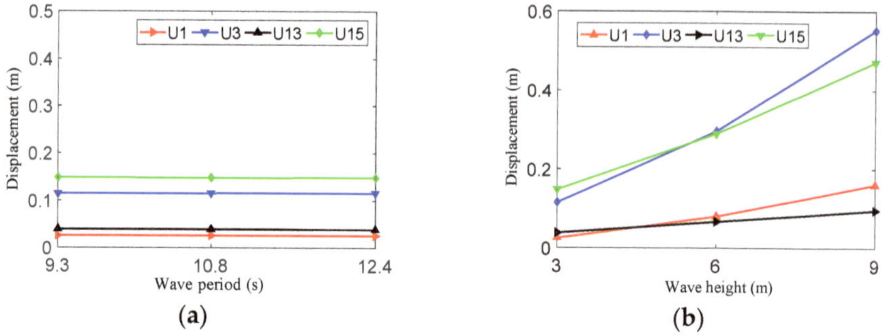

Figure 18. Comparison of peak displacement values with different wave periods and wave heights. (**a**). H = 3 m. (**b**). T = 12.4 s.

The structural displacement curves at different attack angles were compared (Figure 19). The displacement responses of the fixed aquaculture platform are quite different at different attack angles.

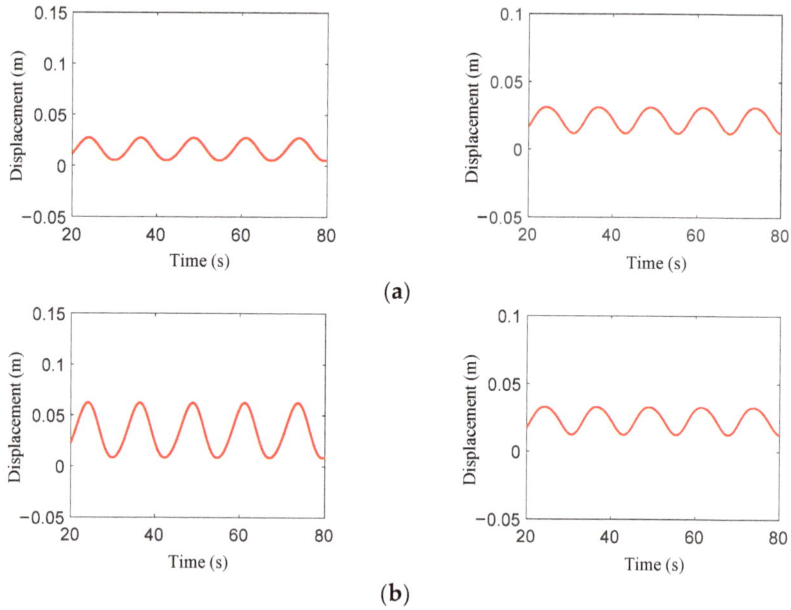

Figure 19. Curves of displacement for U1 and U3 under different attack angles (H = 3 m, T = 12.4 s). (**a**). Comparison of time history curves of U1 at attack angles of 45° (**left**) and 90° (**right**). (**b**). Comparison of time history curves of U3 at attack angles of 45° (**left**) and 90° (**right**).

5. Discussion

5.1. The Influence of Attack Angle on the Dynamic Response of the Structure

The attack angle has a significant influence on the displacement of each position (Figure 20). The displacement of each measuring point of the fixed culture platform is the largest under load in the direction of 0°. The displacement values are smaller under 90° and 45° attack angles and are similar. Most of the members of the fixed aquaculture platform are perpendicular to the load direction at 0°, and the projection area of the net on the plane perpendicular to the load direction is larger than other attack angles, resulting in larger displacement of the structure. At other attack angles, the acting area of the structure is reduced and the displacement is smaller.

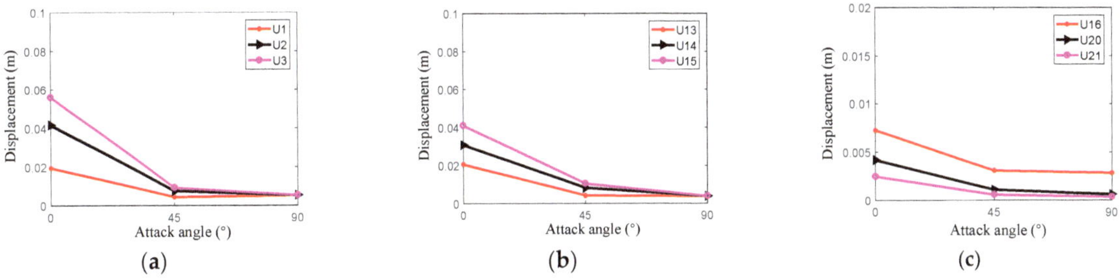

Figure 20. Displacement peak curves of measuring points under different attack angles. (**a**). U1, U2, and U3. (**b**). U13, U14, and U15. (**c**). U13, U14, and U15.

The peak acceleration of the measuring points on the cage is of an order of magnitude equal to the negative quadratic power of 10. The acceleration values reach the maximum at 0° (Figure 21). This is because the major axis of the cage is along the Y axis, and the force area is the largest when the attack angle is 0°. The acceleration responses of most points are very small and almost equal at 90°. Because the force area of the cage is the smallest at 90°, the peak acceleration is the smallest. On the other hand, the acceleration responses at measuring points J1 and J13, which are located at the end of the cage, first decrease and then increase with the increase in attack angle and reach the minimum value when the attack angle is 45°. Therefore, it can be considered that the attack angle of 0° is the most unfavorable direction of load for acceleration.

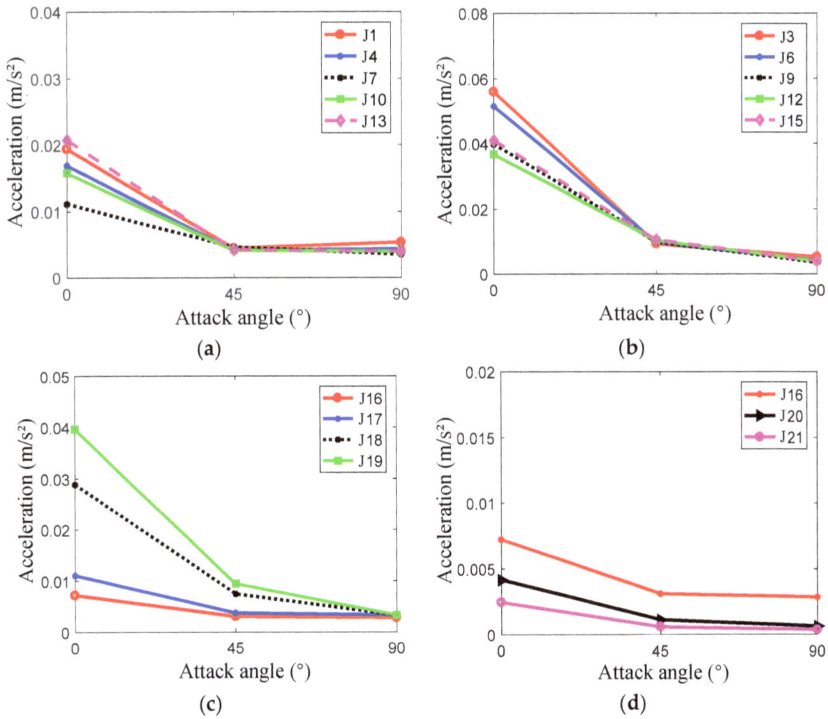

Figure 21. Acceleration peak curves of measuring points under different attack angles. (**a**). Measuring points J1, J4, J7, J10, and J13. (**b**). Measuring points J3, J6, J9, J12, and J15. (**c**). Measuring points J16, J17, J18, and J19. (**d**). Measuring points J16, J20, and J21.

The variation curves of the peak strain at each position at different attack angles are shown in Figure 22. The maximum strain response measuring point is S10 at the attack angle of 0°, S4 at the attack angle of 45°, and S1 at the attack angle of 90°. These measuring points are all at the end of the net cage. Because the force is transferred to both ends, the bending moments at the end are greater and the stress and strain values are also larger. The strain response peak value at each point generally decreases with the increase in attack angle. This is because, when the attack angle is 0°, most of the members of the fixed aquaculture platform are perpendicular to the load direction, and the force area is larger than other attack angles, resulting in larger strain.

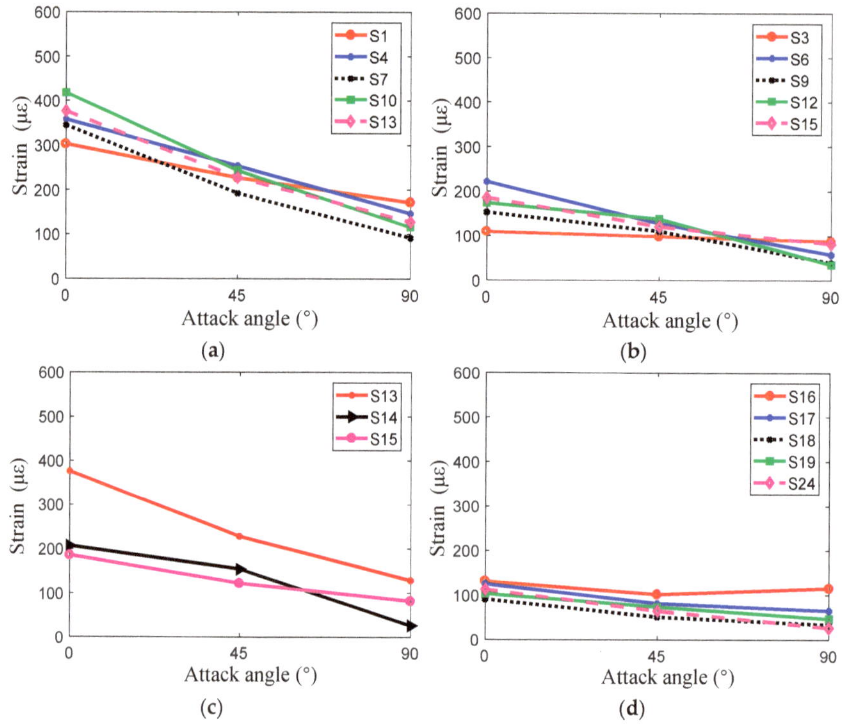

Figure 22. Strain peak curves of measuring points under different attack angles. (**a**). Measuring points S1, S4, S7, S10, and S13. (**b**). Measuring points S3, S6, S9, S12, and S15. (**c**). Measuring points S13, S14, and S15. (**d**). Measuring points S16, S17, S18, S19, and S24.

5.2. The Influence of Water Depth on the Dynamic Response of the Structure

The water level is also an important factor affecting the hydrodynamic response. The extreme water level does great damage to the structure and can cause serious economic losses. Various water levels are designed according to the hydrological data, as shown in Table 3.

Table 3. Different water levels.

Water Level No.	Water Level	Value (m)
A	Extremely high water level	41.73
B	Designed high water level	39.04
C	Normal water level	33.64
D	Designed low water level	30.14
E	Extremely low water level	28.52

Under the extremely high water level, the acceleration values are very small. The acceleration values of measuring points J16 and J17 show little difference at each water level. The acceleration values decrease with the increase in water level (Figure 23). This is because more parts of the structure are submerged into the water, and the influence of water damping thus becomes greater, leading to reduced acceleration values. The acceleration of measuring point J24 is obviously larger than other positions at the central axis of the cage because measuring point 24 is located in the middle of the cage where there is no diagonal bar around it to provide support; thus, its peak acceleration is larger. Along the circumferential direction of the net cage, the peak acceleration of J3 is largest and that of J9 is smallest. Variation in the peak acceleration of J9 with water level is the smallest. At different water levels, the shape of the acceleration frequency spectrum curves is basically the same, and they all have multiple peaks (Figure 24). The frequency corresponding to the maximum peak is about 3 Hz.

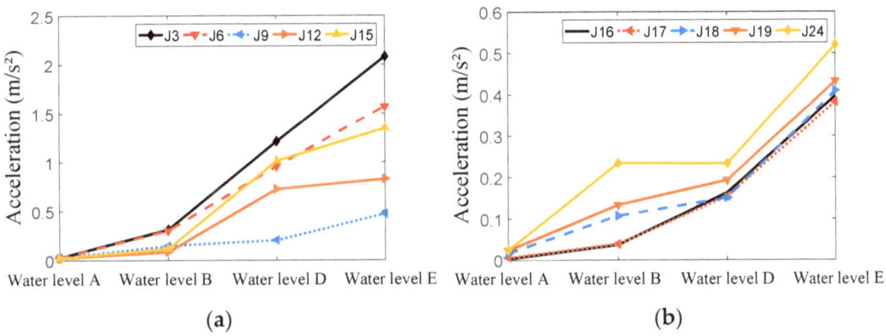

Figure 23. Comparison curves of peak acceleration at different water levels. (**a**). Measuring points J3, J6, J9, J12, and J15. (**b**). Measuring points J16, J17, J18, J19, and J24.

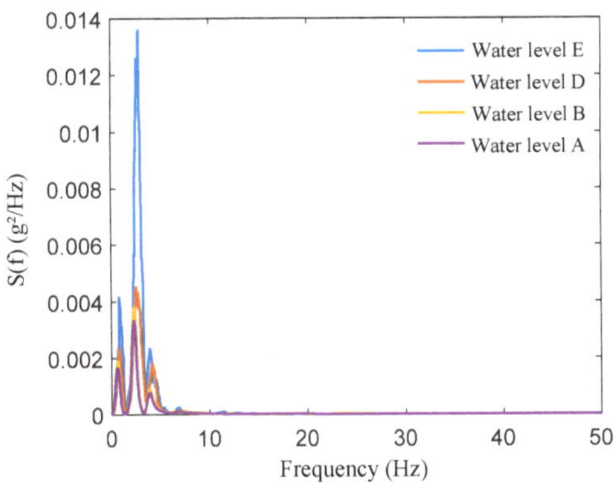

Figure 24. Frequency spectrum curves for J16.

According to Figure 25, the displacement peak values of measuring points U1, U3, U4, and U6 located in the upper part of the cage decrease at first, increase, and then decrease with the increase in water depth. This is because the load only acts on the lower part of the structure at an extremely low water level, and the peak displacement decreases due to the action of water damping when the water level rises. When the water level reaches the designed high water level, the wave load mainly acts on the upper part of the cage and the flow velocity at the water surface is the largest, with the displacement of the measuring point thus reaching the maximum value. When the water level reaches the extremely high water level, the structure is completely submerged in water and the effect of damping is significant, and displacement thus becomes smaller. The displacement peak values of measuring points U10 and U13 at the end of the cage decrease with the increase in water depth. The displacement peak values of measuring points U7, U9, U12, and U15 increase at first and then decrease with the increase in water level.

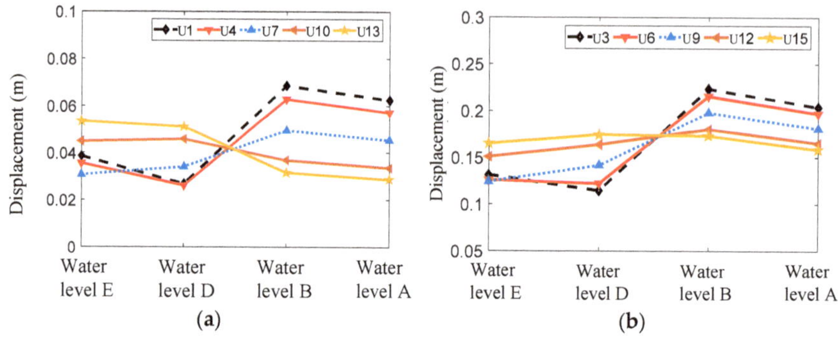

Figure 25. Comparison curves of peak displacement in relation to height under different water levels. (**a**). Measuring points U1, U4, U7, U10, and U13. (**b**). Measuring points U3, U6, U9, U12, and U15.

For the outer frame of the net cage, the strain values decrease with the increase in water level (Figure 26a–c). For the central axis of the net cage, the strain response values also decrease with the increase in water depth (Figure 26d).

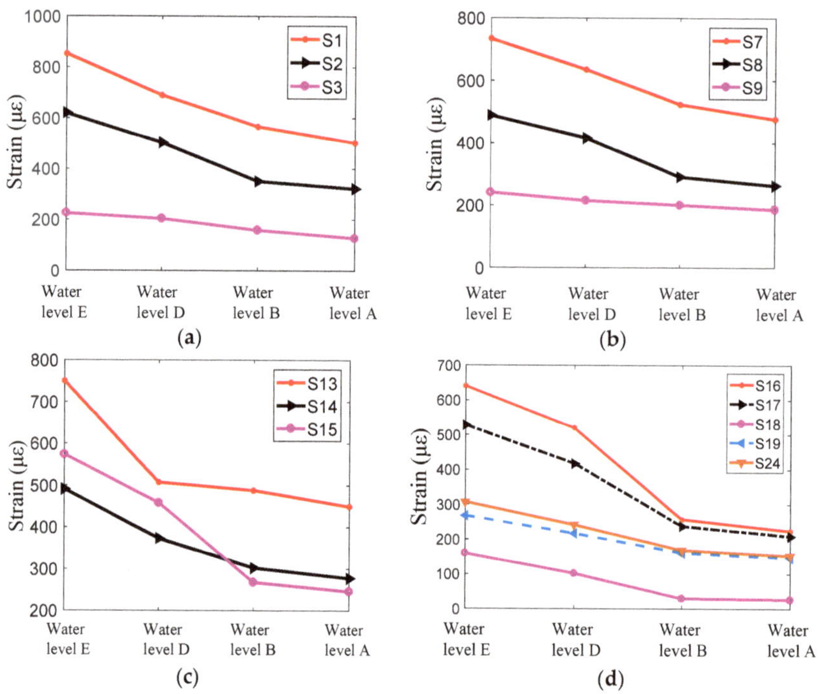

Figure 26. Comparison curves of peak strain response in relation to longitudinal measuring points. (**a**). Measuring points S1, S2, and S3. (**b**). Measuring points S7, S8, and S9. (**c**). Measuring points S13, S14, and S15. (**d**). Measuring points S16, S17, S18, S19, and S24.

In addition, the strain response peak values at measuring points S1, S7, and S13 are much larger than those in the middle of the outer frame of the cage. On the central axis of the cage, the strain response peak value decreases and then increases from the end to the middle position, and the strain response peak value at S18 is the smallest. This is because S16 is the junction of the cage's central axis and the oblique brace of the tower, while S17 is the junction of the cage's central axis and the radial support of the net cage, both of which are places of stress concentration; thus, the strain is larger.

5.3. The Influence of Cage Rotation on the Dynamic Response of the Structure

When cleaning nets in traditional cages, it is necessary to clean manually or after the net has been dismantled, which is time-consuming and laborious. For the fixed aquaculture platform proposed in this paper, the cage can be rotated around the horizontal axis. With rotation of the cage, underwater nets and frames leak out of the water surface to remove biofouling. When the cage rotates, its hydrodynamic response is more complex.

In this part of the numerical simulation, the side support of the structure was removed and only the net cage was calculated. The cage rotates under the action of the environmental load after releasing the rotation constraint around the Y axis.

Whether the cage is fixed or the cage is rotated, the peak acceleration response increases gradually from the end of the cage to the middle of the cage. When the cage is fixed, the peak acceleration changes indistinctly. When the cage rotates, the peak acceleration response of J17, J18, and J19 also changes indistinctly; however, the acceleration of J24 increases greatly (Figure 27). The peak acceleration response of measuring points J17, J18, and J19 is more than 40% higher when the cage is rotated than when the cage is fixed. At J24, the peak acceleration response is 93% higher when the cage is rotated than when the cage is fixed.

Under the two conditions, the acceleration of J24 is transformed using a Fourier transform, and the shape of the curve is basically the same. The frequency corresponding to the peak value is about 0.7 Hz. However, when the cage is rotated, the peak value of the acceleration spectrum curve increases by 25% compared to when the cage is fixed (Figure 28).

Whether the cage is fixed or the cage is rotated, the displacement increases gradually from the end to the middle of the cage. The difference in displacement response is small between the two conditions (Figure 29). The structural frame is made of steel, and the weight of nets cannot be ignored. Therefore, the weight of the structure is very large. In addition, when the cage rotates in water, the water resistance is also significant. Thus, values of displacement at measuring points change indistinctively compared with those without rotation.

Whether the cage is rotated or fixed, the strain response peak value at the measuring point from the end of the cage to the middle decreases at first and then increases, and the strain response peak value is the minimum at measurement point S18. The strain response peak value of each position when the cage is rotating is slightly larger than that when the cage is fixed (Figure 30).

The forces acting on the cage when the cage rotated were compared with those when the cage was fixed (Figure 31). The rotation of the cage has a tiny influence on the force.

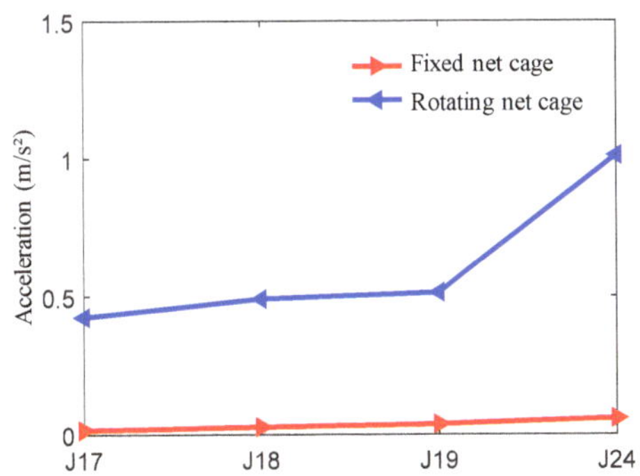

Figure 27. Comparison curve of peak acceleration response.

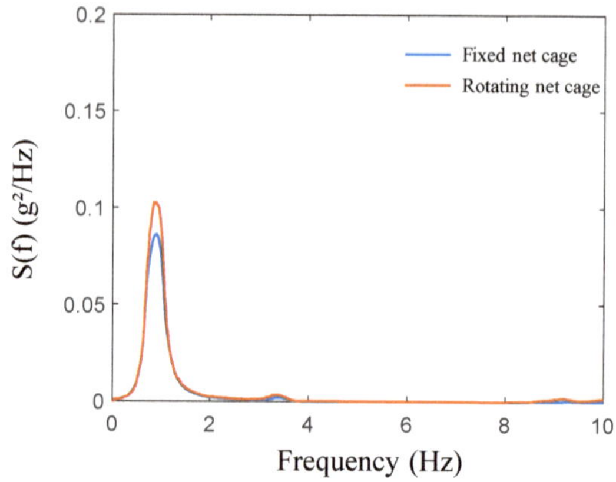

Figure 28. Acceleration frequency domain curve.

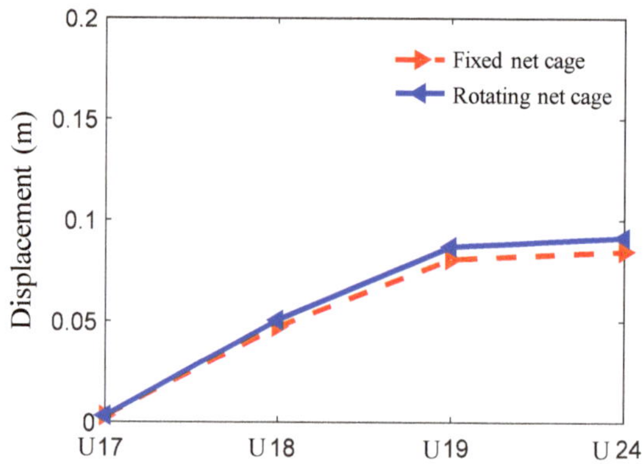

Figure 29. Comparison of peak displacement response between rotatable and fixed net cages.

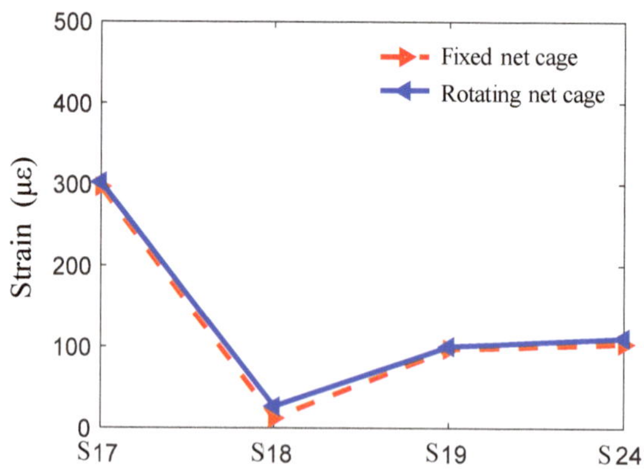

Figure 30. Comparison curve of peak strain response.

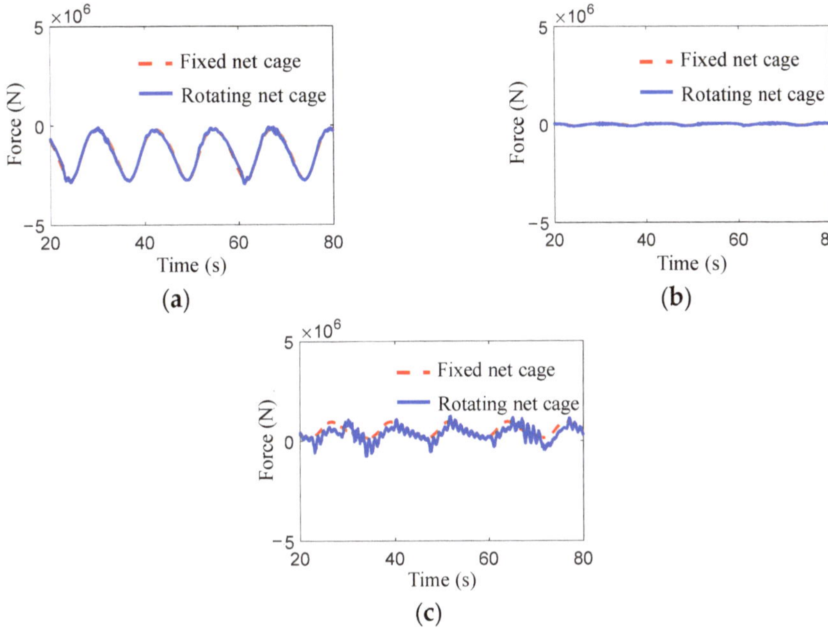

Figure 31. Comparison of force curves acting on the cage. (**a**). The force component along the X direction. (**b**). The force component along the Y direction. (**c**). The force component along the Z direction.

The biofouling on nets often jams the mesh and has many adverse effects on the cultured fish and the safety of facilities. It is recommended to pay attention to regular monitoring of biofouling and ensure timely cleaning to avoid extremely unfavorable conditions for aquaculture production. In addition, materials with antifouling properties should be preferred for nets. For the main structure, antifouling treatment should be carried out to ensure the safety of the cage. Due to the diversity of biofouling in the ocean, any antifouling measures are unlikely to completely eliminate the phenomenon of biofouling. Cleaning biofouling is still a necessary measure in aquaculture production. Therefore, this study proposes a fixed aquaculture platform with a horizontal cylindrical cage which can rotate the net below the water surface into the air, making cleaning more convenient. It can reduce the costs of cleaning biofouling. In addition, compared with the hydrodynamic response of the cage when it is fixed, we find that rotation of the cage driven by waves and currents does not significantly change the load acting on the structure or structural strain.

6. Conclusions

In this study, a novel fixed aquaculture platform is presented in which the net cage can rotate around the horizontal axis to aid in the cleaning of biofouling. The hydrodynamic characteristics of the aquaculture platform were studied numerically. The finite element model of both the structural frame and net were established. The results of strain, acceleration, and displacement under the combined actions of waves and currents at three typical attack angles were calculated. The effects of attack angle, water level, and cage rotation on the hydrodynamic response were discussed. The main conclusions are summarized as follows:

(1) The strain, acceleration, and displacement results of the structure increase with the increase in wave height; however, the change with wave period is not obvious. The acceleration of the structure decreases at first and then increases with the decrease in height. Displacement at the net cage decreases with the increase in height. The strain decreases gradually from the ends of the net cage to the middle of the net cage.

(2) The hydrodynamic response results of the structure are the largest when the attack angle is 0°. Thus, 0° is the most disadvantageous angle to the structure.
(3) The acceleration of each measuring point of the structure increases with the increase in water depth. The largest value of displacement is mostly distributed in the designed low water level and the designed high water level. For the strain response, most of the peak values decrease with the increase in water level, and the strain is largest at the extremely low water level.
(4) When the cage rotates, the acceleration of the cage is larger than when the cage is fixed. However, values of displacement, strain, and the force acting on the cage change indistinctively compared with those without rotation.

This fixed aquaculture platform makes it easier to clean biofouling, and the rotation of the cage does not pose a greater challenge to the safety of the structure. Therefore, this type of structure is more favorable and worth popularizing. There are many aspects that can be further studied about this fixed aquaculture platform, such as the influence of the structure's size and the form of support on the hydrodynamic response of the structure. It is expected to provide a reference for structural design and the formulation of regulatory requirements.

Author Contributions: Conceptualization, C.B.; methodology, K.S. and C.B.; software, S.G.; validation, Z.J.; formal analysis, S.G.; investigation, C.B. and Z.J.; resources, K.S.; data curation, C.B.; writing—original draft preparation, S.G., K.S. and C.B.; writing—review and editing, Z.J. and B.W.; supervision, C.B.; project administration, K.S. and C.B.; funding acquisition, B.W. All authors have read and agreed to the published version of the manuscript.

Funding: This research was funded by the National Natural Science Foundation of China (project nos. 31972843, 52101334, 52071301 and 51939002).

Institutional Review Board Statement: Not applicable.

Informed Consent Statement: Not applicable.

Data Availability Statement: Data will be made available on request.

Acknowledgments: The authors acknowledge technical support from Power China Huadong Engineering Corporation Limited.

Conflicts of Interest: The authors declare no conflict of interest.

References

1. Fitridge, I.; Dempster, T.; Guenther, J.; de Nys, R. The impact and control of biofouling in marine aquaculture: A review. *Biofouling* **2012**, *28*, 649–669. [CrossRef] [PubMed]
2. Bi, C.W.; Chen, Q.P.; Zhao, Y.P.; Su, H.; Wang, X.Y. Experimental investigation on the hydrodynamic performance of plane nets fouled by hydroids in waves. *Ocean Eng.* **2020**, *213*, 107839. [CrossRef]
3. Bi, C.W.; Zhao, Y.P.; Dong, G.H.; Wu, Z.M.; Zhang, Y.; Xu, T.J. Drag on and flow through the hydroid-fouled nets in currents. *Ocean Eng.* **2018**, *161*, 195–204. [CrossRef]
4. Gansel, L.C.; Plew, D.R.; Endresen, P.C.; Olsen, A.I.; Misimi, E.; Guenther, J.; Jensen, Ø. Drag of clean and fouled net panels—Measurements and parameterization of fouling. *PLoS ONE* **2015**, *10*, 0131051. [CrossRef]
5. Pica, D.; Bloecher, N.; Dell'Anno, A.; Bellucci, A.; Pinto, T.; Pola, L.; Puce, S. Dynamics of a biofouling community in finfish aquaculture: A case study from the South Adriatic Sea. *Biofouling* **2019**, *35*, 696–709. [CrossRef]
6. Ali, A.M. Dynamic Behavior of Jacket Type Offshore Structure. *Jordan J. Civ. Eng.* **2012**, *6*, 418–435.
7. Raheem, S.E.A. Nonlinear behaviour of steel fixed offshore platform under environmental loads. *Ships Offshore Struct.* **2016**, *11*, 1–15.
8. Shi, W.; Park, H.; Chung, C.; Baek, J.; Kim, Y.; Kim, C. Load analysis and comparison of different jacket foundations. *Renew. Energy* **2013**, *54*, 201–210. [CrossRef]
9. Shi, W.; Park, H.; Han, J.; Na, S.; Kim, C. A study on the effect of different modeling parameters on the dynamic response of a jacket-type offshore wind turbine in the Korean Southwest Sea. *Renew. Energy* **2013**, *58*, 50–59. [CrossRef]
10. Terro, M.J.; Abdel-Rohman, M. Wave Induced Forces in Offshore Structures Using Linear and Nonlinear Forms of Morison's equation. *J. Vib. Control* **2007**, *13*, 139–157. [CrossRef]
11. Bi, C.W.; Zhao, Y.P.; Dong, G.H. Experimental study on the effects of farmed fish on the hydrodynamic characteristics of the net cage. *Aquaculture* **2020**, *524*, 735239. [CrossRef]
12. Bi, C.W.; Zhao, Y.P.; Dong, G.H.; Zheng, Y.N.; Gui, F.K. A numerical analysis on the hydrodynamic characteristics of net cages using coupled fluid–structure interaction model. *Aquac. Eng.* **2014**, *59*, 1–12. [CrossRef]

13. Gansel, L.C.; Rackebrandt, S.; Oppedal, F.; McClimans, T.A. Flow fields inside stocked fish cages and the near environment. *J. Offshore Mech. Arct. Eng.* **2014**, *136*, 031201. [CrossRef]
14. Patursson, Ø.; Swift, M.R.; Tsukrov, I.; Simonsen, K.; Baldwin, K.; Fredriksson, D.W.; Celikkol, B. Development of a porous media model with application to flow through and around a net panel. *Ocean Eng.* **2010**, *37*, 314–324. [CrossRef]
15. Winthereig-Rasmussen, H.; Simonsen, K.; Patursson, Ø. Flow through fish farming sea cages: Comparing computational fluid dynamics simulations with scaled and full-scale experimental data. *Ocean Eng.* **2016**, *124*, 21–31. [CrossRef]
16. Bi, C.W.; Zhao, Y.P.; Dong, G.H.; Cui, Y.; Gui, F.K. Experimental and numerical investigation on the damping effect of net cages in waves. *J. Fluids Struct.* **2015**, *55*, 122–138. [CrossRef]
17. Lader, P.F.; Olsen, A.; Jensen, A.; Sveen, J.K.; Fredheim, A.; Enerhaug, B. Experimental investigation of the interaction between waves and net structures—Damping mechanism. *Aquac. Eng.* **2007**, *37*, 100–114. [CrossRef]
18. Martin, T.; Tsarau, A.; Bihs, H. A numerical framework for modelling the dynamics of open ocean aquaculture structures in viscous fluids. *Appl. Ocean Res.* **2021**, *106*, 102410. [CrossRef]
19. DeCew, J.; Tsukrov, I.; Risso, A.; Swift, M.R.; Celikkol, B. Modeling of dynamic behavior of a single-point moored submersible fish cage under currents. *Aquac. Eng.* **2010**, *43*, 38–45. [CrossRef]
20. Dong, G.H.; Xu, T.J.; Zhao, Y.P.; Li, Y.C.; Gui, F.K. Numerical simulation of hydrodynamic behavior of gravity cage in irregular waves. *Aquac. Eng.* **2010**, *42*, 90–101. [CrossRef]
21. Huang, C.C.; Tang, H.J.; Liu, J.Y. Dynamical analysis of net cage structures for marine aquaculture: Numerical simulation and model testing. *Aquac. Eng.* **2006**, *35*, 258–270. [CrossRef]
22. Kim, T.; Lee, J.; Fredriksson, D.W.; DeCew, J.; Drach, A.; Moon, K. Engineering analysis of a submersible abalone aquaculture cage system for deployment in exposed marine environments. *Aquac. Eng.* **2014**, *63*, 72–88. [CrossRef]
23. Lader, P.F.; Enerhaug, B. Experimental investigation of forces and geometry of a net cage in uniform flow. *IEEE J. Ocean Eng.* **2005**, *30*, 79–84. [CrossRef]
24. Liu, H.F.; Bi, C.W.; Zhao, Y.P. Experimental and numerical study of the hydrodynamic characteristics of a semisubmersible aquaculture facility in waves. *Ocean Eng.* **2020**, *214*, 107714. [CrossRef]
25. Tsukrov, I.; Drach, A.; DeCew, J.; Swift, M.R.; Celikkol, B. Characterization of geometry and normal drag coefficients of copper nets. *Ocean Eng.* **2011**, *38*, 1979–1988. [CrossRef]
26. Zhao, Y.P.; Li, Y.C.; Dong, G.H.; Gui, F.K.; Teng, B. A numerical study on dynamic properties of the gravity cage in combined wave-current flow. *Ocean Eng.* **2007**, *34*, 2350–2363. [CrossRef]
27. Zhao, Y.P.; Chen, Q.P.; Bi, C.W.; Cui, Y. Experimental investigation on hydrodynamic coefficients of a column-stabilized fish cage in waves. *J. Mar. Sci. Eng.* **2019**, *7*, 418. [CrossRef]
28. Zhao, Y.P.; Bi, C.W.; Chen, Q.P. Numerical investigation of nonlinear wave loads on a trestle-netting enclosure aquaculture facility. *Ocean Eng.* **2022**, *257*, 111610. [CrossRef]
29. Morison, J.R.; Johnson, J.W.; Schaaf, S.A. The force exerted by surface waves on piles. *J. Pet. Technol.* **1950**, *2*, 149–154. [CrossRef]
30. Wilson, J.F. *Dynamics of offshore Structures*; John Wiley & Sons, Inc.: New York, NY, USA, 1984.
31. Chakrabarti, S.K. *Hydrodynamics of Offshore Structures*; Computational Mechanics Publications: Southampton, UK; SpringerVerlag: Heidelberg, Germany, 1987.
32. Fenton, J.D. A Fifth-Order Stokes Theory for Steady Waves. *J. Waterw. Port Coast. Ocean Eng.* **1985**, *111*, 216–234. [CrossRef]
33. Zhao, S.X.; Bi, C.W.; Zhang, D.L.; Yu, H.F. Hydrodynamic response analysis of a 10,000-ton offshore electrical platform in waves using a modified finite element model. *Ocean Eng.* **2021**, *233*, 109194. [CrossRef]

Journal of
Marine Science and Engineering

MDPI

Article

Identification of Large Yellow Croaker under Variable Conditions Based on the Cycle Generative Adversarial Network and Transfer Learning

Shijing Liu [1,2,*], Cheng Qian [1], Xueying Tu [1], Haojun Zheng [3], Lin Zhu [1,*], Huang Liu [1] and Jun Chen [1]

1 Fishery Machinery and Instrument Research Institute, Chinese Academy of Fishery Sciences, Shanghai 200092, China; qiancheng@fmiri.ac.cn (C.Q.); tuxueying@fmiri.ac.cn (X.T.); liuhuang@fmiri.ac.cn (H.L.); chenjun@fmiri.ac.cn (J.C.)
2 Sanya Oceanographic Insitution, Ocean University of China, Sanya 572011, China
3 School of Navigation and Naval Architecture, Dalian Ocean University, Dalian 116023, China; zhenghaojun2023@163.com
* Correspondence: liushijing@fmiri.ac.cn (S.L.); zhulin@fmiri.ac.cn (L.Z.); Tel.: +86-156-1867-6160 (S.L.)

Abstract: Variable-condition fish recognition is a type of cross-scene and cross-camera fish re-identification (re-ID) technology. Due to the difference in the domain distribution of fish images collected under different culture conditions, the available training data cannot be effectively used for the new identification method. To solve these problems, we proposed a new method for identifying large yellow croaker based on the CycleGAN (cycle generative adversarial network) and transfer learning. This method constructs source sample sets and target sample sets by acquiring large yellow croaker images in controllable scenes and actual farming conditions, respectively. The CycleGAN was used as the basic framework for image transformation from the source domain to the target domain to realize data amplification in the target domain. In particular, IDF (identity foreground loss) was used to optimize identity loss judgment criteria, and MMD (maximum mean discrepancy) was used to narrow the distribution between the source domain and target domain. Finally, transfer learning was carried out with the expanded samples to realize the identification of large yellow croaker under varying conditions. The experimental results showed that the proposed method achieved good identification results in both the controlled scene and the actual culture scene, with an average recognition accuracy of 96.9% and 94%, respectively. These provide effective technical support for the next steps in fish behavior tracking and phenotype measurement.

Keywords: fish recognition; CycleGAN; foreground mask loss; maximum mean discrepancy; transfer learning

Citation: Liu, S.; Qian, C.; Tu, X.; Zheng, H.; Zhu, L.; Liu, H.; Chen, J. Identification of Large Yellow Croaker under Variable Conditions Based on the Cycle Generative Adversarial Network and Transfer Learning. *J. Mar. Sci. Eng.* **2023**, *11*, 1461. https://doi.org/10.3390/jmse11071461

Academic Editor: Alexei M. Orlov

Received: 27 June 2023
Revised: 20 July 2023
Accepted: 21 July 2023
Published: 22 July 2023

1. Introduction

The large yellow croaker (*Larimichthys crocea*) is marine migratory fish of the Pacific Northwest [1]. In recent years, due to its high economic value, large yellow croaker has become one of the most commercially valuable marine fishery species in China's aquaculture production [2]. Accurate identification of large yellow croaker under variable conditions is of great significance to improve the ability of the high-throughput detection of fish phenotypes in genetic breeding and aquaculture production [3]. Affected by differences in sampling methods [4,5], illumination [6], and the farming environment [7,8], the images obtained in different farming scenarios have different a domain distribution, which limits the effect of data interoperability and increases the difficulty associated with the industrial application of identification technology. In recent years, with the development of transfer learning [9,10] and person re-ID [11,12], a possible solution for the accurate identification of fish targets under variable working conditions has been provided.

With the progress of information technology such as artificial intelligence and deep learning, the identification technology of production objects, diseases, and behaviors in the agricultural field has been continuously developed and has been widely used in different fields of the industry [13,14]. However, compared with static objects such as rice and plants [15,16], and large land-based animals such as cattle and sheep [17,18], the development of underwater freestyle moving-target recognition technology is slow, and relevant studies have mostly focused on application scenarios where specific working conditions and training data are easy to obtain [19]. To solve this problem, transfer learning technology has been introduced into the field of fish identification. For example, Zhang et al. [6] proposed a transfer learning method based on a residual network to realize unconstrained swimming fish identification. Yuan et al. [20] used a metric learning network based on a residual structure to realize 5-way, 15-shot fish target recognition, and the recognition accuracy was higher than 90%. The method based on small samples and transfer learning can effectively improve the accuracy of fish identification. However, the application of this method has certain limitations when targeting unconstrained swimming fish under obvious actual farming conditions with relatively different backgrounds and postures. This is mainly due to: (1) the variation in sampling device and scene leading to a difference in domain distribution between the target domain and source domain, which results in the ineffective use of available training data in the new recognition domain; and (2) the change in fish swimming posture leading to the dispersion of target features, and a single data source cannot cover all feature spaces, which reduces the adaptability of the algorithm to different features.

Re-identification (Re-ID) is a technology that unifies images from different source domains into the feature space of the target domain through image domain-to-domain conversion to achieve data enhancement. It is mainly used to solve the limitations of supervised methods in the application of real scenes, and has made significant progress in the field of pedestrian re-identification. For example, Wang et al. [21] used attribute features to transfer the model to an unlabeled dataset; Deng et al. [22] embedded a twin network into the CycleGAN [23] to realize image transmission from the source domain to the target domain. Ye et al. proposed RACE (robust anchor embedding) [24] and DGM (dynamic graph co-matching) [25] to solve the video-based unsupervised person re-identification problem. Tang et al. [26] used the CycleGAN and MMD methods to strengthen the retention of pedestrian identity information and narrow the distribution of the domain.

Inspired by this image domain transfer method, we proposed a large yellow croaker recognition method based on cyclic adversarial networks and transfer learning. In this study, large yellow croaker images were collected as source samples in controlled scenes in a specific environment to provide basic image samples for fish recognition in different scenes. The large yellow croaker in the scene to be identified was collected as the target domain sample. The CycleGAN was adopted as the basic model for the transfer from the source domain to the target domain. The foreground mask self-evaluation method [27] was used to optimize the evaluation effect of the model's identity loss. MMD was introduced as the loss function to improve the model's ability to adapt to the distribution of pull and push domains. Then, the expanded sample was used for transfer learning to reduce the influence of uneven distribution of transfer learning sample data on the recognition accuracy, and realize the recognition of a free-swimming large yellow croaker. Finally, the ablation test and comparison test were used to verify the effectiveness of the proposed method.

2. Proposed Method

2.1. Method Overview

In this study, the ReID method was mainly used to unify the style of fish images obtained from different scenes, increase the number of target samples to be identified in the application scene and improve the adaptability of the algorithm. Therefore, in order to preserve identity information and extract the distribution between different domains, we embedded a foreground mask loss and a MMD layer in the CycleGAN to enable image domain-to-domain transfer. In addition, transfer learning has been shown to have advantages in feature reuse, but due to the uneven distribution of pre-trained samples, the performance of the models varies significantly in different recognition tasks. Therefore, in order to increase the recognition ability of the fish features of the transfer learning pre-trained model, we optimized the knowledge transfer process by expanding the fish dataset. The overall framework of the proposed method is shown in Table 1.

Table 1. The results of fish ablation test to source and target. We evaluated the source and target with protocol recall, specificity and mAP (%).

Method	Target to Source			Source to Target		
	Recall	**Specificity**	**mAP**	**Recall**	**Specificity**	**mAP**
No Transfer	52	86	56.2	24.2	90.9	57.6
Direct Transfer	47.5	100	73.8	30.3	100	65.2
CycleGAN	57.5	100	78.8	18.2	97.5	61.7
CycleGAN + IDF	60	100	80	12.1	97.5	58.9
CycleGAN + MMD	65	95	81	24.3	100	65.8
CycleGAN + IDF + MMD	77.5	100	88.7	69.5	97.5	85

As shown in Figure 1, in our algorithm, source domains (specific breeding scenarios) and target domains (ship farming scenarios) were input into the CycleGAN to generate false target domains and false source domains. In a large-scale water mass, due to the relatively sparse spatial distribution of fish, it is easy to obtain background images without fish, and common foreground extraction algorithms can segment the foreground and background more accurately. Therefore, in the conversion process, using the difference in foreground image changes to calculate identity loss, the false source domain identity information could be pulled to the target domain identity information. At the same time, the distribution of the false target domain will be pulled towards the target domain. After the translation was completed, the tagged fish source domain image was transferred to the target domain image to realize the expansion of the fish sample set in the culture scene. Finally, the transfer model was trained with the expanded data to further improve the target recognition accuracy.

2.2. *CycleGAN-Based Translation*

The CycleGAN is an image transformation model based on the generative adversarial network(s), which consists of two pairs of generators and discriminators. G is the mapping function from the source domain to the target domain, and $\hat{G}$ is the mapping function from the target domain to the source domain. D_S and D_T are style discriminators for the source and target domains, respectively. S and T represent the source and destination domains, respectively. The CycleGAN mainly realizes the image conversion of two different domains by minimizing the loss function, so as to realize the multi-modal conversion between domains. The objective function consists of three parts: adversarial loss, cycle consistency loss and identity loss. The purpose of the adversarial loss function is to make the generated image indistinguishable from the real image of the target domain. The adversarial loss is used to maximize the probability of the discriminator to output the image to the generator, which is used to improve the quality of the converted image and make it more realistic. Applying adversarial loss to the two mapping functions, the objectives are expressed as:

$$\mathcal{L}_{GAN}G, D_T, S, T = \mathbb{E}_{t\sim P_{data}t}[logD_T(t)] + \mathbb{E}_{s\sim P_{data}s}[\log(1 - D_T Gs)] \tag{1}$$

$$\mathcal{L}_{GAN}\hat{G}, D_s, T, S = \mathbb{E}_{s\sim P_{data}s}[logD_T(s)] + \mathbb{E}_{t\sim P_{data}t}\left[\log(1 - D_S\hat{G}t)\right] \tag{2}$$

where s and t are the source domain image and the target domain image, respectively. Since the full diversity of the target domain cannot be captured using adversarial loss alone, the generator may produce a limited or repetitive output, and the correct mapping from a single input s to the desired output t cannot be guaranteed. Therefore, the CycleGAN uses cycle consistency loss so that the learned mapping function has periodic consistency. The cycle consistency loss improves the generator's ability to generate images that retain the original image by minimizing the difference between the original input image and the cyclic production image, thereby improving the accuracy of image conversion. The cycle consistency loss is expressed as:

$$\mathcal{L}_{cyc}G,\hat{G} = \mathbb{E}_{s\sim P_{data}s}\left[\left\|\hat{G}(G(s)) - s\right\|_1\right] + \mathbb{E}_{t\sim P_{data}t}\left[\left\|G(\hat{G}(t)) - t\right\|_1\right] \tag{3}$$

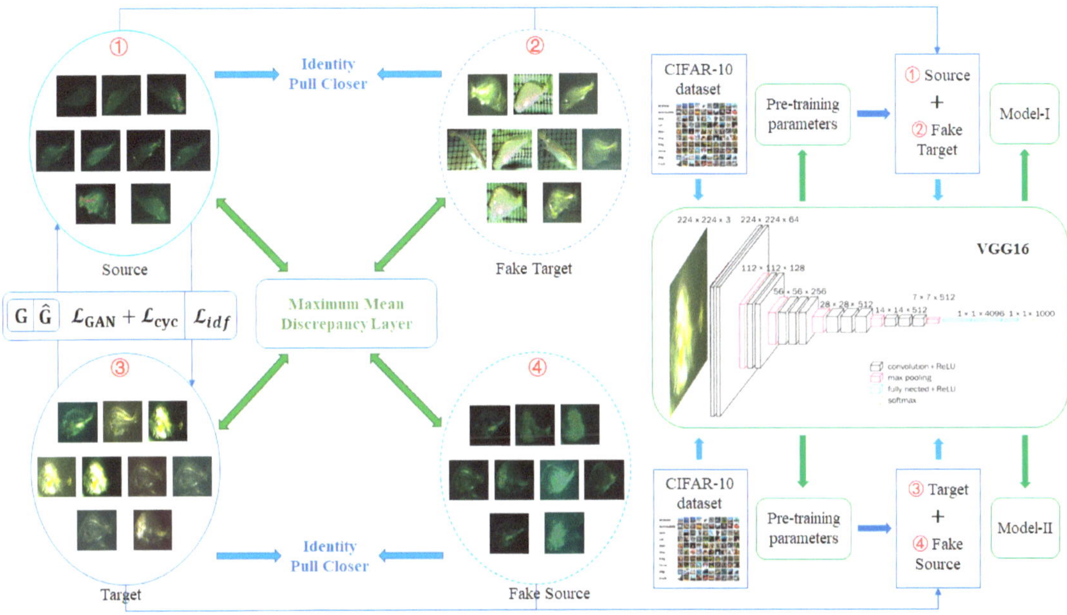

Figure 1. The framework for our method. There were two components of our method (i.e., a data transfer layer and a knowledge transfer layer). The data transfer part was mainly used to achieve sample expansion in the target domain, including CycleGAN, IDF, and maximum mean discrepancy. CycleGAN was mainly used to transfer images from the source domain to the target domain. IDF restricts CycleGAN to retaining fish identity information during the transfer process. The maximum mean discrepancy was used to narrow the distribution between the source and destination domains during transmission. Knowledge transfer was mainly used to improve the ability of the model to recognize the characteristics of fish, and the amplified data were mainly used to increase the effect of the transfer model on knowledge transfer.

2.3. Identity Foreground Loss

As part of the CycleGAN loss function, the identity loss forces the generator to not change the characteristics of the input image, but to maintain its own characteristics. The CycleGAN identity loss only uses global characteristics to count identity loss, and does not consider the impact of background noise on identity information, which leads to the mis-generation of identity during style transfer. However, under actual farming conditions, phenomena such as light absorption, scattering and diffraction caused by water turbidity reduce the feature difference between foreground fish and background noise in the image. This results in unclear identification of the fish after style conversion, which increases the risk of misidentification. To solve this problem and ensure the correct identification of fish as much as possible, we introduced the foreground constraint into the identity loss to evaluate the changes in fish before and after migration. Due to the large volume of aquaculture water and the relatively dispersed distribution of fish under actual farming conditions, the background difference method could be easily used to obtain fish foreground images [27]. Therefore, the fish foreground images were used as the constraint conditions, and Formula (4) was used to calculate the loss of fish identity information.

$$\begin{aligned}\mathcal{L}_{IDF} = & \mathbb{E}_{s\sim P_{data}s}[\|(G(s)-s)\odot M(s)\|_2] \\ & +\mathbb{E}_{t\sim P_{data}t}[\|(\hat{G}(t)-t)\odot M(t)\|_2]\end{aligned} \tag{4}$$

where $M(s)$ and $M(t)$ represent the foreground mask of the fish image with a specific pose, and $\odot$ represents the same or logical operation.

2.4. Maximum Mean Discrepancy

For large yellow croaker images collected under different working conditions, the CycleGAN only transferred the background style of each image from the source domain to the target domain, ignoring the intra-domain distribution differences. The distribution difference provides different reference features for target recognition, which is very important in the task of target recognition with variable characteristics. Maximum mean discrepancy is mainly used to evaluate whether the distribution of two datasets is similar, and in the field of style transfer, it is mainly used to minimize the distribution difference between two networks. Therefore, the maximum mean discrepancy was used to measure the distribution difference between different sampling scenarios to solve the problem of fish sample enhancement.

$$\mathcal{L}_{MMD} = \left[\frac{1}{m(m-1)}\sum_{i\neq j}^{m} k(s_i,s_j) + \frac{1}{n(n-1)}\sum_{i\neq j}^{m} k(t_i,t_j) - \frac{2}{mn}\sum_{i,j=1}^{m,n} k(s_i,t_j)\right]^{\frac{1}{2}} \tag{5}$$

where k is the kernel function, m and n are the number of samples in the source and target domains, respectively, and i and j represent the coordinates of samples in specific domains. As is shown in Formula (6), the Gaussian kernel function was chosen in this paper to calculate the inner product between feature graphs.

$$\mathrm{k}(s,s') = exp\left(-\frac{\|s-s'\|}{2\sigma^2}\right) \tag{6}$$

2.5. Full Objective Function

By combining the CycleGAN, foreground mask loss and maximum mean discrepancy, we could achieve the full objective of CGAN-TM as:

$$\mathcal{L} = \mathcal{L}_{GAN}G, D_T, S, T + \mathcal{L}_{GAN}\hat{G}, D_s, T, S + \lambda_1\mathcal{L}_{cyc}G, \hat{G} + \lambda_2\mathcal{L}_{IDF} + \lambda_3\mathcal{L}_{MMD} \tag{7}$$

The λ_2 and λ_3 control the weights of foreground mask loss and maximum mean discrepancy during the translation process, respectively. Detailed analysis of the parameter sensitivity is presented in Section 4.7.

2.6. Transfer Learning

According to the actual farming conditions, it is difficult to construct a sufficient field sample set according to the change in the farming environment, so the identification of large yellow croaker becomes a small-sample recognition situation. In order to simplify the complexity of the model integration application, this study adopted VGG-16 as the basic transfer learning framework, and used the CIFAR-10 dataset (open dataset, 10 categories, 60,000 images) to pre-train the model. The new training samples composed of the original small-sample data and the migrated data were used to optimize the pre-trained model parameters, and the optimized model was used to realize fish target recognition.

3. Experiments

3.1. Datasets and Evaluation Protocol

In order to evaluate the effectiveness of the method proposed in this paper, we constructed two image sample datasets: source domain and target domain. The source domain samples were collected in a recirculating aquaculture system with a controlled sampling environment, and the target domain samples were collected in an actual farming environment on an aquaculture ship. We took the source domain and the target domain as the identification scenes and verified each one.

Source area image: A total of 360 large yellow croakers with different specifications were placed in the temporary rearing tank. An underwater camera was used and the underwater depth of the camera was 40 cm. The camera was parallel to the water's surface during sampling, and the sampling was continuous for 24 h. A total of 600 images of large yellow croakers in different swimming states were selected to construct a source sample set, including 480 large yellow croaker images for training and 120 images for testing.

Target area image: We selected the "Guoxin 1" aquaculture ship, No. 1 warehouse, to collect the actual farmed fish images. The warehouse is 15 m deep and 8 m in diameter, with a total of approximately 10,000 large yellow croakers. In order to avoid the impact of fish and the influence of circulating water during sampling, a sliding rail was used to conduct continuous sampling at a depth of 4 m underwater for 1 h. A total of 300 images of large yellow croakers were obtained, of which 240 images were used for training and the remaining 60 images were used for testing.

We used VGG-16 as the core framework to verify the effect of fish image transfer and the effectiveness of transfer learning in different domains. We used recall, specificity and the mean average precision (mAP) to evaluate the performance of data transfer on the source domain and target domain. Meanwhile, we selected the recall and mean average precision (mAP) to evaluate transfer learning effects.

3.2. Implementation Details

Our method was implemented using the Pytorch framework. For the CycleGAN, we used foreground mask loss instead of the identity loss function. We calculate the MMD losses using five Gaussian cores with different σ values (0.25, 0.5, 1, 2, 4) and trained them with the CycleGAN. In Equation (7), λ_1, λ_2 and λ_3 were set as 10, 5 and (0.6, 0.8), respectively.

In order to reduce the complexity of the model framework, VGG-16, which is consistent with the CycleGAN, was selected as the transfer learning backbone network and pre-trained on the CIFAR-10 dataset. We used SGD to optimize the model, and the SGD momentum parameter set to 0.9, the weight attenuation parameter was set to 0.0005, and the learning rate was set to 0.0002. In the transfer learning stage, the original data, the generated fake data and the amplified data were used for transfer learning. Due to the small number of model parameters, freezing specific convolution layers had no obvious effect on reducing the training time, so all weight parameters were updated for transfer learning. We set the learning rate of the full connection layer to 0.01, the output dimensions to 2, the batch_size to 16, and the epoch to 60.

The GPU used was RTX A5000, the system used was Windows10, and the Pytorch version used was 1.0. Several randomly selected generated images are shown in Figure 2.

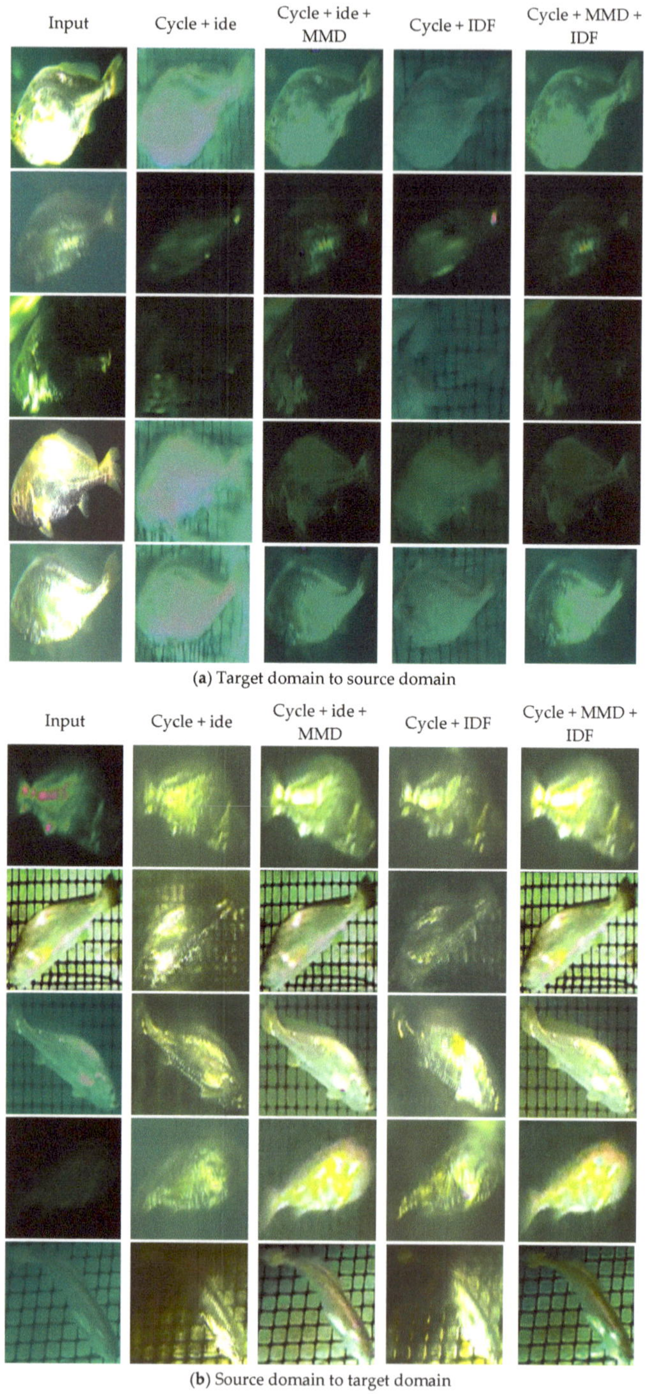

Figure 2. Source domain and target domain ablation test images. From left to right: input, CycleGAN, CycleGAN+ maximum mean discrepancy, CycleGAN+ foreground mask loss, CycleGAN+ maximum mean discrepancy + foreground mask loss (our complete method).

4. Evaluation

The main goal of fish data domain transfer was to expand the training samples, while the goal of transfer learning was to improve the fish recognition rate of a specific sample number. In order to verify the validity of this algorithm, we verified the effect of domain transfer from the source domain to the target domain, and from the target domain to the source domain.

4.1. Performance of Direct Transfer

Due to the insufficient number of samples, the model demonstrated poor performance in the source domain and the target domain. As is shown in Table 1, the recall rate of 52% and 24.24% and the mAP of 56.17% and 57.58%, in the source domain and the target domain, respectively, were achieved. However, in order to expand the number of samples, the performance of the model was slightly improved when the source domain and the target domain were directly migrated. For example, the recall rate of the data migrated to the target domain was 30.3%. Furthermore, due to the poor quality of the data in the target domain, the recall rate decreased by 4.5% after direct transfer to the source domain, and the performance decreased significantly. The main reason for this was that the source domain and target domain samples were collected under different settings, resulting in different domain distributions.

4.2. Effectiveness of the CycleGAN

As the source and target datasets are often collected in different environments, the CycleGAN is able to efficiently generate images with similar styles of datasets. Therefore, we used the CycleGAN to transfer the source domain and target domain image styles to each other, obtain fake source and fake target data. We combined the fake data with the original training data for training. As is shown in Table 1, after adding pseudo-training samples, the recall rate and mAP value of the model in the source domain increased by 10% and 5%, respectively. However, the model recall dropped to 18.2 percent in the target domain, and mAP dropped to 61.65 percent. This was mainly due to the poor quality of the target domain samples and the unsupervised transmission process of the CycleGAN, so the generated images contained a lot of noise and did not take into account the distribution of different datasets.

4.3. Necessity of Identity Foreground Loss

In order to enhance the transfer effect of fish feature information, we introduced identity foreground loss (IDF) into the CycleGAN. As is shown in Figure 2, by supervising the process of identity transfer, IDF reduces the interference of similar background features on the foreground transfer and eliminates the noise in the process of image generation. Finally, it improves the performance of the transfer model in the task of fish sample expansion. As is shown in Table 1, CycleGAN + IDF can increase the source domain recall rate to 60% and the mAP value to 80%. However, the target domain recall rate dropped to 12.1 percent and the mAP value dropped to 58.9 percent. As can be seen from Figure 2, due to the poor image quality of the target domain, the difference between the foreground and background was reduced. However, the CycleGAN + IDF was only concerned with the image difference between two different domains, but did not take into account the image difference between a specific domain, which reduced the transfer effect from the source domain to the target domain, resulting in an obvious loss of fish features in the generated images.

4.4. Importance of Maximum Mean Discrepancy

We embedded the MMD into the CycleGAN with IDF, trying to narrow the distribution by reducing the maximum mean discrepancy between the foreground in different domains. As can be seen from Table 1, the recall rate and mAP value of the model increased to 65% and 81%, respectively, after the transfer of the target domain to the source domain. Furthermore, the increase was 24.25% and 65.75%, respectively, after the transfer of the source domain to the target domain. The results show that embedding MMD loss in the CycleGAN can successfully minimize the distribution differences between different foreground samples, which makes fish target feature extraction more efficient in different datasets. However, it can be seen from Figure 2 that the image generated by adding only MMD loss function for low-quality image data transfer still had local identity feature loss.

4.5. Practicability of Our Method

We verified the practicability of the proposed method by migrating from the source domain to the target domain and from the target domain to the source domain. Obviously, with the CycleGAN, IDF, MMD, the recall rate and mAP accuracy of the final identification results were the highest, reaching 77.5%, 88.75%, 69.5%, and 84.95%, respectively. These results increased by 30%, 15%, 39.2%, and 19.8%, respectively, compared with direct transfer. Since only 300 samples of the target set were selected, the results further prove the practicability of the proposed method in terms of its practical application.

4.6. Parameter Sensitivity

In this study, three parameters, λ_1, λ_2 and λ_3, control the relative importance of three target losses. We evaluated their influence on the mutual transfer between the source and target domains. λ_1 is the original parameter in the CycleGAN, and parameter 10 has been proved to be the optimal choice in the literature [23,26]. In this study, the foreground mask loss was used to optimize the identity loss function in the CycleGAN, so λ_2 could learn from the original parameters. λ_3 is a key parameter controlling MMD loss weight, so this section mainly compared the sensitivity of λ_2 and λ_3; the comparison results are shown in Tables 2 and 3. It is clear that both the foreground identity loss and MMD loss have been proven to be effective compared to the case of $\lambda_2 = 0$ and $\lambda_3 = 0$. From Table 3, we can see that foreground identity loss was positive when the target domain was transferred to the source domain. However, due to the poor image quality of the target domain, the features of the target to be recognized were not obvious. When the image was transferred from the source domain to the target domain, the transfer effect was poor. As can be seen from Table 4, when the weight was small, MMD loss had a significant impact on the recognition effect, and when the weight was large, the recognition effect changed slowly with the weight. Therefore, for different datasets, the values of λ_2 and λ_3 should be carefully selected due to the difference in data quality and domain distribution.

Table 2. The recall and mAP (%) results of different λ_2 values on source and target. λ_1 and λ_3 and are fixed at 10 and 0, respectively.

λ_2	Target to Source		Source to Target	
	Recall	**mAP**	**Recall**	**mAP**
0	57.5	78.75	18.2	61.65
2.5	58.5	79.5	12	58.8
5	60	80	12.1	58.9
7.5	58.75	78.3	11.6	58.3
10	58	70.75	11.2	58.1

Table 3. The recall and mAP (%) results of different λ_3 values on source and target. λ_1 and λ_2 and are fixed at 10 and 5, respectively.

λ_3	Target to Source		Source to Target	
	Recall	**mAP**	**Recall**	**mAP**
0	60	80	12.1	58.9
0.2	65.3	84	30.3	65.2

Table 3. *Cont.*

λ_3	Target to Source		Source to Target	
	Recall	**mAP**	**Recall**	**mAP**
0.4	70.5	85.5	54.2	77.3
0.6	77.5	88.7	60.1	80.9
0.8	77	88.4	69.5	85
1	76	88.1	69.2	84.8

Table 4. Comparison with the state-of-the-art unsupervised methods for source and target. Recall and mAP(%) were selected as the metric protocols. First results are annotated by bold type.

Methods	**Target to Source**		**Source to Target**	
	Recall	**mAP**	**Recall**	**mAP**
PTGAN [27]	60	80	12.1	58.9
CamStyle [28]	71.79	80.3	33.12	70.62
StarGAN [29]	82.05	88.39	24.03	66.95
Our Method	77.5	88.7	69.5	85

4.7. Comparison with State-of-the-Art Methods

We compared the proposed method with state-of-the-art methods, including inter-domain comparative transfer [27] and multi-domain joint transfer [28,29], etc. The experimental results are shown in Table 4. PTGAN (person transfer generative adversarial network) mainly considers domain differences between datasets without considering identity information loss caused by intra-domain deformation. This is similar to the method of only considering IDF loss function in the ablation experiment in this study, resulting in poor performance. CamStyle (camera style) uses label smooth regularization (LSR) to reduce the overfitting risk caused by noisy generated samples, and achieves a good effect in the target domain. However, since the loss of identity difference is not considered, the feature loss of transfer samples seriously reduces the performance of source domain recognition. StarGAN uses the mask vector to optimize the feature differences in different datasets and improve the algorithm's transfer effect among features. However, in the field of underwater free-swimming fish recognition, especially the transfer learning when the features of acquired fish images are seriously lost, the algorithm's transfer recognition effect is poor. In the target domain, recall and mAP were 24.03% and 66.95%, respectively. After the destination domain was transferred to the source domain, the mAP reached 88.39%. Compared with the above methods, this study preserved the identity information in the transmission process by introducing IDF loss, thus eliminating the background noise to a certain extent. Meanwhile, the MMD layer was adopted to learn the distribution of unlabeled datasets, thus successfully reducing the distribution difference between different foreground samples.

4.8. Effectiveness of Transfer Learning

From Table 5, we can see that the recognition accuracy of the original data was higher than that of the fake data, and the recognition accuracy of the amplified data was the highest, with recall reaching 96.5% and 87%, respectively. Overall, the recognition accuracy of fish in the source domain was higher than that in the target domain. On the whole, the fish identification accuracy in the source area was higher than that in the target area. This was mainly due to the low image quality of the target domain, resulting in more loss of identity information during data migration. By comparing the overall recognition accuracy and fish recognition accuracy, we found that, although the recall of false source domain data was low, the mAP value was high. This proves that the background recognition rate was high. It was further demonstrated that data migration effectively distinguished between background and foreground features. On the whole, the transfer learning method effectively improved the target recognition accuracy. The recognition accuracy after the amplification of the source domain and the target domain reached 96.9% and 94%, respectively, which reflects the effectiveness of the combination of data amplification and transfer learning.

Table 5. The results of transfer learning test to source and target. We evaluated the source and target with protocol recall and mAP(%).

Training Data	Recall	mAP
Source	80.3	89.6
F (Target)	36.4	44.3
Source + F (Target)	96.5	96.9
Target	79.2	85
F (Source)	31.8	70.3
Target + F (Source)	87	94

5. Conclusions

In this paper, we proposed an improved CycleGAN and transfer learning method to recognize the large yellow croaker (*Larimichthys crocea*) in a factory ship farming scene. There are still many problems associated with a variable scene recognition task (e.g., the distribution of different datasets cannot be pulled closer during the translation process and a large number of learning samples are difficult to obtain under production conditions). To solve the first problem, we introduced the foreground ID loss and maximum mean discrepancy into the CycleGAN framework. Meanwhile, to enhance the practicality of the technology, we used transfer learning to improve recognition accuracy. We conducted extensive experiments and the results have validated the effectiveness of our method. When compared with state-of-the-art methods, the improved CycleGAN method can achieve competitive performance with a simple framework, and the final test results show that the data amplification method of domain transfer can improve the recognition accuracy of small-sample transfer learning.

Author Contributions: Conceptualization, S.L., H.L. and J.C.; data curation, H.Z.; formal analysis, C.Q. and H.Z.; funding acquisition, S.L.; methodology, S.L.; project administration, S.L. and X.T.; resources, C.Q.; software, C.Q. and H.Z.; writing—original draft, S.L.; writing—review and editing, S.L. and L.Z. All authors have read and agreed to the published version of the manuscript.

Funding: This study was supported by Central Public-interest Scientific Institution Basal Research Fund, CAFS(NO. 2022XT06) and the earmarked fund for CARS 48.

Institutional Review Board Statement: This study complied with the regulations and guidelines established by the Animal Care and Use Committee of Fishery Machinery and Instrument Research Institute, Chinese Academy of Fishery Sciences (FMIRI-AWE-2022-001, approved on 30 September 2022).

Informed Consent Statement: Not applicable.

Data Availability Statement: Not applicable.

Acknowledgments: The authors would like to express their thanks to L.Z. of the Fishery Machinery and Instrument Research Institute, Chinese Academy of Fishery Sciences, for reviewing this article. The authors are thankful for the financial support received from the Central Public-interest Scientific Institution Basal Research Fund, CAFS(NO. 2022XT06) and the earmarked fund for CARS 48.

Conflicts of Interest: The authors declare no conflict of interest.

References

1. Wu, Y.; Yu, X.; Suo, N.; Bai, H.; Ke, Q.; Chen, J.; Pan, Y.; Zheng, W.; Xu, P. Thermal tolerance, safety margins and acclimation capacity assessments reveal the climate vulnerability of large yellow croaker aquaculture. *Aquaculture* **2022**, *561*, 738665. [CrossRef]
2. Bai, Y.; Wang, J.; Zhao, J.; Ke, Q.; Qu, A.; Deng, Y.; Zeng, J.; Gong, J.; Chen, J.; Pan, Y.; et al. Genomic selection for visceral white-nodules diseases resistance in large yellow croaker. *Aquaculture* **2022**, *559*, 738421. [CrossRef]
3. Sandford, M.; Castillo, G.; Hung, T.C. A review of fish identification methods applied on small fish. *Rev. Aquac.* **2020**, *12*, 542–554. [CrossRef]
4. Alaba, S.Y.; Nabi, M.M.; Shah, C.; Prior, J.; Campbell, M.D.; Wallace, F.; Ball, E.B.; Moorhead, R. Class-aware fish species recognition using deep learning for an imbalanced dataset. *Sensors* **2022**, *22*, 8268. [CrossRef] [PubMed]
5. Chang, C.C.; Ubina, N.A.; Cheng, S.C.; Lan, H.Y.; Chen, K.C.; Huang, C.C. A Two-Mode Underwater Smart Sensor Object for Precision Aquaculture Based on AIoT Technology. *Sensors* **2022**, *22*, 7603. [CrossRef] [PubMed]
6. Hsiao, Y.H.; Chen, C.C.; Lin, S.I.; Lin, F.P. Real-world underwater fish recognition and identification, using sparse representation. *Ecol. Inform.* **2014**, *23*, 13–21. [CrossRef]
7. Zhang, Z.; Du, X.; Jin, L.; Wang, S.; Wang, L.; Liu, X. Large-scale underwater fish recognition via deep adversarial learning. *Knowl. Inf. Syst.* **2022**, *64*, 353–379. [CrossRef]
8. Liang, J.M.; Mishra, S.; Cheng, Y.L. Applying Image Recognition and Tracking Methods for Fish Physiology Detection Based on a Visual Sensor. *Sensors* **2022**, *22*, 5545. [CrossRef] [PubMed]
9. Zhang, S.; Liu, W.; Zhu, Y.; Han, W.; Huang, Y.; Li, J. Research on fish identification in tropical waters under unconstrained environment based on transfer learning. *Earth Sci. Inform.* **2022**, *15*, 1155–1166. [CrossRef]
10. Xu, X.; Li, W.; Duan, Q. Transfer learning and SE-ResNet152 networks-based for small-scale unbalanced fish species identification. *Comput. Electron. Agric.* **2021**, *180*, 105878. [CrossRef]
11. Saghafi, M.A.; Hussain, A.; Zaman, H.B.; Saad, M.H.M. Review of person re-identification techniques. *IET Comput. Vis.* **2014**, *8*, 455–474. [CrossRef]
12. Huang, N.; Liu, J.; Miao, Y.; Zhang, Q.; Han, J. Deep learning for visible-infrared cross-modality person re-identification: A comprehensive review. *Inf. Fusion* **2022**, *91*, 396–411. [CrossRef]
13. Shruthi, U.; Nagaveni, V.; Raghavendra, B.K. A review on machine learning classification techniques for plant disease detection. In Proceedings of the 2019 5th International Conference on Advanced Computing & Communication Systems (ICACCS), Coimbatore, India, 15–16 March 2019; pp. 281–284.
14. Mahmud, M.S.; Zahid, A.; Das, A.K.; Muzammil, M.; Khan, M.U. A systematic literature review on deep learning applica-tions for precision cattle farming. *Comput. Electron. Agric.* **2021**, *187*, 106313. [CrossRef]
15. Duong, H.T.; Hoang, V.T. Dimensionality reduction based on feature selection for rice varieties recognition. In Proceedings of the 2019 4th International Conference on Information Technology (InCIT), Bangkok, Thailand, 24–25 October 2019; pp. 199–284.
16. Chen, J.; Chen, W.; Zeb, A.; Yang, S.; Zhang, D. Lightweight inception networks for the recognition and detection of rice plant diseases. *IEEE Sens. J.* **2022**, *22*, 14628–14638. [CrossRef]
17. Zhang, C.; Zhang, H.; Tian, F.; Zhou, Y.; Zhao, S.; Du, X. Research on sheep face recognition algorithm based on improved AlexNet model. *Neural Comput. Appl.* **2023**, *35*, 1–9. [CrossRef]
18. Peng, Y.; Kondo, N.; Fujiura, T.; Suzuki, T.; Ouma, S.; Yoshioka, H.; Itoyama, E. Dam behavior patterns in Japanese black beef cattle prior to calving: Automated detection using LSTM-RNN. *Comput. Electron. Agric.* **2020**, *169*, 105178. [CrossRef]
19. Barbedo, J. A Review on the Use of Computer Vision and Artificial Intelligence for Fish Recognition, Monitoring, and Management. *Fishes* **2022**, *7*, 335. [CrossRef]
20. Yuan, P.; Song, J.; Xu, H. Fish Image Recognition Based on Residual Network and Few-shot Learning. *Trans. Chin. Soc. Agric. Mach.* **2022**, *53*, 282–290.
21. Wang, J.; Zhu, X.; Gong, S.; Li, W. Transferable joint attribute-identity deep learning for unsupervised person re-identification. In Proceedings of the IEEE Conference on Computer Vision and Pattern Recognition, Salt Lake City, UT, USA, 18–22 June 2018; pp. 2275–2284.
22. Deng, W.; Zheng, L.; Ye, Q.; Kang, G.; Yang, Y.; Jiao, J. Image-image domain adaptation with preserved self-similarity and domain-dissimilarity for person re-identification. In Proceedings of the IEEE Conference on Computer Vision and Pattern Recognition, Salt Lake City, UT, USA, 18–22 June 2018; pp. 994–1003.
23. Zhu, J.Y.; Park, T.; Isola, P.; Efros, A.A. Unpaired image-to-image translation using cycle-consistent adversarial networks. In Proceedings of the IEEE International Conference on Computer Vision, Venice, Italy, 22–29 October 2017; pp. 2223–2232.
24. Ye, M.; Lan, X.; Yuen, P.C. Robust anchor embedding for unsupervised video person re-identification in the wild. In Proceedings of the European Conference on Computer Vision (ECCV), Munich, Germany, 8–14 September 2018; pp. 170–186.
25. Ye, M.; Li, J.; Ma, A.J.; Zheng, L.; Yuen, P.C. Dynamic graph co-matching for unsupervised video-based person re-identification. *IEEE Trans. Image Process.* **2019**, *28*, 2976–2990. [CrossRef] [PubMed]
26. Tang, Y.; Yang, X.; Wang, N.; Song, B.; Gao, X. CGAN-TM: A novel domain-to-domain transferring method for person re-identification. *IEEE Trans. Image Process.* **2020**, *29*, 5641–5651. [CrossRef] [PubMed]
27. Wei, L.; Zhang, S.; Gao, W.; Tian, Q. Person transfer gan to bridge domain gap for person re-identification. In Proceedings of the IEEE Conference on Computer Vision and Pattern Recognition, Salt Lake City, UT, USA, 18–22 June 2018; pp. 79–88.

28. Zhong, Z.; Zheng, L.; Zheng, Z.; Li, S.; Yang, Y. Camera style adaptation for person re-identification. In Proceedings of the IEEE Conference on Computer Vision and Pattern Recognition, Salt Lake City, UT, USA, 18–22 June 2018; pp. 5157–5166.
29. Choi, Y.; Choi, M.; Kim, M.; Ha, J.W.; Kim, S.; Choo, J. Stargan: Unified generative adversarial networks for multi-domain image-to-image translation. In Proceedings of the IEEE Conference on Computer Vision and Pattern Recognition, Salt Lake City, UT, USA, 18–22 June 2018; pp. 8789–8797.

Journal of
Marine Science
and Engineering

MDPI

Article

Sloshing Response of an Aquaculture Vessel: An Experimental Study

Yanwu Tao [1], Renqing Zhu [1,*], Jiayang Gu [2], Qi Wei [1], Fangxin Hu [1], Xiaosen Xu [2], Zhongyu Zhang [2] and Zhiyu Li [3]

1 School of Naval Architecture and Ocean Engineering, Jiangsu University of Science and Technology, Zhenjiang 212003, China; tywjkd@126.com (Y.T.)
2 Marine Equipment and Technology Institute, Jiangsu University of Science and Technology, Zhenjiang 212003, China
3 Marine Design & Research Institute of China, Shanghai 200023, China
* Correspondence: zjcyzrq@163.com

Abstract: The sloshing response is crucial to the design and operation of aquaculture vessels and affects the safety of the culture equipment and the efficiency of the culture operation. A 1/50 scaled model was utilized to investigate the coupled sloshing response characteristics of a novel aquaculture vessel in a wave basin. Two wave directions (beam and head wave) and two filling levels (81.5% and 47.4%) are taken into account. The time-domain and frequency-domain characteristics of the sloshing response under the linear regular wave and extreme operational sea state were investigated using regular wave tests and irregular wave tests, respectively. The sloshing mechanism in the aquaculture tanks is complicated, due to the coupling effect between external waves, ship motion, and internal sloshing. In linear regular waves, the wave frequency mode dominates the sloshing response, which is larger under beam wave conditions than under head wave conditions and larger under half load conditions than full load conditions. The irregular wave test results confirmed the regular wave test conclusions, but the sloshing response has stronger nonlinearity, higher natural modes appeared, and the amplitude of the higher natural modes is also relatively larger.

Keywords: aquaculture vessel; model test; sloshing; coupling effect; higher natural mode; regular wave; irregular wave

Citation: Tao, Y.; Zhu, R.; Gu, J.; Wei, Q.; Hu, F.; Xu, X.; Zhang, Z.; Li, Z. Sloshing Response of an Aquaculture Vessel: An Experimental Study. *J. Mar. Sci. Eng.* **2023**, *11*, 2122. https://doi.org/10.3390/jmse11112122

Academic Editor: Spyros A. Mavrakos

Received: 7 October 2023
Revised: 26 October 2023
Accepted: 2 November 2023
Published: 6 November 2023

1. Introduction

With wild fish catches already approaching their maximum and human demand for mariculture increasing [1], mariculture production has more than doubled in the last 20 years and is predicted to more than double again in the next 30 years [2]. New farming techniques are needed to achieve sustainable aquaculture development. Researchers have proposed various forms of offshore aquaculture equipment concepts, including improved versions of traditional net pens [3], offshore closed aquaculture systems [4], ship-like structures [5], as well as integration on multi-purpose platforms [6], and energy production units. In these concepts, offshore closed aquaculture systems can improve the culture efficiency by controlling the culture environment (water temperature, NO_X, etc.), isolating parasites and reducing environmental impacts through excreta and bait residue filtering. A novel form of farm vessel for deep-sea aquaculture with numerous closed aquaculture tanks is being developed and promoted by Chinese researchers [7]. The vessel has a total aquaculture volume of 100,000 m^3 and is able to sail autonomously to adapt to the harsher sea environment.

In harsh seas, violent sloshing responses may occur in the aquaculture tanks of aquaculture vessels. Severe sloshing will affect the cultural operations in the tanks, as well as the fish's growth and survival. In comparison to ordinary liquid cargo ships, aquaculture vessels must pay more attention to the sloshing problem, and there is minimal literature on the subject. The sloshing problem is a classical issue in the fields of both land and sea liquid transport shipping, storage, and aerospace, and it has been intensively and extensively studied by a large number of scholars [8]. The study of the sloshing problem is mainly based on the theoretical method based on the potential flow theory [9], the experimental method based on the scaled model [10,11], and the numerical method based on RANS [12,13], MPS [14,15], and SPH [16], etc. For the aquaculture vessel with closed tanks, the coupling effect of hull motion and sloshing as well as the sloshing suppression methods are of qualitative concern.

The theoretical analysis of the problem of the coupled action of hull motion and sloshing is mostly solved by separating the internal and external domains, in which the external domain of the hull is solved by the linear potential flow theory, and the internal liquid tank domain is dealt with by two methods: one is based on the viscous method of calculating the sloshing loads based on the viscous method such as the linear potential flow method [17–19], the multimodal method [20], the finite difference method [21], and the fourth-order Runge–Kutta method [22], and the other one is based on the viscous method [23–25]. With the development of computer technology and computational fluid dynamics, scholars have widely adopted the RANS method to calculate the sloshing in the liquid tanks and the coupling between the hull's motion and the liquid tanks [26,27]. For experimental studies, Zhao et al. [28] investigated the response of a single liquid tank under the white noise wave. Kim et al. [29] and Zhao et al. [30] conducted tests on ships with two tanks arranged fore and aft, and Igbadumhe et al. [31] and Li et al. [32] investigated the coupled motions and the sloshing response of an FPSO and an aquaculture vessel with left and right double rows of tank arrangements.

Although the sloshing effect can be used on the anti-roll tanks [33] and tuned liquid damper [34] to improve the motion performance, the sloshing effect is more noteworthy for its adverse effects on the free surface's stability and the impact load on the structure. To suppress the sloshing response, scholars have designed various methods to decrease the free surface area or utilize obstacles to dissipate the kinetic energy of inner water [35], such as arranging trusses and protruding structures on the bulkhead [36], or fitting fixed [37] or floating [38,39] horizontal or vertical structures inside the tanks. These sloshing suppression methods are widely used on conventional vessels, but these protruding structures may affect fishing operations and even cause potential fish damage. Considering the suitability for fishing, some scholars have explored the sloshing suppression method in aquaculture vessels. Cui et al. [40] studied the sloshing response of an aquaculture tank with an inclined top using a numerical simulation, and Gao et al. [41] investigated the sloshing effect of arranging intermittent vertical cylinders and continuous vertical flat plates on the top of the tank and the effect on the velocity field. Wiegerink et al. [42] designed an annular sloshing suppression structure with a rectangular cross-section for a cylindrical closed aquaculture platform and validated its suppression effect by experimental and numerical methods.

This paper focuses on analyzing the sloshing response of a novel aquaculture vessel using the experimental method, and it is structured as follows. In Section 2, the model test scheme and the calibration of the sensors are described. In Section 3, the frequency-domain sloshing response characteristics under linear regular waves are investigated. The wave frequency mode and higher natural mode of the sloshing response are identified, and the effects of the position of the aquaculture tank and the walkway on the sloshing response are analyzed. In Section 4, the complex sloshing response under extreme operating conditions is analyzed based on irregular wave tests.

2. Vessel Description and Experimental Setup

The design operation area of the aquaculture vessel is the Yellow Sea and South China Sea of China, and the design operation depth is 100 m–500 m. The aquaculture vessel is designed to extract deep water and maintain the proper flow speed and temperature in the tanks by the recirculating water systems for the culture of Atlantic salmon and pseudosciaena crocea. The aquaculture vessel has a two-propeller propulsion system that allows it to cruise autonomously or relocate during typhoons. To investigate the hydrodynamic and sloshing performance of the aquaculture vessel, model tests at a scale (λ) of 1:50 were carried out.

2.1. Description of the Vessel

The aquaculture vessel is arranged in a double row of tanks, with longitudinal and transverse bulkheads separating the hull into several near-square aquaculture tanks. The side view of the main hull is shown in Figure 1. Aquaculture tanks No.1 to No.14 (in Figure 1) are standard tanks with an aquaculture volume of 5300 m^3, and No.15 to No.18 are non-standard tanks. Compared with the conventional liquid carriers, the aquaculture vessel has more aquaculture tanks, the longitudinal and transverse dimensions of the aquaculture tanks are close to each other, and small equipment cabins are arranged at the transverse and longitudinal intervals of the aquaculture tanks. Economically, the construction cost of an aquaculture vessel is close to that of a conventional liquid cargo vessel of the same displacement. The main particulars of the vessel are presented in Table 1.

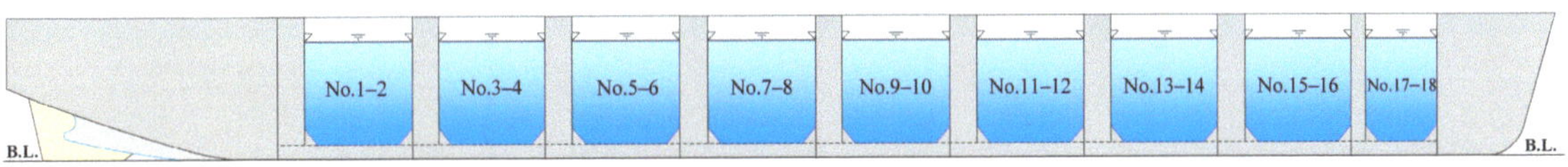

Figure 1. Side view of the aquaculture vessel.

Table 1. Main particulars of the aquaculture vessel.

Designation	Signal	Unit	Value	
			Full Scale	Model
Length overall	L_{OA}	m	258.20	5.164
Length between perpendiculars	L_{PP}	m	250.56	5.011
Breadth	B	m	44.00	0.880
Depth	D	m	22.80	0.456

The aquaculture vessel consists of two typical loading conditions, full load and half load, which correspond to a filling level (water depth/tank height) of 81.5% and 47.4% in the aquaculture tanks, respectively. Since sufficient space needs to be retained in the upper part of the aquaculture tanks for personnel operations, the design full load condition of the aquaculture vessel has a lower filling level than that defined for a conventional liquid cargo vessel (95% or more). The mass and moment of inertia parameters of the vessel for the two filling levels are shown in Table 2.

Table 2. Mass and moment of inertia parameters [43].

Designation	Signal	Unit	Full Load		Half Load	
			Full Scale	Model	Full Scale	Model
Draft	d	m	14.00	0.28	10.80	0.216
Displacement	Δ	t	138,971	1.085	105,326	0.822
Center of gravity above BL	V_{CG}	m	11.78	0.236	10.34	0.207
Center of gravity from AP	L_{CG}	m	125.31	2.506	126.01	2.520
Roll radius of gyration	k_{xx}	m	13.68	0.274	14.39	0.288
Pitch radius of gyration	k_{yy}	m	61.53	1.231	64.84	1.297
Yaw radius of gyration	k_{zz}	m	62.33	1.247	65.63	1.313

The aquaculture tanks have chamfers on the bottom and side walls, and the vertical walls are fitted with walkways to enable personnel culture operation inside the tanks. The transverse section of the vessel at a standard aquaculture tank is shown in Figure 2.

The standard tanks have a width of 19.0 m, a length of 17.84 m, and a water depth of 16.74 m for the full load condition and 9.74 m for the half load condition, respectively. The depth-to-length ratio (h/l) of the aquaculture tank, i.e., the ratio of the water depth (h) to the length in the sloshing direction (l), is an important factor affecting the sloshing characteristics. For transverse sloshing, the h/l was 0.51 and 0.89 for full load and half load conditions, respectively; for longitudinal sloshing, it was 0.55 and 0.95 for full load and half load conditions.

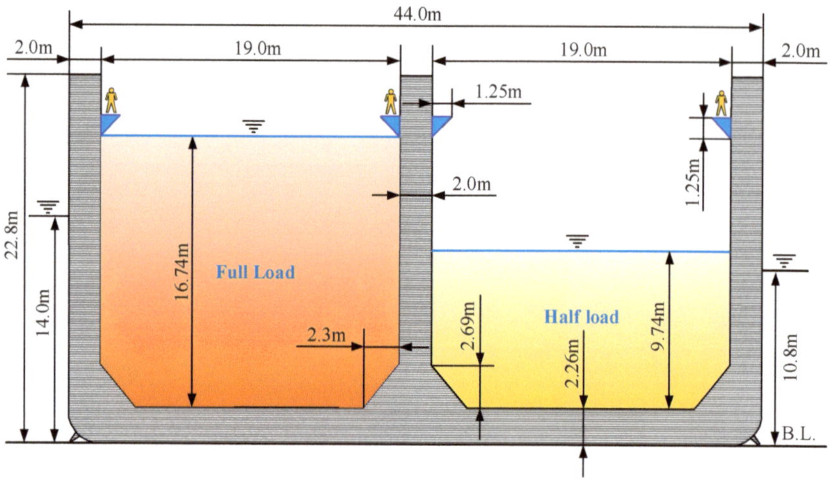

Figure 2. Transverse section of the aquaculture vessel at a standard aquaculture tank.

2.2. Facility and Test Model

Model tests were performed in the Special Vehicle Research Institute at the AVIC (Aviation Industry Corporation) of China. The basin has an overall length of 60 m, width of 60 m, and depth of 5 m. The wave generators are capable of making waves with wavelengths ranging from 0.5 m to 15 m, wave heights between 0.05 m and 0.5 m, and maximum significant wave heights of 0.3 m for long-crested irregular waves. The main body of the aquaculture vessel model is composed of multi-layer board and fiber reinforced plastics, and the internal aquaculture tanks are made of 6 mm thick plexiglass. The aquaculture vessel model and inner tankers are shown in Figures 3 and 4. The support structure is specially designed to maintain the counterweight block in the test, and the position of the counterweight unit can be precisely modulated by screws. The model geometric tolerances and mass distribution have been verified in the previous work [43], which satisfies the requirements of the ITTC Recommended Procedures and Guidelines [44].

Figure 3. Aquaculture vessel model in the basin.

Figure 4. Aquaculture tank model.

Four horizontal mooring cables consisting of soft springs and thin wire ropes were used to prevent the model from drifting off. The stiffness of the spring is 88 N/m, and the pre-tension is 45 N. The cable does not slacken during testing, and the natural period of the surge and sway motion caused by the mooring cables is much larger than the natural period of the wave frequency motion (heave, roll, and pitch). A single anchor chain mooring is used for the aquaculture vessel during the aquaculture operation, which has instability in beam wave conditions. To study the extreme sloshing response of the aquaculture vessel under the beam wave, the same mooring arrangement is still used for the irregular wave test.

2.3. Sensor Arrangement and Calibration

In the model test, a set of gyroscope units was installed at the hull's center of gravity to measure the roll and pitch motion; a mini camera was installed in the open space at the stern of the hull to record the sloshing phenomenon of the aft aquaculture tanks (No.1 and No.2); and 10 wave gauges (WG) were set up at multiple aquaculture tanks to measure the wave elevation in the typical position. The arrangement of the camera and all the sensors is shown in Figure 5.

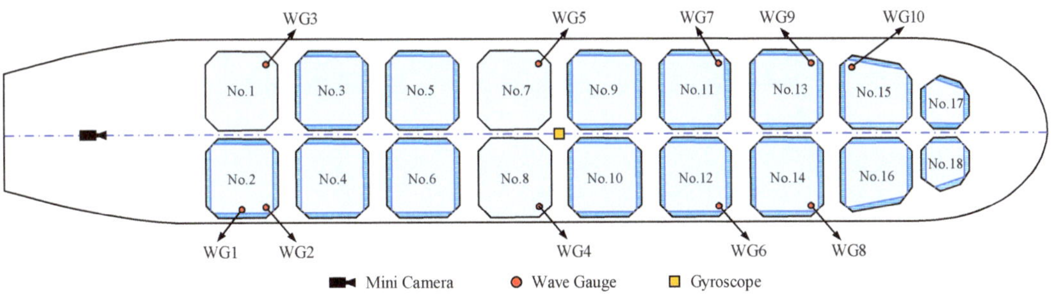

Figure 5. Arrangement of the camera and all the sensors.

The wave gauge WG1–WG3 arrangement at the hull stern is shown in Figure 6. The still free surface of the aquaculture tank under full load conditions was at the lower edge of the inclined brace plate of the walkway. To reduce the impact of the walkway on the sensors, the capacitance filaments of the wave gauges were placed at a specific distance from the walkway's outer edge, and all wave gauges were placed at a distance (d_1 in Figure 6) of 3.5 cm from the tank walls they were closest to. Therefore, at full scale, the distances from WG1 and WG2 to the center of the aquaculture tank are 7.75 m and 6.67 m, respectively. Tank No.1 in the stern, as well as Nos. 7 and 8 in the amidships, does not install the walkway and aims to evaluate the sloshing response unaffected by the walkway. WG1 is situated in the center of the aquaculture tank's sidewall and mainly measures the pure transverse sloshing response; the remaining WGs are located at the aquaculture tanks' corners and can measure both longitudinal and transverse sloshing responses.

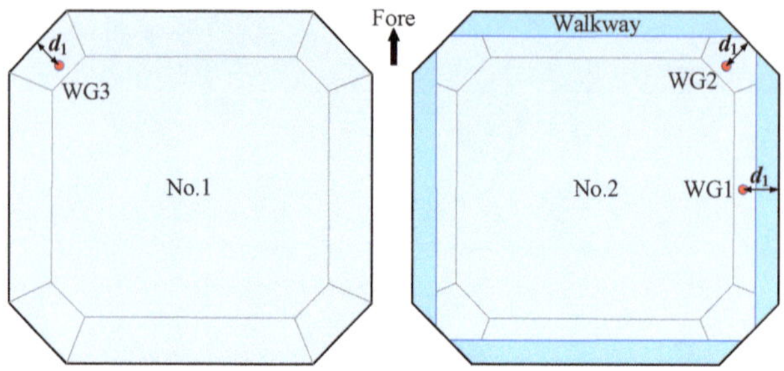

Figure 6. Arrangement of WG1–WG3 at the hull stern.

The wave gauge has a range of 50 cm and a sampling frequency of 100 Hz, with an accuracy of 0.15%. Before the model test, the acquisition unit and gyroscope system are calibrated at a special testing institution. The gyroscope system is sampled at 1.25 kHz. The camera records at a frame rate of 60 fps and 4 K resolution. The parameters and accuracy of the main sensors are summarized in Table 3.

Table 3. Parameters and accuracy of main sensors.

Instrument	Sensor Type	Measuring Range	Accuracy
Data acquisition unit	PCM-006	0–5 V	0.01 V
Gyroscope	IMU610H	−90°–90°	0.05°
Wave gauge	YWH200-D	50cm	0.15%
Electronic hanging scale	OCS-3T	2000 kg	0.5 kg
Electronic platform scale	MTC002C	100 kg	0.01 kg

3. Sloshing Response under Regular Waves

When a regular wave test is used for the study, the wave parameters should consider the coupling effect between the external wave, hull motion, and internal sloshing. In this chapter, the parameters of the regular waves (Section 3.1) are first determined considering the aquaculture tanks' natural sloshing properties, as well as the motion performance of the vessel and the basin's wave-making capacity. Following that, the sloshing response under the beam wave and head wave is studied. Due to the significant difference in the magnitude of the sloshing response for the two wave headings, the beam wave condition (Section 3.2) and the head wave condition (Section 3.3) are investigated independently.

3.1. Regular Wave Parameters

If the chamfers on the bottom and sides of the aquaculture tank are neglected and the tank can be considered as a cuboid, then its n-th natural period of sloshing response can be given by the formula

$$T_n = \frac{2\pi}{\sqrt{n\pi g \tanh(n\pi h/l)/l}}, n = 1, 2, 3, \ldots \quad (1)$$

where T_n is the n-th natural period, g is the gravity acceleration, h is the water depth, and l is the width of the tank, respectively. The first four natural frequencies and periods of transverse and longitudinal sloshing at full scale are shown in Table 4. The n-th natural frequencies of transverse and longitudinal sloshing are denoted as $f_{\mathrm{T}n}$ and $f_{\mathrm{L}n}$, and the n-th natural period of transverse and longitudinal sloshing are denoted as $T_{\mathrm{T}n}$ and $T_{\mathrm{L}n}$, respectively. It can be seen that the first four natural periods of the sloshing range from 2.39 s to 4.95 s.

Table 4. Theoretical natural frequencies and periods of sloshing in full scale.

Mode Order n	Load Condition	Transverse Sloshing		Longitudinal Sloshing	
		Frequency f_{Tn} (Hz)	Period T_{Tn} (s)	Frequency f_{Ln} (Hz)	Period T_{Ln} (s)
1	Full load	0.202	4.953	0.209	4.794
	Half load	0.195	5.134	0.203	4.938
2	Full load	0.287	3.489	0.296	3.380
	Half load	0.286	3.494	0.296	3.384
3	Full load Half load	0.351	2.848	0.362	2.760
4	Full load Half load	0.405	2.467	0.418	2.390

The wavelength range of the regular wave should be at least 0.5 L_{pp} to 2.0 L_{pp}, and the wave steepness should be around 1/50, according to the ITTC method [45]. The minimum wave height of the wavemaker for regular waves is 0.05 m. If the wave period is close to the sloshing natural period, the wave steepness must be less than 1/15, and the wavemaker cannot generate continuous waves with a specified period and wave height. Combining factors such as the basin's wave-making capacity and model scale, the wave height of the regular wave was finally chosen as 50 mm, the wavelength range was 2 m–10.125 m, the wavelength to vessel length ratio was 0.4–2.025, and the corresponding wave period was 8 s–18 s at full scale. The wave parameters of the regular wave test are shown in Table 5.

Table 5. Wave parameters of regular wave model test.

Model.			Full Scale		λ_w/L_{pp}	H/λ_w
H_m (m)	λ_m (m)	T_m (s)	λ_s (m)	T_s (s)		
0.05	2.000	1.131	100.00	8.0	0.400	1/40
	3.125	1.414	156.30	10.0	0.625	1/63
	3.781	1.556	189.06	11.0	0.756	1/76
	4.500	1.697	225.00	12.0	0.900	1/90
	4.883	1.768	244.14	12.5	0.977	1/98
	5.281	1.838	264.06	13.0	1.056	1/106
	6.125	1.980	306.25	14.0	1.225	1/123
	7.031	2.121	351.56	15.0	1.406	1/141
	8.000	2.263	400.00	16.0	1.600	1/160
	10.125	2.546	506.25	18.0	2.025	1/203

3.2. Beam Wave Condition

Since the vessel has multiple aquaculture tanks, the time history and spectral analyses of the sloshing response at typical locations were first performed, and then, the effects of location and walkways on localized sloshing were evaluated by considering the differences between the in-vessel and in-tank locations.

3.2.1. Time and Frequency Domain Response

Based on the Froude scaling law [46], the conversion relations for the main physical quantities are shown in Table 6. The subscripts s and m denote the full scale and model scale, respectively.

Table 6. Froude scaling law for the involved physical quantities.

Physical Quantity	Model	Full Scale
Time	t_m	$t_s = t_m\sqrt{\lambda}$
Wave period	T_m	$T_s = T_m\sqrt{\lambda}$
Frequency	f_m	$f_s = f_m/\sqrt{\lambda}$
Sloshing amplitude	ζ_m	$\zeta_s = \zeta_m\,\lambda$

The wave gauge WG1 mainly measures transverse sloshing and is less affected by longitudinal sloshing, so the data of WG1 was chosen as a typical sloshing response to analyze. The time history and response spectrum of the sloshing response at WG1 in the full load condition under beam waves is shown in Figure 7. The incident wave frequency (f_W) is shown in Figure 7b as a red background region, and the first natural frequency (f_{T1}) of transverse sloshing is shown as a gray background region. When the wave period is less than 14 s, the sloshing response time history is nearly sinusoidal, and the sloshing response is dominated by the wave frequency response. However, when the wave period is greater than or equal to 14 s, the sloshing response shows a specific higher natural mode, and the higher natural mode is mainly caused by the first natural mode. The truncation of the data has some effect on the peak position of the spectrum, combined with the wave period uncertainty during the model test, resulting in the peak period in the sloshing spectrum deviating from the theoretical value.

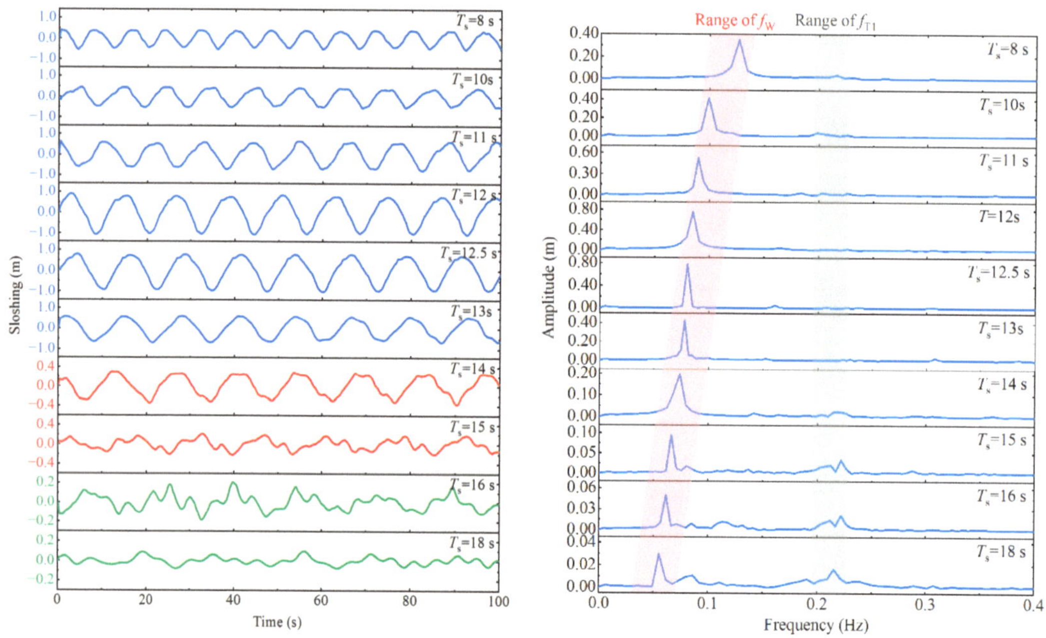

(**a**) (**b**)

Figure 7. Sloshing time history and response spectrum under full load and regular beam wave condition. (**a**) Sloshing time history; (**b**) Response spectrum.

For data processing, the initial steady-state interval was intercepted for the sloshing responses of all wave gauges, and band-pass filtering was used to extract the sloshing responses corresponding to the different modes. The wave-frequency mode of the sloshing has a truncation frequency of 0.9–1.1 f_W, and similarly, the first natural mode cutoff ranges from 0.9 to 1.1 f_{T1}. The sloshing RAO is defined as the ratio of the sloshing amplitude to the wave height (H_m) of the external incident wave. The sloshing RAOs of the wave frequency mode and the first natural mode of all wave gauges under the full load condition at beam sea are shown in Figure 8. The roll natural period at the full load condition is around 12.2 s. According to Figure 8a, the sloshing RAO of wave frequency mode reaches its maximum value when the wave period is close to the roll natural period. There are certain variances in the sloshing response at different wave gauges, and to represent the discretization of the sloshing RAOs in various wave gauges, the dimensionless standard deviation of the sloshing RAOs (denote as $\hat{\sigma}(\text{RAO})$) is defined as the ratio of the RAOs' standard deviation to their mean value for all wave gauges. When the wave period is smaller than 15 s, the $\hat{\sigma}$(RAO) is smaller than 6%, and when the wave period is greater than or equal to 15 s, it ranges from 13.5% to 22.5%. Overall, when the wave period is close to the roll natural period, the $\hat{\sigma}$(RAO) is minimized; from another perspective, the larger the sloshing RAO, the smaller its dimensionless dispersion. In Section 3.2.2, more details of the variations in sloshing response among the wave gauges are investigated.

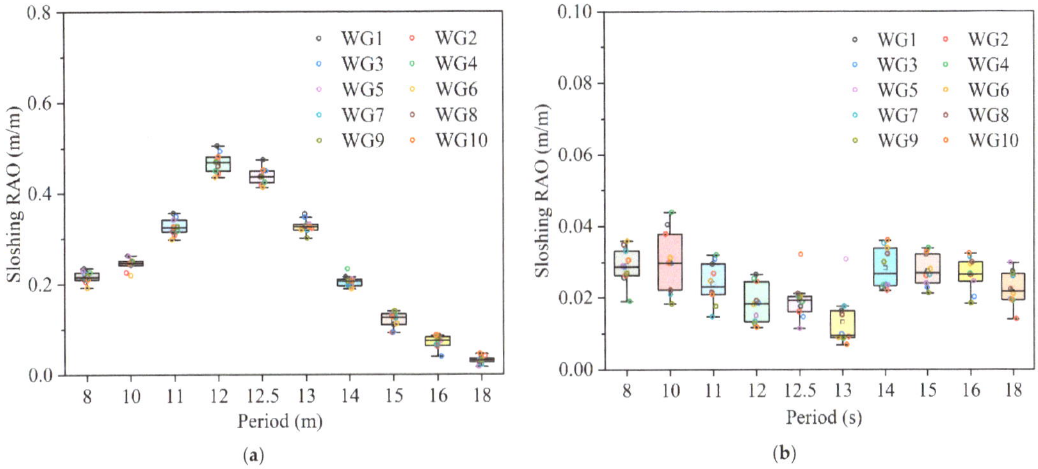

Figure 8. Sloshing RAOs under full load and beam wave condition. (**a**) Wave frequency mode; (**b**) First natural mode.

The first natural mode of the sloshing RAO has a smaller amplitude near the roll natural period (12 s–13 s). In particular, when the wave period is 10 s, the wave period is about twice the first natural period, and a larger first natural mode is excited. The ratio of the first natural mode to the wave frequency mode increases as the wave period shifts away from the roll natural period.

The time history and response spectrum of the sloshing response at WG1 in the half load condition under beam waves are shown in Figure 9. As can be seen from Figure 9a, the sloshing responses for all wave periods, except for wave periods of 8 s and 12 s, exhibit a distinct multi-frequency superposition mode, which can be confirmed by the sloshing response spectrum in Figure 9b. For all wave periods, the sloshing response showed a certain doubling frequency mode ($2f_W$), especially at wave periods of 15 s–18 s, the ratio of the doubling frequency mode ($2f_W$) to the wave frequency (f_W) mode increased from 0.5 to 0.9. When the wave period is 10 s, it triggers a significant first natural frequency (f_{T1}) mode, with an amplitude approximately twice that of the wave frequency mode. The first natural mode of the sloshing response is progressively larger than the wave frequency mode when the period is greater than or equal to 13 s.

The sloshing RAOs of the wave frequency mode and the first natural mode of all wave gauges under the half load condition at beam sea are shown in Figure 10. The roll natural period in the half load condition is around 11.6 s. From Figure 10a, it can be seen that the sloshing RAO of the wave frequency mode in the half load condition is maximum when the wave period is close to the roll natural period, which is similar to that of the full load condition. There is a certain variation in the sloshing RAO at different wave gauges, specifically, the $\hat{\sigma}$(RAO) is less than 11% for wave periods shorter than 15 s, and 21.2–24.7% for wave periods of 15 s to 18 s.

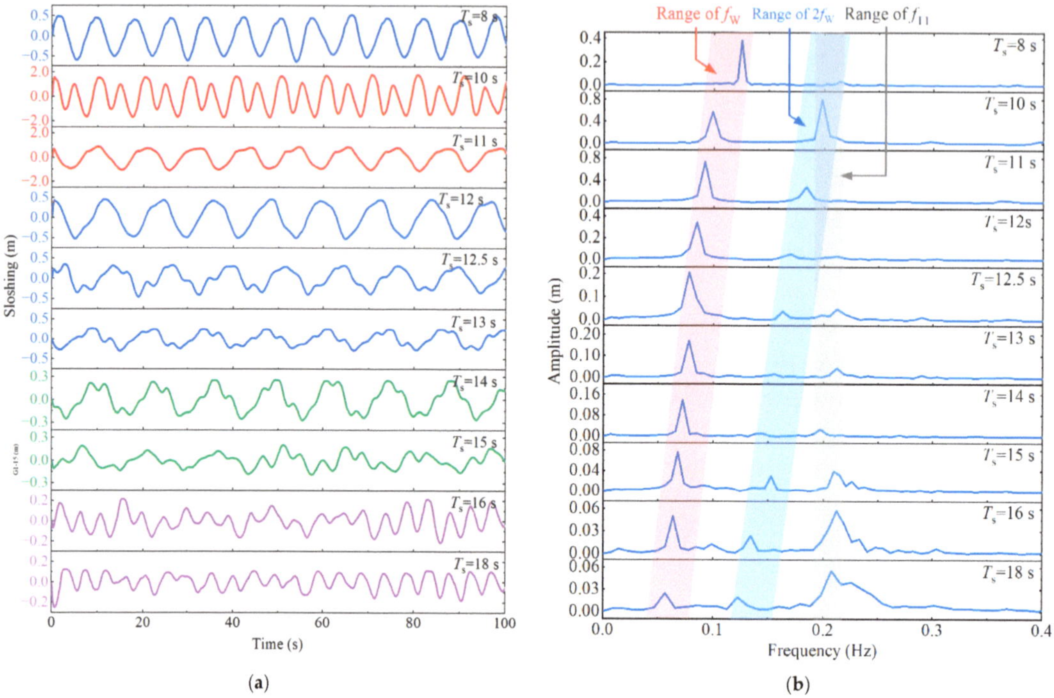

Figure 9. Sloshing response and response spectrum under half load and regular beam wave condition. (**a**) Sloshing time history; (**b**) Response spectrum.

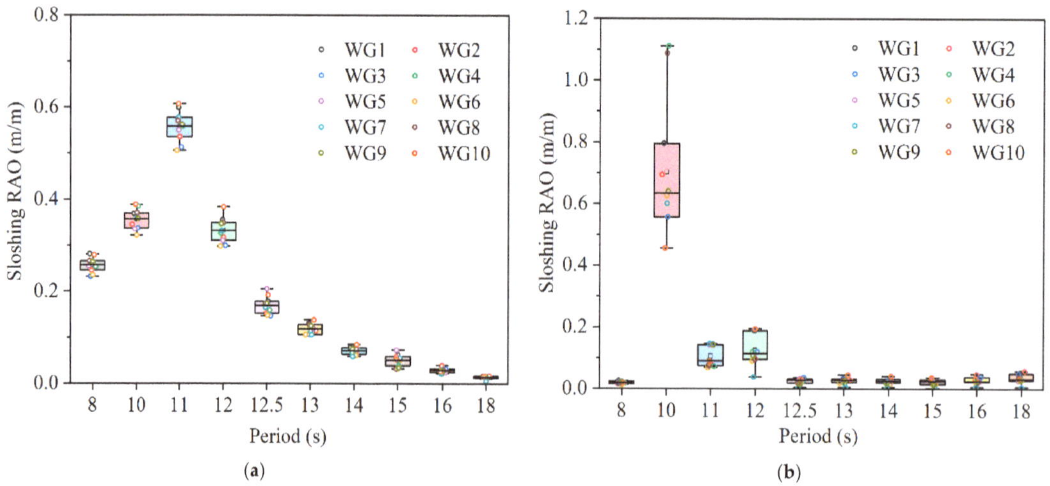

Figure 10. Sloshing RAO under half load and beam wave condition. (**a**) Wave frequency mode; (**b**) First natural mode.

Figure 10b shows that the first natural mode with a wave period of 10 s is significant compared with other wave periods in the half load condition. The mean value of the first natural mode is close to 1.9 times that of the wave frequency mode, and close to 3 times at some wave gauges, which mainly resulted from the wave period being close to 2 times the first natural period. Except for wave periods around 10 s, both the first natural mode

(f_{T1}) and the doubling frequency mode ($2f_W$) are larger under long-period waves ($T_s \geq 15$ s) and their amplitudes are bigger than the wave frequency mode for some wave periods. Comparatively, for the beam wave condition, the wave frequency mode, the first natural mode, and the doubling frequency mode are larger in the half load condition than those in the full load condition.

3.2.2. The Effect of Tank Position and Walkway

Due to the various distances from each aquaculture tank to the center of the vessel, the acceleration of each tank varies somewhat, which may result in different sloshing responses in different aquaculture tanks. Furthermore, the aquaculture tanks are fitted with walkways, which can affect the free surface shape within the tank. In this section, the effect of the aquaculture tanks' location in the vessel and the location within the aquaculture tank on the sloshing are specifically investigated.

The vessel has a double-row arrangement of aquaculture tanks, with different longitudinal and transverse positions of the tanks in the hull. Figure 11 shows the sloshing response of the wave gauges at typical locations under full load conditions with a wave period of 12 s, and walkways are installed in all the aquaculture tanks where the wave gauges are installed. In particular, Figure 11a shows the sloshing response of three wave gauges (WG2, WG6, and WG8) at different longitudinal positions, and Figure 11b shows the sloshing response of wave gauges (WG8 and WG9) at different transverse positions. The sloshing amplitudes of WG2, WG6, and WG8 are 1.30 m, 1.28 m, and 1.36 m, respectively, and the relative difference at WG2 and WG8 is 6.2%. The sloshing amplitudes of WG8 and WG9 were 1.36 m and 1.39 m, respectively, with a relative difference of 2.2%.

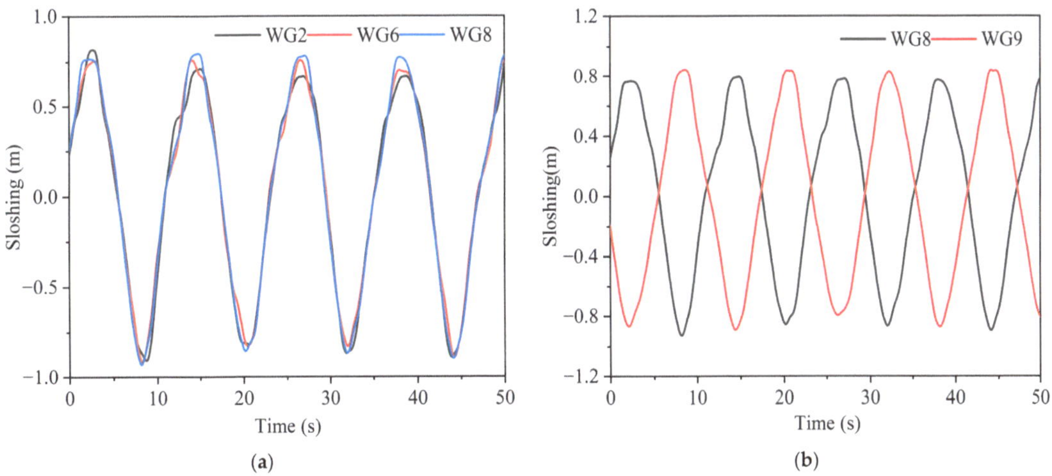

Figure 11. Sloshing response at different locations under full load condition. (**a**) Different longitudinal positions; (**b**) Different transverse positions.

Figure 12 shows the sloshing response of WG1 at the center of the side walkway and WG2 at the walkway connection in the tank No.2. The sloshing response amplitude of WG1 is 14.6% larger than that of WG2, and in addition, WG1 has a localized peak just before the maximum, which is caused by the reflective effect of the inclined brace plate of the walkway.

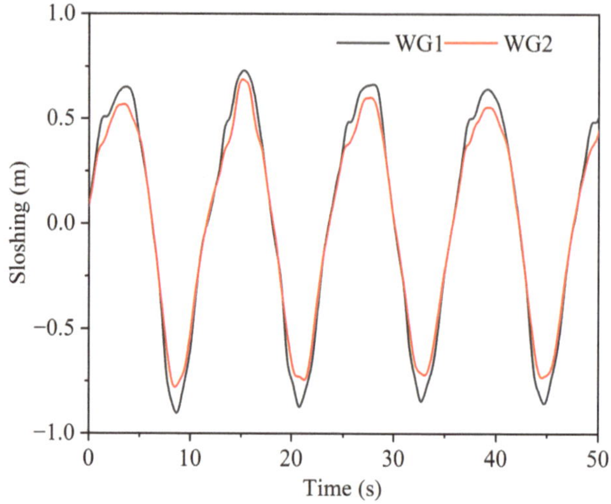

Figure 12. Sloshing response at different locations within the tank.

In general, the sloshing response of WG1 is about 15% larger than the other wave gauges in the beam wave condition, based on an analysis of the amplitudes of all the wave gauges. Meanwhile, the distribution of the sloshing amplitude at different wave height gauges shows a certain randomness, and the difference in sloshing response between the other wave gauges (WG2–WG10) is less than 10%.

3.3. Head Wave Condition

WG2 is located in the tank No.2 where the walkway is installed, and its longitudinal sloshing is representative; therefore, the sloshing response of WG2 is analyzed under head wave conditions. The time history and response spectrum of the sloshing response at WG2 in the full load condition under head waves are shown in Figure 13.

In the head wave condition, the sloshing response is dominated by the wave frequency mode for both loading conditions. As shown in Figure 13, the doubling frequency mode exists ($2f_W$) when the wave period is 12 s–13 s for the full load condition, and the first natural mode occurs when the wave period is large (T_s = 15 s–18 s). From Figure 14, it can be found that a significant doubling frequency mode ($2f_W$) occurs in the half load condition, with periods of 10 s–12.5 s, while the first natural mode occurs at periods of 16 s–18 s. In contrast, for the same regular wave excitation, the sloshing amplitude in the head wave condition is smaller than that of the beam wave condition, and its nonlinearity is weaker likewise.

The sloshing RAOs of the wave frequency mode in the full load and half load conditions under the heading wave condition are shown in Figure 15. Since WG1 mainly measures the transverse sloshing, the sloshing response of this wave gauge is not analyzed in the head wave condition. Furthermore, several wave gauge data were removed since the sloshing amplitude in some wave periods was smaller than 1 mm, which causes challenges to the measurement and data processing. It can be seen that the sloshing RAO of the wave frequency mode for the full and half load conditions shows a maximum at the wave period of 11 s–13 s. The peak sloshing RAO in the head wave condition does not occur in the range of the pitch natural period. The sloshing response measured at the different wave gauges was relatively dispersive, with a $\hat{\sigma}$(RAO) of 31–47%, of which the dispersion was slightly larger for the half load condition than for the full load condition. In the regular wave tests, the longitudinal sloshing response in the head wave condition is at a small level, and its amplitude is about 1/10 of that of the corresponding beam wave condition, so the sloshing response at different tanks in the head wave condition was not analyzed specifically.

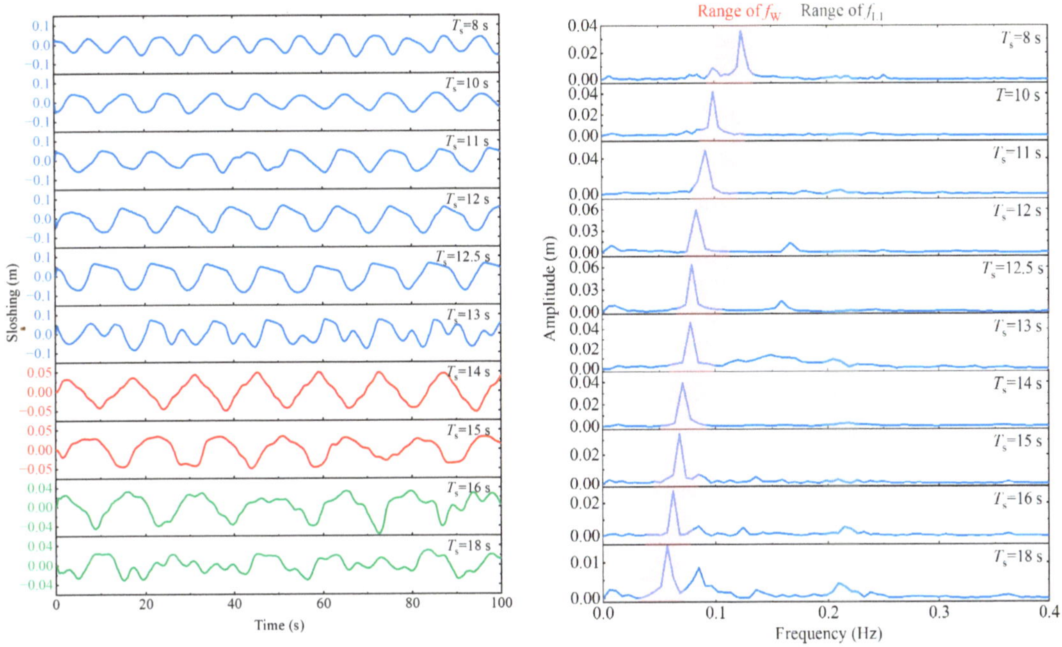

(a) (b)

Figure 13. Sloshing response and response spectrum under full load and regular head wave condition. (**a**) Sloshing response; (**b**) Response spectrum.

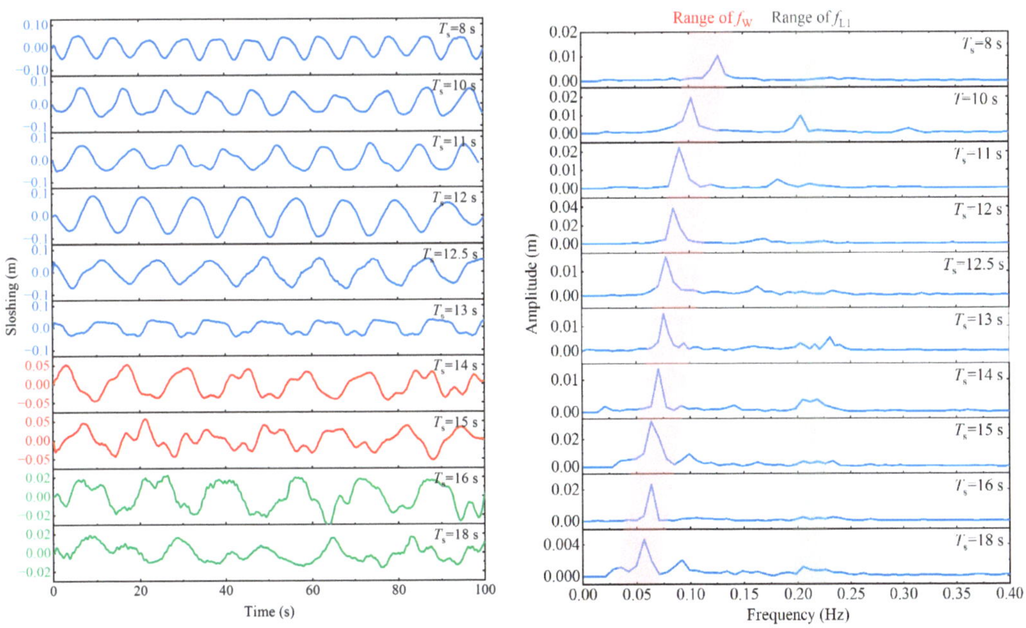

(a) (b)

Figure 14. Sloshing response and response spectrum under half load and regular head wave condition. (**a**) Sloshing response; (**b**) Response spectrum.

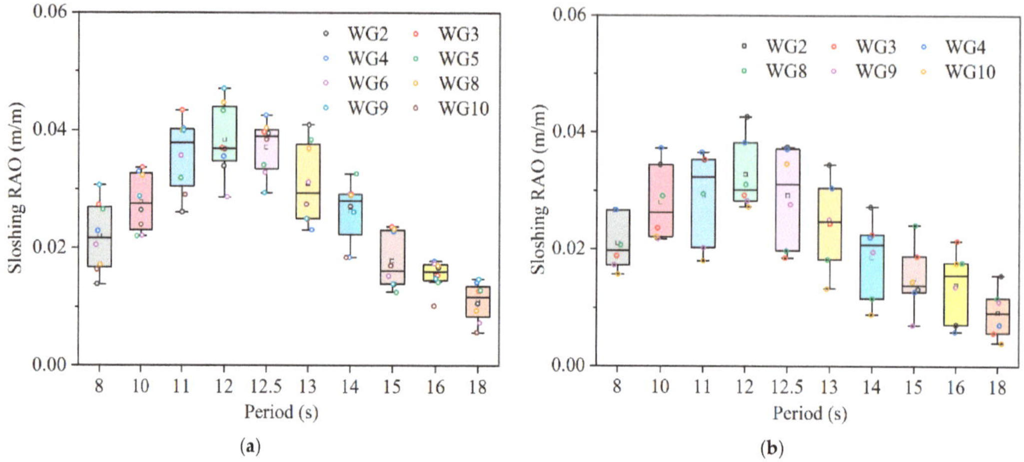

Figure 15. Wave frequency sloshing RAO under head wave condition. (**a**) Full load; (**b**) Half load.

4. Sloshing Response under Irregular Waves

The previous section investigated the response characteristics of the aquaculture vessel under the excitation of linear regular waves with a specific period range, whereas the waves encountered during the actual operation are extremely irregular, and this chapter investigates the sloshing response of the aquaculture vessel under the designed extreme operating sea conditions. Similar to the regular wave response study, this chapter considers two typical loading conditions, full and half load, and two typical wave directions, beam and heading sea.

4.1. Irregular Wave Condition

The design limit operating sea state of the aquaculture vessels is a significant wave height ($H_{1/3}$) of 5.8 m, with a peak period (T_p) of 12 s. The wave spectrum is selected as the JONSWAP spectrum [47].

$$S(\omega) = \frac{5}{16}(1 - 0.287\ln(\gamma)) \cdot \frac{H_{1/3}^2}{\omega^5}\omega_p^4 \exp\left(-\frac{5}{4}\left(\frac{\omega}{\omega_p}\right)^{-4}\right) \cdot \gamma^{\exp\left(-0.5\left(\frac{\omega-\omega_p}{\sigma\cdot\omega_p}\right)\right)} \tag{2}$$

where $\omega_p = 2\pi/T_p$ is the angular spectral peak frequency, γ is the non-dimensional peak shape parameter, σ is the spectral width parameter, and $\sigma = \begin{cases} 0.07 \text{ for } \omega \le \omega_p \\ 0.09 \text{ for } \omega > \omega_p \end{cases}$.

To minimize the effect of wave reflection in a single long-duration irregular wave test, multiple short-duration tests with different random seed numbers were re-ran, which makes the total effective duration of the irregular wave longer than 3 h at full scale. The irregular wave was calibrated before the test, and the significant wave height varied from the design value by a maximum of 2.5%, with a maximum variation of the spectral peak period of 6.56%.

Since there are large differences in the amplitude and nonlinearity of the sloshing response under the beam and head seas, the sloshing response of the two wave directions is analyzed independently.

4.2. Beam Wave Condition

The sloshing time history and response spectrum of wave gauges WG1–WG3 at the stern of the aquaculture vessel under full load are shown in Figure 16. The maximum and minimum sloshing amplitude of WG1 are 1.72 m and −1.66 m, respectively, and the range of the sloshing response (deviation between maximum and minimum) is 3.38 m, which is larger than 2.90 m of WG2 and 3.15 m of WG3. In the extreme operational sea state, the sloshing response was dominated by the wave frequency mode, but a significant first to third natural mode appeared. The amplitude of first natural mode is about 1/3 of the wave frequency mode, and for high natural modes, the amplitude of the modes decreases as the order of the natural modes increases.

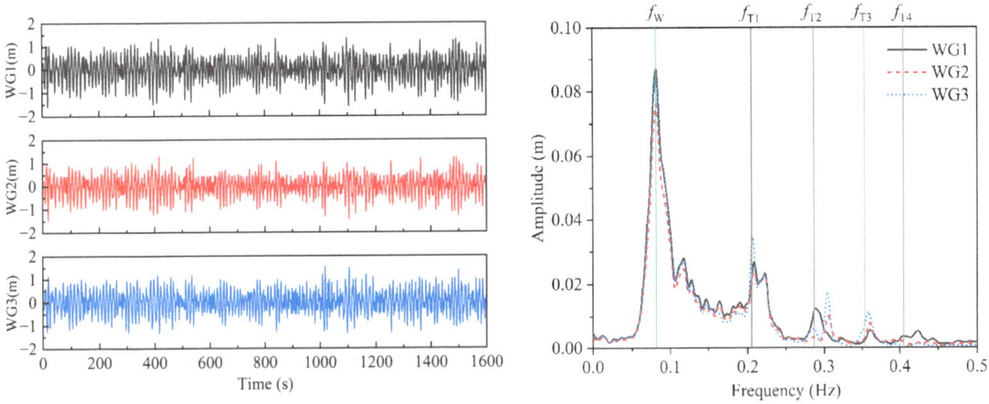

(a) (b)

Figure 16. Sloshing response and response spectrum under full load and irregular beam wave condition. (**a**) Sloshing response; (**b**) Response spectrum.

The typical free surface of the No.1 aquaculture tank without a walkway under the beam wave and full load condition is shown in Figure 17, and the upwelling process at the walkway connection of the tank No.2 is shown in Figure 18. The free surface in tank No.1 showed a large elevation at the side bulkheads, and the free surface was nearly in a planar state. In this condition, the sloshing response can be considered as a quasi-static adjustment process of the static water surface in the tank in response to the hull's heeling. Figure 18 demonstrates a typical upwelling and receding process at the walkway connection. The superposition of transverse and longitudinal sloshing at the side chamfer of the tank walls and a narrowed angled walkway support plate can cause greater wave upwelling. Both transverse and longitudinal sloshing are superimposed at the corners of the aquaculture tanks, and the narrowing of the upper part of the two adjacent walkways accelerates the upwelling, with jets appearing in the upper gaps (Figure 18IV,V). For the entire test, free liquid level observations revealed that there was no impact on the roofs of the aquaculture tank Nos.1–2.

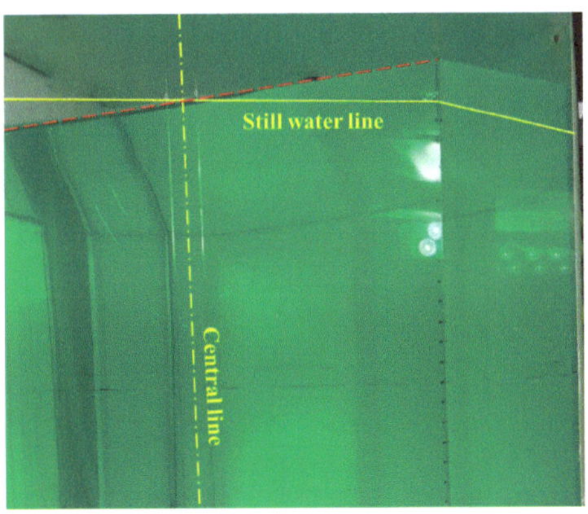

Figure 17. Free surface of tank No.1. Red line: transient free surface at the tank wall.

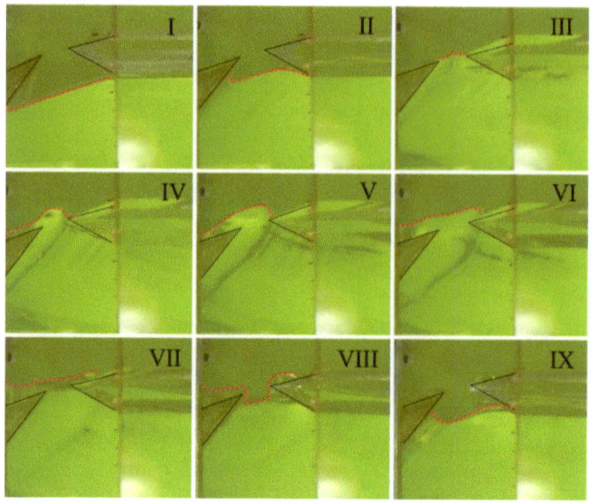

Figure 18. Upwelling process at the walkway connection of tank No.2. The black line is the end of the walkway and the red line is the transient free surface at the tank wall. The free surface begins at the design waterline (**I**), gradually rises to the top of the walkway (**II–III**), further causing a localized upsurging swell at the walkway connection (**IV–V**), and then stays briefly on the walkway (**VI–VII**) and finally begins to fall (**VIII–IX**).

The sloshing time history and response spectrum of WG1–WG3 under the half load and beam wave condition are shown in Figure 19. The maximum and minimum sloshing amplitude of WG1 are 2.27 m and −2.35 m, respectively, and the range of the sloshing response is 4.62 m, which is larger than 4.18 m of WG2 and 3.66 m of WG3. Significant higher natural modes were observed in the aquaculture tank at the half load condition, particularly the amplitude of first and second natural modes (f_{T1} and f_{T2}) were close to the wave frequency mode, and this large second natural mode was not found in the regular wave test. Since the sloshing response in this state is relatively violent, the sloshing is presented as a three-dimensional (3D) pattern, which further leads to notable first and second natural modes (f_{L1} and f_{L2}) of longitudinal sloshing of WG2 and WG3 as well. On the other hand, in the regular wave test, the doubling frequency mode can be observed for a specific period range of waves, but it is not significant in the irregular response spectrum.

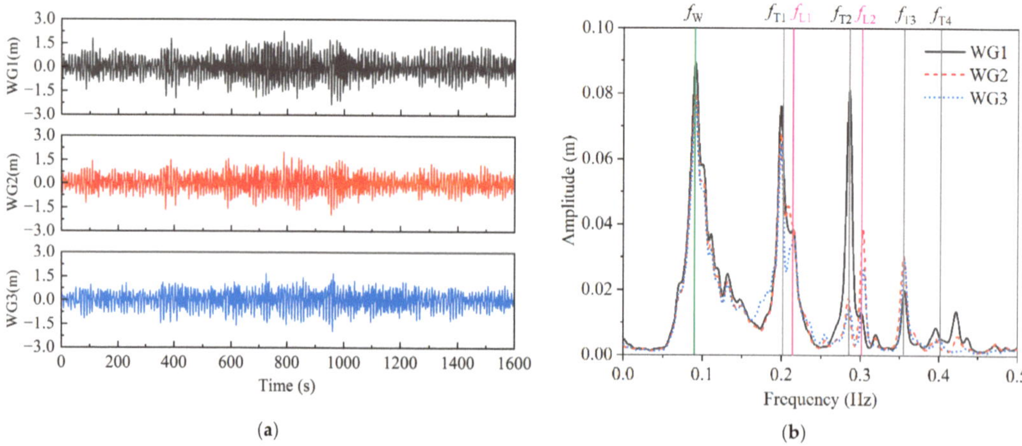

Figure 19. Sloshing response and response spectrum under half load and irregular beam irregular wave condition. (**a**) Sloshing response; (**b**) Response spectrum.

When comparing the spectrum of the starting stable stage and the later stage of the sloshing response in the regular wave test, it was found that the later stage sloshing response showed a higher natural mode, and the amplitude of the higher natural mode was also larger than that of the starting stable stage. The duration of a single regular wave test is about 45 s–80 s, while a single irregular wave test is about 300 s, which is much longer than that of a regular wave test. On the other hand, the wave height of the irregular wave test is also much larger than that of the regular wave, so the irregular wave test is more likely to evolve to higher natural modes, and the higher natural modes are also excited for a sufficient time, and the amplitude will also increase.

By observing the waveshape in the irregular wave test, it is found that at the start stage when the vessel just encountered the wave, the sloshing response is dominated by the wave frequency mode and first natural mode, and with the growth of the encounter time, the second natural mode appeared, and finally, mixed higher natural modes appeared and formed the 3D standing waves, and the typical evolution of the free surface is shown in Figure 20.

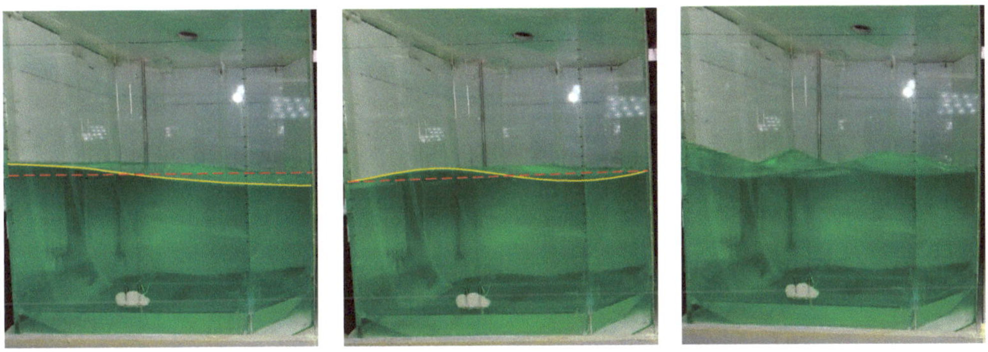

(**a**) (**b**) (**c**)

Figure 20. Typical evolution of the free surface under half load and beam wave condition. (**a**) First natural mode; (**b**) Second natural mode; (**c**) 3D standing waves. Red lines: the still water line; Yellow lines: the transient free surface at the tank wall.

4.3. Head Wave Condition

The sloshing time history and response spectrum of wave gauges WG1–WG3 under the half load and beam wave condition are shown in Figure 21. WG3 has the largest sloshing amplitude with maximum and minimum values of 0.17 m and −0.25 m, respectively. The sloshing amplitude of WG2 is slightly smaller than WG3, whereas the sloshing amplitude of WG1 is much smaller than the other two wave gauges. The wave frequency mode dominates the longitudinal sloshing in the full load condition, and the amplitude of the first natural mode of WG2 and WG3 is around 1/4 that of the wave frequency mode.

The sloshing time history and response spectrum of WG1–WG3 under the half load and head wave condition are shown in Figure 22. WG2 has the largest sloshing amplitude with maximum and minimum values of 0.22 m and −0.25 m, respectively. The sloshing amplitude of WG3 is slightly smaller than WG2. Similar to the full load condition, the wave frequency mode dominates the longitudinal sloshing in the half load condition. The sloshing response spectrum has a significant first natural frequency of the longitudinal sloshing (f_{L1}) of WG2 and WG3. The first and second natural modes (f_{L1} and f_{L2}) in the half load condition are bigger than those in the full load condition.

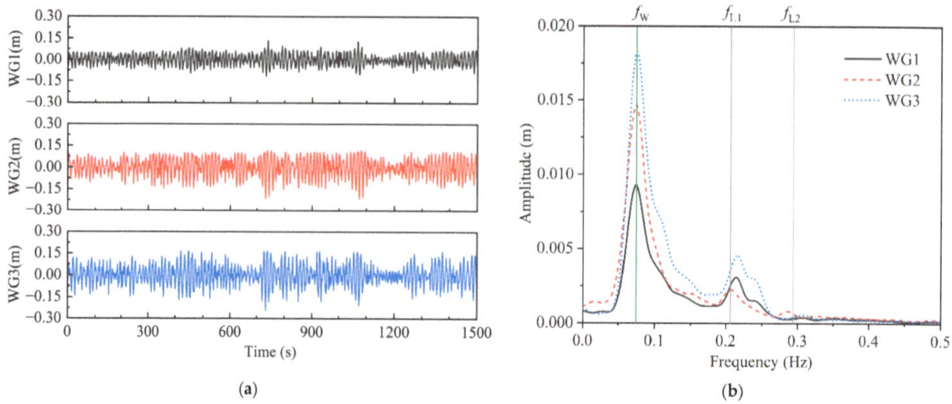

Figure 21. Sloshing response and response spectrum under full load and irregular head irregular wave condition. (**a**) Sloshing response; (**b**) Response spectrum.

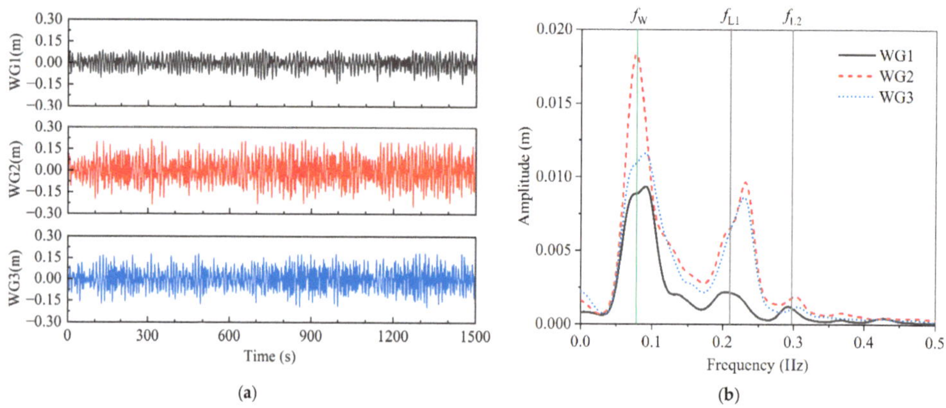

Figure 22. Sloshing response and response spectrum under half load and irregular head wave condition. (**a**) Sloshing response; (**b**) Response spectrum.

Typical free liquid surface shapes for the full and half load conditions are shown in Figure 23, where the free surface is plate-like for the full load condition and a weaker first natural mode appears on the half load condition.

(a) (b)

Figure 23. Typical free surface shape under head wave condition. (**a**) Full load; (**b**) Half load.

5. Conclusions

The sloshing response of an aquaculture vessel coupled with external waves is investigated in this study using the experimental method, with two typical filling levels and two wave directions. The frequency domain response characteristics of the sloshing under linear regular waves are analyzed in the regular wave test, and the statistical and frequency domain characteristics of the aquaculture tanks under the limiting sea state are investigated in the irregular wave test. The main conclusions are as follows:

1. In regular wave conditions, the sloshing response is dominated by the wave frequency mode on the whole, except for the case of the wave period of 10 s under the beam wave and half load condition. For the beam wave condition, the wave frequency sloshing has a maximum value when the wave period is close to the roll natural period, meanwhile, the peak of wave frequency sloshing mode in the half load condition is slightly larger than that in the full load condition. For the heading wave condition, the wave frequency sloshing mode is larger when the wave period is about 10 s to 13 s, and at this time, the wave frequency sloshing mode of the half load condition is close to that of the full load condition. The double-row tank arrangement of the vessel can reduce the breadth of the aquaculture tank, so that the first natural period of the tank deviates from the roll and pitch natural period of the hull, and the first resonance phenomenon can be better avoided. This arrangement concept is a useful scheme for the design of similar aquaculture equipment.
2. In regular wave conditions, with a wave period of 10 s, there is a significant first natural mode since the wave period is almost twice the first natural period. Particularly in the beam wave and half load condition, the amplitude of the first natural mode is around three times the wave frequency mode. In the extreme operational sea state, two times the first natural period is in the main energy range of the irregular waves, which likewise causes a more significant first natural response.
3. In the extreme sea state, the sloshing amplitude in the beam wave condition is about 7–10 times that of the head wave condition at the same filling level. The sloshing amplitude in the half load condition is 1.4 and 1.15 times that of the full load state for the beam wave and head wave condition, respectively. Therefore, a half load condition should be avoided during the culture operation, and in addition, the designer can enlarge the designed water depth in the aquaculture tank to increase the filling level. Green water occurred on the roof of the walkway in the beam wave conditions but did not impact the roof of the aquaculture tanks. In this case, personnel should not be allowed to enter the tank for culture operations, while the designer needs to pay attention to the impact loads of the walkway. Complex 3D standing waves with first and second natural modes of transverse and longitudinal sloshing were observed under the beam wave and half load conditions. Operationally, aquaculture vessels should choose an appropriate mooring scheme or sail autonomously away from typhoons to avoid harsh beam seas.
4. The nonlinearity of the sloshing response is much stronger in the extreme sea state than in the regular wave at the same filling level and wave direction, and the sloshing response in the extreme sea state has higher natural modes, particularly a significant fourth natural mode in beam wave and half load conditions. The proportion of higher natural modes to wave frequency modes has increased as well. This is owing to the irregular wave test having a long duration, which is more likely to trigger higher natural modes, and the higher natural modes evolved over enough time to increase in amplitude.

In the present study, the sloshing response was studied using the experimental method, as a comparison, the numerical simulations are also being studied, which will be reported later separately. Furthermore, the impact loads caused by sloshing and the flow pattern of the aquaculture tanks are being investigated, both of which are important for the design of the aquaculture vessels.

Author Contributions: Data curation, Y.T., Q.W. and F.H.; funding acquisition, R.Z. and J.G.; investigation, Z.Z. and X.X.; methodology, Z.Z. and X.X.; supervision, R.Z. and J.G.; writing—original draft, Y.T.; Conceptualizing, Z.L. All authors have read and agreed to the published version of the manuscript.

Funding: This work was supported by the National Natural Science Foundation of China [Grant No. 52271318].

Institutional Review Board Statement: Not applicable.

Informed Consent Statement: Not applicable.

Data Availability Statement: Data are contained within the article.

Acknowledgments: The authors acknowledge the meticulous test adjustment and data collection work of Weifa Liu at Guangzhou Salvage Bureau of the Ministry of Transport and Tengjiao Sang at AVIC Special Vehicle Research Institute of China.

Conflicts of Interest: The authors declare no conflict of interest.

Abbreviations

NO_X	Nitrogen oxides
RANS	Reynolds averaged Navier–Stokes
CIP	Constraint interpolation profile
SPH	Smoothed Particle Hydrodynamics
BL	Base line
AP	After perpendicular
WG	Wave gauge
AVIC	Aviation Industry Corporation
FPSO	Floating production storage and offloading unit
ITTC	International Towing Tank Conference
RAO	Response amplitude operator
JONSWAP	Joint North Sea Wave Project

References

1. FAO. *The State of World Fisheries and Aquaculture 2022. Towards Blue Transformation*; FAO: Rome, Italy, 2022; p. 266.
2. DNV. *Oceans' Future to 2050: Marine Aquaculture Forecast*; DNV: Høvik, Norway, 2021; p. 48.
3. Zhao, Y.P.; Guan, C.T.; Bi, C.W.; Liu, H.F.; Cui, Y. Experimental Investigations on Hydrodynamic Responses of a Semi-Submersible Offshore Fish Farm in Waves. *J. Mar. Sci. Eng.* **2019**, *7*, 238. [CrossRef]
4. Shen, Y.G.; Firoozkoohi, R.; Greco, M.; Faltinsen, O.M. Comparative investigation: Closed versus semi-closed vertical cylinder-shaped fish cage in waves. *Ocean Eng.* **2022**, *245*, 110397. [CrossRef]
5. Li, L.; Jiang, Z.; Vangdal Høiland, A.; Chen Ong, M. Numerical Analysis of a Vessel-Shaped Offshore Fish Farm. *J. Offshore Mech. Arct. Eng.* **2018**, *140*, 041201. [CrossRef]
6. Chu, Y.I.; Wang, C.M. Hydrodynamic Response Analysis of Combined Spar Wind Turbine and Fish Cage for Offshore Fish Farms. *Int. J. Struct. Stab. Dyn.* **2020**, *20*, 2050104. [CrossRef]
7. Li, Z.; Guo, X.; Cui, M. Numerical investigation of flow characteristics in a rearing tank aboard an aquaculture vessel. *Aquac. Eng.* **2022**, *98*, 102272. [CrossRef]
8. Zheng, J.H.; Xue, M.A.; Dou, P.; He, Y.M. A review on liquid sloshing hydrodynamics. *J. Hydrodyn.* **2021**, *33*, 1089–1104. [CrossRef]
9. Faltinsen, O.M.; Rognebakke, O.F.; Timokha, A.N. Resonant three-dimensional nonlinear sloshing in a square-base basin. *J. Fluid Mech.* **2003**, *487*, 1–42. [CrossRef]
10. Lugni, C.; Brocchini, M.; Faltinsen, O.M. Wave impact loads: The role of the flip-through. *Phys. Fluids* **2006**, *18*, 122101. [CrossRef]
11. Souto-Iglesias, A.; Botia-Vera, E.; Martín, A.; Pérez-Arribas, F. A set of canonical problems in sloshing. Part 0: Experimental setup and data processing. *Ocean Eng.* **2011**, *38*, 1823–1830. [CrossRef]
12. Zhu, R.Q.; Wu, Y.S. A Numerical Study on Sloshing Phenomena in a Liquid Tank. *Shipbuild. China* **2002**, *43*, 17–23.
13. Chen, Y.; Xue, M.-A. Numerical Simulation of Liquid Sloshing with Different Filling Levels Using OpenFOAM and Experimental Validation. *Water* **2018**, *10*, 1752. [CrossRef]
14. Sanchez-Mondragon, J.; Felix-Gonzalez, I.; Cruces-Giron, A.R. Turbulence analysis for vertical baffle configurations on prismatic tanks by the MPS method. *Ocean Eng.* **2022**, *264*, 112392. [CrossRef]
15. Zhang, Y.X.; Wan, D.C.; Hino, T. Comparative study of MPS method and level-set method for sloshing flows. *J. Hydrodyn. Ser. B* **2014**, *26*, 577–585. [CrossRef]
16. Cao, X.Y.; Ming, F.R.; Zhang, A.M. Sloshing in a rectangular tank based on SPH simulation. *Appl. Ocean Res.* **2014**, *47*, 241–254. [CrossRef]

17. Rognebakke, O.F.; Faltinsen, O.M. Coupling of Sloshing and Ship Motions. *J. Ship Res.* **2003**, *47*, 208–221. [CrossRef]
18. Newman, J.N. Wave effects on vessels with internal tanks. In Proceedings of the 20th International Workshop on Water Waves and Floating Bodies, Svalbard, Norway, 29 May–1 June 2005.
19. Martic, I.; Degiuli, N.; Catipovic, I. Evaluation of the added resistance and ship motions coupled with sloshing using potential flow theory. *Brodogradnja* **2016**, *67*, 109–122. [CrossRef]
20. Lee, D.Y.; Choi, H.S.; Faltinsen, O.M. A study on the sloshing effect on the motion of 2d boxes in regular waves. *J. Hydrodyn.* **2010**, *22*, 429–434. [CrossRef]
21. Nam, B.-W.; Kim, Y.; Kim, D.-W.; Kim, Y.-S. Experimental and numerical studies on ship motion responses coupled with sloshing in waves. *J. Ship Res.* **2009**, *53*, 68–82. [CrossRef]
22. Chen, B.F.; Chiang, H.W. Complete two-dimensional analysis of sea-wave-induced fully non-linear sloshing fluid in a rigid floating tank. *Ocean Eng.* **2000**, *27*, 953–977. [CrossRef]
23. Zhao, D.; Hu, Z.; Chen, G.; Chen, X.; Feng, X. Coupling analysis between vessel motion and internal nonlinear sloshing for FLNG applications. *J. Fluids Struct.* **2018**, *76*, 431–453. [CrossRef]
24. Su, Y.; Liu, Z.Y. Coupling effects of barge motion and sloshing. *Ocean Eng.* **2017**, *140*, 352–360. [CrossRef]
25. Xiao, K.L.; Chen, Z.G.; Dai, Y. Numerical Study on Coupling Effect of Multi-tank Sloshing and Ship Motion. *Ship Eng.* **2020**, *42*, 50–54. [CrossRef]
26. Zhuang, Y.; Yin, C.; Wan, D. Numerical study on coupling effect of LNG tank sloshing and ship motion in wave. In Proceedings of the The Eleven Asian Computational Fluid Dynamics Conference (ACFD11), Dalian, China, 16–20 September 2016; pp. 432–436.
27. Liu, L.W.; Feng, D.K.; Wang, X.Z.; Zhang, Z.G.; Yu, J.W.; Chen, M.X. Numerical study on the effect of sloshing on ship parametric roll. *Ocean Eng.* **2022**, *247*, 110612. [CrossRef]
28. Zhao, W.; Yang, J.; Tao, L.; White, D. Research on the Coupling Effects Between Ship Motions and Sloshing. In Proceedings of the ASME 2014 33rd International Conference on Ocean, Offshore and Arctic Engineering, San Francisco, CA, USA, 8–13 June 2014.
29. Kim, Y.; Nam, B.W.; Kim, D.W.; Kim, Y.S. Study on coupling effects of ship motion and sloshing. *Ocean Eng.* **2007**, *34*, 2176–2187. [CrossRef]
30. Zhao, W.; Taylor, P.H.; Wolgamot, H.A.; Eatock Taylor, R. Identifying linear and nonlinear coupling between fluid sloshing in tanks, roll of a barge and external free-surface waves. *J. Fluid Mech.* **2018**, *844*, 403–434. [CrossRef]
31. Igbadumhe, J.F.; Bonoli, J.; Dzielski, J.; Furth, M. Coupled FPSO roll motion response and tank sloshing in a pair of two row cargo tanks. *Ocean Eng.* **2023**, *278*, 114273. [CrossRef]
32. Li, H.; Sun, Z.Y.; Han, B.B.; Shao, Y.H.; Deng, B.L. Research on the motion response of aquaculture ship and tank sloshing under rolling response. *Brodogradnja* **2022**, *73*, 1–15. [CrossRef]
33. Neves, M.A.S.; Merino, J.A.; Rodríguez, C.A. A nonlinear model of parametric rolling stabilization by anti-roll tanks. *Ocean Eng.* **2009**, *36*, 1048–1059. [CrossRef]
34. Dou, P.; Xue, M.-A.; Zheng, J.; Zhang, C.; Qian, L. Numerical and experimental study of tuned liquid damper effects on suppressing nonlinear vibration of elastic supporting structural platform. *Nonlinear Dyn.* **2020**, *99*, 2675–2691. [CrossRef]
35. Jithu, S.; Janardhanan, S.; Asok, N. Numerical Study on the Effect of Ball Baffles in Reducing Sloshing Loads in Ship Tanks. In Proceedings of the International Conference on Computational and Experimental Marine Hydrodynamics, Chennai, India, 25–27 November 2018.
36. Kim, T.W.; Kim, S.K.; Park, S.B.; Lee, J.M. Design of Independent Type-B LNG Fuel Tank: Comparative Study between Finite Element Analysis and International Guidance. *Adv. Mater. Sci. Eng.* **2018**, *2018*, 5734172. [CrossRef]
37. Trimulyono, A.; Atthariq, H.; Chrismianto, D.; Samuel, S. Investigation of sloshing in the prismatic tank with vertical and T-shape baffles. *Brodogradnja* **2022**, *73*, 43–58. [CrossRef]
38. Yu, Y.M.; Ma, N.; Fan, S.M.; Gu, X.C. Experimental and numerical studies on sloshing in a membrane-type LNG tank with two floating plates. *Ocean Eng.* **2017**, *129*, 217–227. [CrossRef]
39. Koh, C.G.; Luo, M.; Gao, M.; Bai, W. Modelling of liquid sloshing with constrained floating baffle. *Comput. Struct.* **2013**, *122*, 270–279. [CrossRef]
40. Cui, M.; Li, Z.; Zhang, C.; Guo, X. Statistical investigation into the flow field of closed aquaculture tanks aboard a platform under periodic oscillation. *Ocean Eng.* **2022**, *248*, 110677. [CrossRef]
41. Gao, R.; Cui, M.C.; Wang, Q.W.; Wang, J. Numerical Aanalysis on Roll Fish-suitability of Aquaculture Water Tank. *Ship Eng.* **2020**, *42*, 35–47. [CrossRef]
42. Wiegerink, J.J.; Baldock, T.E.; Callaghan, D.P.; Wang, C.M. Experimental study on hydrodynamic response of a floating rigid fish tank model with slosh suppression blocks. *Ocean Eng.* **2023**, *273*, 113772. [CrossRef]
43. Tao, Y.W.; Zhu, R.Q.; Gu, J.Y.; Li, Z.Y.; Zhang, Z.Y.; Xu, X.S. Experimental and numerical investigation of the hydrodynamic response of an aquaculture vessel. *Ocean Eng.* **2023**, *279*, 114505. [CrossRef]
44. ITTC. Recommended Procedures and Guidelines: Ship Models. 2017, 7.5-01-01-01. Available online: https://www.ittc.info/media/9571/75-01-01-01.pdf (accessed on 6 October 2023).
45. ITTC. Recommended Procedures and Guidelines: Seakeeping Experiments. 2017, 7.5-02-07-02.1. Available online: https://www.ittc.info/media/9705/75-02-07-021.pdf (accessed on 6 October 2023).

46. ITTC. Recommended Procedures and Guidelines: Sloshing Model Tests. 2021, 7.5-02-07-02.7. Available online: https://www.ittc.info/media/9715/75-02-07-027.pdf (accessed on 6 October 2023).
47. DNV. *Environmental Conditions and Environmental Loads*; DNV: Høvik, Norway, 2014; DNV-RP-C205, 182. Available online: https://www.dnv.com/rules-standards/index.html (accessed on 6 October 2023).

Journal of
Marine Science
and Engineering

MDPI

Article

The Photoperiod Significantly Influences the Growth Rate, Digestive Efficiency, Immune Response, and Antioxidant Activities in the Juvenile Scalloped Spiny Lobster (*Panulirus homarus*)

Yinggang Wang [1,2], Rui Yang [1,2], Zhengyi Fu [1,2,3], Zhenhua Ma [1,2,3,*] and Zemin Bai [4,*]

1 Key Laboratory of Efficient Utilization and Processing of Marine Fishery Resources of Hainan Province, Sanya Tropical Fisheries Research Institute, Sanya 572018, China; yinggangwang2024@163.com (Y.W.)
2 South China Sea Fisheries Research Institute, Chinese Academy of Fishery Sciences, Guangzhou 510300, China
3 College of Science and Engineering, Flinders University, Adelaide 5001, Australia
4 Yazhou Bay Agriculture and Aquaculture Co., Ltd., Sanya 572025, China
* Correspondence: zhenhua.ma@scsfri.ac.cn (Z.M.); baizemin@163.com (Z.B.)

Abstract: This study aimed to elucidate the effects of different photoperiods (0 L:24 D, 6 L:18 D, 12 L:12 D, 18 L:6 D, 24 L:0 D, "Light (L) and Dark (D)") on the growth performance and physiological responses of the juvenile scalloped spiny lobster (*Panulirus homarus*). Over a period of 56 days, parameters such as growth rate, digestive enzyme, immune enzyme, and antioxidant enzyme were meticulously evaluated in 90 lobsters subjected to these varying light conditions. The present study found no significant differences in survival rate (SR), molting frequency (MF), and meat yield production (MYP) among the various photoperiod treatments ($p > 0.05$). Notably, the highest weight gain rate (WGR) and specific growth rate (SGR) were observed under a 12 L:12 D photoperiod. In the continuous dark phase (0 L:24 D), pepsin (PEP) activity remained high in gastric tissues, while trypsin (TRYP) and chymotrypsin (CHT) activities reached the highest in hepatopancreas tissues. The α-amylase (AMS) activity in the hepatopancreas was most elevated under 18 L:6 D, and the optimal lipase (LPS) activity was recorded under 12 L:12 D. The activity of acid phosphatase (ACP) in the hepatopancreas was highest in the absence of light (0 L:24 D), whereas the activities of alkaline phosphatase (AKP) and lysozyme (LZM) were most effective under the 12 L:12 D photoperiod. The total antioxidant capacity (T-AOC), along with catalase (CAT) and superoxide dismutase (SOD) activities of the hepatopancreas reached the highest at 12 L:12 D. The highest activity of glutathione peroxidase (GSH-Px) was seen under 18 L:6 D. The concentration of malondialdehyde (MDA), a marker of oxidative stress, was found to be highest under 12 L:12 D. Consequently, this specific photoperiod is essential for achieving optimal growth and maintaining appropriate physiological balance in the scalloped spiny lobster during aquaculture. These findings provide a foundational guideline for establishing the lighting environment in the farming of the juvenile scalloped spiny lobster.

Keywords: lobster *Panulirus homarus*; photoperiod; growth rate; digestive enzymes; non-specific immunity

Citation: Wang, Y.; Yang, R.; Fu, Z.; Ma, Z.; Bai, Z. The Photoperiod Significantly Influences the Growth Rate, Digestive Efficiency, Immune Response, and Antioxidant Activities in the Juvenile Scalloped Spiny Lobster (*Panulirus homarus*). *J. Mar. Sci. Eng.* **2024**, *12*, 389. https://doi.org/10.3390/jmse12030389

Academic Editor: Valerio Zupo

Received: 29 January 2024
Revised: 18 February 2024
Accepted: 21 February 2024
Published: 24 February 2024

1. Introduction

The scalloped spiny lobster (*Panulirus homarus* Linnaeus, 1878), a large warm-water crustacean, inhabits tropical and subtropical coasts [1]. Its distribution and aquaculture are primarily concentrated in the Indo-West Pacific region [2,3]. Recognized as one of the most valuable seafood products [4], the scalloped spiny lobster is highly sought after in international markets due to its exquisite taste and rich nutritional profile [5]. However, with the development of fishery, the natural resources of lobsters have decreased dramatically [4–11]. In order to meet the market demand, larvae are captured from the wild and reared artificially [6–11]. Current research on this species encompasses a range of topics, including breeding practices [6], nutritional analysis [7], capture techniques [8,9], resource assessment [9–11], habitat studies [12], physiological characteristics [13–15], disease management [16,17], and genomics [18–20]. Despite these advancements, the impact of lighting conditions on scalloped spiny lobsters remains relatively underexplored. Presently, there is a notable gap in standardized breeding protocols that account for light conditions.

Light serves as a critical ecological factor within aquatic ecosystems [21]. It exerts a profound influence on various physiological processes and feeding behaviors in aquatic animals [22], modulating growth performance and developmental stages [23–25]. Notably, light impacts the growth and development of diverse species, including fish [26,27], shrimp [28], crabs [29], other crustaceans [22], and turtles [30], along with various physiological responses. As a pivotal environmental element, light is instrumental in regulating the circadian rhythms of these organisms [31]. Light possesses three fundamental characteristics: intensity, spectrum (or light quality), and photoperiod [21]. Among these, the photoperiod is particularly crucial for managing biological rhythms in aquatic life [32]. The diverse array of life on Earth is profoundly influenced by the varying light conditions resulting from the planet's rotation around the Sun [32,33]. To cope with these consistent changes, terrestrial and aquatic organisms have developed intricate internal circadian rhythm systems [33]. Consequently, the photoperiod is recognized as a critical environmental cue impacting these biological rhythms [21]. In crustaceans, the photoperiod can directly or indirectly affect the circadian system, influencing growth, molting, and reproductive processes [34,35]. Optimizing photoperiods in aquaculture can enhance the growth and development rates of aquatic animals, thereby potentially shortening their production cycles [36].

Extensive research illustrates the impact of the photoperiod on vital activities and physiological responses in various aquatic species. In eastern rock lobsters (*Sagmariasus verreauxi*), longer photoperiods (18 L:6 D and 24 L:0 D) significantly improved survival and growth rates during larval metamorphosis [37]. Photoperiods of 6 L:18 D, 12 L:12 D, and 18 L:6 D have been shown to promote early larval growth, molting, and feeding in spiny lobsters (*Jasus edwardsii*) [38]. In juvenile goldfish (*Carassius auratus*), a light duration exceeding 16 h enhanced fat formation, lipolysis, and fatty acid oxidation [33]. Long photoperiods (16 L:8 D) induced stress in blunt snout bream (*Megalobrama amblycephala*), increasing plasma cortisol levels and causing oxidative stress [39]. Similarly, long photoperiods (18 L:6 D) significantly elevated plasma gonadotropins in greater amberjack (*Seriola dumerili*) [40] and affected osmoregulation and hepatic energy metabolism in cultured olive flounder (*Paralichthys olivaceus*), triggering stress responses [41]. Shorter photoperiods (2 L:22 D and 4 L:20 D) enhanced innate immune and antioxidant responses in white leg shrimp (*Penaeus vannamei*), reducing mortality in adult white leg shrimp [34]. Continuous darkness (0 L:24 D) was found to increase reproductive efficiency and sperm production in male narrow-clawed crayfish (*Pontastacus leptodactylus*) [42] and to elevate molt frequency and growth rate in juvenile mud crabs (*Scylla paramamosain*) [29]. Persian sturgeon (*Acipenser persicus*) exhibited the lowest stress levels in a no-light environment (0 L:24 D) [43]. These findings underscore the importance of the photoperiod in influencing the growth and behavior of aquatic animals. Adjusting the photoperiod in aquacultural settings may offer a means to shorten production cycles and enhance overall production efficiency. However, the specific effects of the photoperiod on the growth, development, and physiology of the scalloped spiny lobster remain to be fully determined.

Altered photoperiods have been shown to significantly influence the functioning of digestive and non-specific immune enzymes in aquatic animals. Several studies have suggested that external environmental factors, including light, can modulate the activity of digestive enzymes, which are critical indicators of digestive performance in aquatic species [22]. For instance, a 13 L:11 D photoperiod significantly alters the activities of pepsin (PEP), lipase (LPS), α-amylase (AMS), trypsin (TRYP), and chymotrypsin (CHT) in the tiger shrimp (*Macrobrachium tenellum*) [44]. Total darkness (0 L:24 D) has been reported to maximize LPS and AMS activities in the fairy shrimp (*Branchinecta orientalis*) [45,46], while a prolonged photoperiod (18 L:6 D) affects LPS and TRYP activities in spotted sea bass (*Lateolabrax maculatus*) [23]. Similarly, the highest TRYP and LPS activities in the Chinese soft-shell turtle were observed in complete darkness (0 L:24 D) [30].

Key barrier enzymes like alkaline phosphatase (AKP), acid phosphatase (ACP), and lysozyme (LZM), which protect aquatic animals from pathogenic bacteria, are widely distributed in their organs and tissues, serving as vital indicators of immune levels [47–49]. LZM can release hydrolytic enzymes that break down pathogens, thus playing an immune defense role [48–50]. ACP can catalyze the hydrolysis of organophosphorus; participate in phagocytosis, nodules, and envelope formation; regulate the activity of immune cells; help the immune system to better recognize and remove foreign substances; and maintain health [47,49]. The role of AKP in the immune system is mainly reflected in promoting the activation and proliferation of immune cells and enhancing the intercellular immune response [47]. It can enhance immunity by promoting the activity of immune cells, helping cells to recognize and resist [47,50,51]. Optimal ACP and LZM activities were recorded in freshwater shrimp (*Macrobrachium rosenbergii*) under complete darkness [50] and in tiger puffer (*Takifugu rubripes*) larvae under a long photoperiod (16 L:8 D) [51]. In brown frogs (*Rana dybowskii*), the best LZM activity was noted under light-free conditions [52]. The activity level of total antioxidant capacity (T-AOC) reflects the aquatic organisms' antioxidant enzymes and non-enzymatic antioxidants, crucial for stress resistance and overall health and growth [53]. Reactive oxygen species (ROS) overproduction can impair physiological functions and cause oxidative damage to essential biomolecules [39]. Enzymes like superoxide dismutase (SOD), catalase (CAT), and glutathione peroxidase (GSH-Px) form the core of the organism's antioxidant system, scavenging ROS and playing a pivotal role in their elimination [54]. Malondialdehyde (MDA), a marker of ROS levels and a byproduct of lipid peroxidation, can indicate cellular and tissue damage [35]. In white leg shrimp, MDA levels increased with longer light durations [35], while blunt snout bream exhibited higher SOD, CAT, and GSH activities with light durations exceeding 12 h [39]. Additionally, abalones showed a gradual increase in T-AOC, SOD, GSH-Px activity, and GSH contents with increasing light duration [55]. These findings underscore the significance of the photoperiod in determining the activities of digestive and immune enzymes in aquatic animals. Regulating the photoperiod can, thus, enhance digestive and immune functions in these species. However, the specific impacts of the photoperiod on the digestive, immunological, and antioxidant enzymes in the scalloped spiny lobster remain to be elucidated.

The primary objective of this study was to examine the physiological responses of the scalloped spiny lobster to various photoperiods (0 L:24 D, 6 L:18 D, 12 L:12 D, 18 L:6 D, 24 L:0 D). Specifically, the research focused on assessing the impact of these photoperiods on the lobster's digestive performance, immune function, and antioxidant capacity. The outcomes of this study aim to establish an optimal photoperiod for the scalloped spiny lobster aquaculture. This would not only contribute to formulating a standardized system for regulating light duration in lobster farming but also optimize the production cycle and enhance the economic efficiency of this valuable seafood resource.

2. Materials and Methods

2.1. Experimental Materials

Scallop spiny lobsters are lobsters that have been farmed for six months after being caught in the wild. The experiment was conducted in the circulating water system of an indoor factory in Lingshui Experimental Station, Sanya Tropical Fisheries Research Institute (Sanya, China). For the experiment, ninety healthy lobsters of uniform size were selected, each averaging a weight of 171.28 ± 23.12 g. Initially, six samples were randomly placed in fifteen 15-L fiberglass canisters for a week of domestication under experimental photoperiodic conditions. Feeding was conducted every evening at 18:00 using the satiation method, where 8 to 12% of the lobsters' body mass in chilled crab bait was provided. Subsequently, at approximately 8:00 am the following day, tank maintenance involved cleaning the bottom of the tanks to remove residual feed and feces.

2.2. Experimental Design

The scalloped spiny lobsters were accurately weighed and allocated into fifteen 15-L experimental tanks, part of a recirculating water system. Each treatment group consisted of three replicates, with six lobsters per replicate. The experiment utilized natural seawater, which was sedimented, sand-filtered, and continuously aerated for 24 h. To ensure optimal water quality, a daily 100% water change was implemented. Monitored water quality parameters included the following: temperature between 28 and 32 °C, salinity from 28 to 32‰, pH levels ranging from 7.5 to 8.5, dissolved oxygen concentration of at least 7.0 mg/L, ammonia nitrogen concentration below 0.02 mg/L, and nitrite levels not exceeding 0.02 mg/L. Based on previous studies [21,56], five photoperiod groups were established: 24 L:0 D, 18 L:6 D, 12 L:12 D, 6 L:18 D, and 0 L:24 D (L: light, D: dark). Photoperiods were controlled using a programmable timer (Gongniu Group Co., Ltd., Ningbo, China). The experiment spanned 56 days, with a maintained light intensity of 100 lx and a spectrum range of white light (λ = 400–770 nm) (Opple Lighting Co., Ltd., Shanghai, China). Shade cloths were employed to achieve total darkness in each experimental group, with lights scheduled to turn on daily at 6:30 am.

2.3. Calculation of Growth Performance

Growth indicators such as weight gain rate (WGR), specific growth rate (SGR), survival rate (SR), molting frequency (MF), hepatopancreatic index (HSI), and meat yield production (MYP) of the scalloped spiny lobster were measured and calculated using the following equations:

$$\mathrm{WGR} = (\mathrm{W_t} - \mathrm{W_0})/\mathrm{W_0} \times 100\%$$
$$\mathrm{SGR} = (\ln \mathrm{W_t} - \ln \mathrm{W_0})/\mathrm{t} \times 100\%$$
$$\mathrm{SR} = \mathrm{N_f}/\mathrm{N_i} \times 100\%$$
$$\mathrm{MF} = (\mathrm{N_m}/\mathrm{N_s}) \times 100\%$$
$$\mathrm{HSI} = (\mathrm{W_g}/\mathrm{W_0}) \times 100\%$$
$$\mathrm{MYP} = (\mathrm{W_S}/\mathrm{W_t}) \times 100\%$$

where WGR is weight gain rate (%), SGR is specific growth rate (%/d), SR is survival rate (%), MF is molting frequency (%), HSI is hepatopancreas index (%), and MYP is meat yield production (%). W_0 is the initial body mass of the experimental lobster (g), W_t is the final body mass of the experimental lobster (g), t is the experimental time (d), N_f is the initial number of experimental lobsters (only), N_i is the final number of experimental lobsters (only), N_m is the number of molted experimental lobsters in a single bucket (only), N_s is the total number of experimental lobsters in a single bucket (only), W_s is the net meat weight (g), and W_t is the final body mass of the experimental lobster (g).

2.4. Sample Collection and Processing

At the end of the experiment, lobsters were anesthetized in an ice bath for two minutes. Before dissection, their body surfaces were carefully dried using absorbent paper. The dissection was conducted using sterilized tools and performed in a consistently cold environment to preserve tissue integrity. Appropriate amounts of lobster stomach and hepatopancreas tissues were collected in 2 mL freezing tubes, frozen in liquid nitrogen, and stored at −80 °C in a refrigerator. To minimize sampling error, samples from each treatment group were stored separately, and then samples with different duplicates from the same treatment group were mixed and extracted.

For tissue analysis, a 10% homogenized tissue solution was prepared. Specified amounts of tissue samples were weighed and added to a pre-cooled homogenization medium, followed by thorough grinding. The resulting tissue homogenate was then transferred to a high-speed centrifuge, maintained at 4 °C, and centrifuged at 3500 rpm for 10 min. Post-centrifugation, the supernatant of the homogenized tissue was collected and stored at −80 °C for subsequent analysis.

Prior to enzyme activity assays, the supernatant was diluted to the optimal concentration as per the requirements of each specific enzyme assay. Enzyme activities were then quantitatively determined according to the standardized procedures provided in the respective assay kits.

2.5. Determination of Indicators of Enzyme Activity

All biochemical assays were conducted using kits provided by the Nanjing Jiancheng Institute of Biological Engineering (Nanjing, China), strictly following the manufacturer's instructions. For the analysis of digestive enzymes and related indicators, gastric tissues were specifically harvested to measure PEP activity and total protein (TP) content. Similarly, hepatopancreatic tissues were collected for a comprehensive assessment of various enzymes and biochemical markers. These included LPS, AMS, TRYP, and CHT activities, alongside assays for ACP, AKP, and LZM activities. Additionally, key oxidative stress markers and antioxidant enzymes were quantified in hepatopancreas, including MDA content, SOD, CAT, GSH-Px, and T-AOC. Concurrently, protein TP was also performed on these samples.

The protein content was determined using the Coomassie Brilliant Blue method with bovine serum protein as the standard used in the protein quantitative kit (Catalog No. A045-4, Nanjing, China), incubated at 37 °C for 30 min at 562 nm wavelength, and the protein concentration was measured using microplate colorimetry. A PEP detection kit (catalog No. A080-1-1, Nanjing, China) was used to determine the activity of PEP in animal tissue samples. PEP can hydrolyze protein to produce phenol-containing amino acids, and phenol reagents can be reduced to blue substances by phenol-containing amino acids. The absorbance value at the zero setting point of distilled water at a 37 °C water bath for 20 min was determined via colorimetry. The unit of activity is defined as follows: 1 μg tyrosine generated by decomposing protein per milligram at 37 °C per minute is equivalent to 1 unit of enzyme activity (U/mgprot). A TRYP assay kit (Catalog No. A080-2-2, Nanjing, China) was used to determine the activity of TRYP in the animal tissue samples. TRYP can catalyze the hydrolysis of the ester chain of ethyl arginine and increase its absorbance at 253 nm. The activity of the enzyme can be calculated according to the change in absorbance. The activity unit was defined as follows: under the condition of pH 8.0, 37 °C, the absorbance change of 0.003 per minute by TRYP contained in each milliliter of serum (pulp) is an enzyme activity unit (U/mgprot). The AMS assay kit (catalog No. C016-1-1, Nanjing, China) was used to determine the activity of AMS in animal tissue samples. AMS can hydrolyze starch to produce glucose, maltose, and dextrin. When the concentration of a substrate was known and excessive, iodine solution was added to combine with unhydrolyzed starch to form a blue complex. The amount of hydrolyzed starch was calculated according to the depth of blue, so as to calculate the activity of AMS. The unit of activity was defined as follows: each milligram of protein in the tissue reacted with the substrate at 37 °C for 30 min, and hydrolyzed 10 mg of starch was defined as 1 unit of amylase activity (U/mgprot). A CHT assay kit (Catalog No. A080-3-1, Nanjing, China) was used to determine CHT activity in animal tissue samples. Using casein as a substrate, CHT hydrolyzed protein to produce phenol-containing amino acids, phenol reagents were reduced to blue substances by phenol-containing amino acids, and CHT activity was determined via colorimetry. The unit of activity was defined as follows: 1μg amino acid generated by decomposing protein per milligram at 37 °C per minute was equivalent to 1 unit of enzyme activity (U/mgprot). The activity of LPS in animal tissue samples was determined with an LPS detection kit (catalog No. A054-2-1, Nanjing, China). 1, 2-o-dilaurin-racemic glycerol-3-valerate-(6-methylhalide) ester +H_2O→1, 2-o-dilaurin-racemic glycerol + valerate-(6-methylhalide) ester, valerate-(6-methylhalide) ester→valerate + 6-methylhalide (color development), at 580 nm wavelength. The activity of LPS was determined according to the production rate of the red product. The activity unit was defined as follows: at 37 °C, each gram of hiprotein reacted with the substrate in this reaction system for 1 min, and each consumption of 1 μmol of the substrate was an enzyme activity unit (U/gprot).

An ACP assay kit (Catalog No. A060-2-2, Nanjing, China) was used to determine the activity of ACP in the animal tissue samples. ACP decomposes disodium phenyl phosphate to produce free phenol and phosphoric acid. Phenol reacts with 4-amino-antipyrine in alkaline solution to oxidize red quinone derivatives by potassium ferricyanide. The activity of enzyme was measured according to the red intensity. The unit of activity was defined as follows: 100 mL of serum or liquid at 37 °C with the matrix for 30 min to produce 1 mg of phenol as 1 Gold unit/mgprot. An AKP assay kit (Catalog No. A059-2-2, Nanjing, China) was used to determine AKP activity in the animal tissue samples. AKP decomposed phenylene disodium phosphate to produce free phenol and phosphoric acid. Phenol reacted with 4-amino-antipyrine in alkaline solution to oxidize red quinone derivatives by potassium ferricyanide. The activity of enzyme was measured according to the red intensity. The activity unit was defined as one Gold unit/mgprot for 1 mg of phenol produced per gram of hiprotein interacting with the matrix at 37 °C for 15 min. The LZM assay kit (Catalog No. A050-1-1, Nanjing, China) was used to determine the activity of LZM in the animal tissue samples. LZM can hydrolyze peptidoglycan on the cell wall of bacteria, resulting in bacterial lysation with decreased concentration and increased transmittance, so the content of LZM was estimated according to the change in transmittance. The activity unit was defined as follows: accurate water bath for 15 min per milliliter of liquid at 37 °C, ice water bath for 3 min below 0 °C, removed tube-by-tube, poured into the light diameter of a 1 cm colorimetric dish, at a 530 nm wavelength, double steaming water regulation light transmission rate of 100%, and the colorimetric determination of the light value of each tube is a unit of vitality (U/mL).

A T-AOC assay kit (Catalog No. A015-1, Nanjing, China) was used to determine the activity of T-AOC in the animal tissue samples. Under the action of appropriate oxidants, ABTS is oxidized to green $ABTS^+$, and in the presence of antioxidants, the production of $ABTS^{\cdot+}$ is inhibited. The T-AOC of the samples was determined by measuring the absorbance of $ABTS^+$ at 405 nm. The activity unit was defined as follows: reaction at room temperature for 6 min, wavelength of 405 nm, and absorbance (OD) value of the reaction system could be directly used as T-AOC activity unit (mM). A CAT test kit (Catalog No.:A007-1-1, Nanjing, China) was used to determine the activity of CAT in the animal tissue samples. The decomposition reaction of H_2O_2 by CAT could be quickly stopped by adding ammonium molybdate. The remaining H_2O_2 reacted with ammonium molybdate to produce a light-yellow complex. The activity of CAT was calculated by measuring its change at 405 nm. The unit is defined as the decomposition of 1 μmol of H_2O_2 per milligram of histone per second as one unit of activity (U/gHb). The GSH-Px activity in the tissues was measured with a GSH-Px determination kit (Catalog No. A005-1, Nanjing, China). The GSH-Px activity was expressed by the consumption rate of GSH in the enzymatic reaction, while the more stable yellow substance formed by GSH and dithiodinitrobenzoic acid was determined through colorimetry to calculate the GSH-Px activity. Through the colorimetric method, a 1 cm optical path cuvette was used at a 412 nm wavelength, the distilled water was adjusted to zero, the absorbance value was measured, and its activity was calculated. The activity unit U indicates that the GSH concentration in the reaction system is reduced by 1% per milligram of protein per minute by deducting a non-enzymatic reaction in $\mu mol \cdot L^{-1}$. The SOD test kit (Catalog No. A001-3, Nanjing, China) was used to measure the activity of the SOD in the animal tissue samples. The activity of the SOD was determined using the xanthine oxidase method. The absorbance value was measured at the wavelength of 550 nm through colorimetry to calculate its activity. The activity unit was defined as follows: when the SOD inhibition rate reached 50% per milligram of tissue protein in 1 mL of the reaction solution, the corresponding amount of SOD was 1 SOD activity unit ($U \cdot mgprot^{-1}$). The MDA determination kit (Catalog No. A003-1, Nanjing, China) was used to measure the content of MDA in the animal tissues. The MDA was condensed with thiobarbituric acid to form a red substance, and MDA was determined through colorimetry at 532 nm. All samples were processed in triplicate.

2.6. Data Analysis

All collected data underwent a normality test to confirm their distribution patterns. Subsequent statistical analyses were conducted using one-way ANOVA, followed by

Duncan's multiple range test to identify significant differences among the groups. The levels of significance were set at $p < 0.05$ for significant differences and $p < 0.01$ for highly significant differences. These analyses were performed using the SPSS 26.0 (SPSS, Chicago, IL, USA) statistical software package. For data presentation, all values were expressed as mean ± standard deviation (mean ± SD). Graphical representations of the data were generated using Origin 2022 (OriginLab Corporation, Northampton, MA, USA) software.

3. Results

3.1. Growth Performance

According to the data in Table 1 and the ANOVA results in Table 2, the WGR and SGR were significantly higher in lobsters exposed to a 12 h light and 12 h dark cycle (12 L:12 D) compared to those under 6 L:18 D and continuous darkness (0 L:24 D) conditions. While longer photoperiods (24 L:0 D and 18 L:6 D) facilitated faster growth than shorter ones (6 L:18 D and 0 L:24 D), the differences between these longer photoperiod groups were not statistically significant. Notably, the HSI was significantly greater in the 6 L:18 D group than in other experimental conditions. However, the photoperiod did not significantly influence ($p > 0.05$) the SR, MF, and MYP of the scalloped spiny lobsters, as detailed in Table 1.

Table 1. Influence of photoperiod on growth performance of *Panulirus homarus*.

Group	WGR (%)	SGR (%/d)	SR (%)	MF (%)	HSI (%)	MYP (%)
24 L:0 D	25.09 ± 4.80 [ab]	0.40 ± 0.07 [ab]	100.00 ± 0.00	88.89 ± 31.43	3.49 ± 0.19 [b]	23.14 ± 1.17
18 L:6 D	23.21 ± 5.36 [ab]	0.37 ± 0.08 [ab]	100.00 ± 0.00	88.89 ± 15.71	3.41 ± 0.22 [b]	23.32 ± 1.63
12 L:12 D	32.24 ± 4.35 [a]	0.50 ± 0.06 [a]	100.00 ± 0.00	88.89 ± 31.43	3.48 ± 0.19 [b]	24.16 ± 0.80
6 L:18 D	19.79 ± 2.24 [b]	0.32 ± 0.03 [b]	100.00 ± 0.00	100.00 ± 0.00	4.09 ± 0.23 [a]	23.76 ± 0.52
0 L:24 D	21.51 ± 2.96 [b]	0.35 ± 0.04 [b]	100.00 ± 0.00	111.11 ± 15.71	3.48 ± 0.26 [b]	24.60 ± 1.71

Note: Different lowercase values within the same column are significant ($p < 0.05$). WGR is weight gain rate (%), SGR is specific growth rate (%/d), SR is survival rate (%), MF is molting frequency (%), HSI is hepatopancreas index (%), MYP is meat yield production (%).

Table 2. The ANOVA results of WGR, SGR, SR, MF, his, and MYP.

Item	df	MS	F	*p*
WGR (%)	4	69.73	2.75	0.09
SGR (%/d)	4	0.01	2.71	0.09
SR (%)	4	0.00		
MF (%)	4	0.09	0.40	0.80
HSI (%)	4	0.24	3.33	0.06
MYP (%)	4	1.08	0.46	0.77

3.2. Digestive Properties

According to the results of the variance analysis shown in Figure 1 and Table 3, PEP activity was found to be significantly higher ($p < 0.05$) in the 0 L:24 D (10.71 ± 0.56 U·mgprot^{-1}), 6 L:18 D (9.49 ± 0.56 U·mgprot^{-1}), and 18 L:6 D (10.23 ± 0.99 U·mgprot^{-1}) photoperiod groups compared to the 24 L:0 D (7.04 ± 0.35 U·mgprot^{-1}) and 12 L:12 D (7.47 ± 0.43 U·mgprot^{-1}) groups, as shown in Figure 1A. Similarly, TRYP and CHT activities were significantly elevated ($p < 0.05$) in the 0 L:24 D (1173.83 ± 536.76 U·mgprot^{-1} and 3.15 ± 0.18 U·mgprot^{-1}) group relative to the other experimental groups, as depicted in Figure 1B,D, respectively. For AMS activity, the highest levels were observed in the 24 L:0 D (11.70 ± 2.15 U·mgprot^{-1}) and 18 L:6 D (13.86 ± 2.77 U·mgprot^{-1}) groups, significantly surpassing those in other groups ($p < 0.05$), as indicated in Figure 1C. Additionally, LPS activity was significantly higher ($p < 0.05$) in the 0 L:24 D (0.70 ± 0.09 U·gprot^{-1}) and 12 L:12 D (1.07 ± 0.18 U·gprot^{-1}) photoperiod groups compared to the rest, as illustrated in Figure 1E.

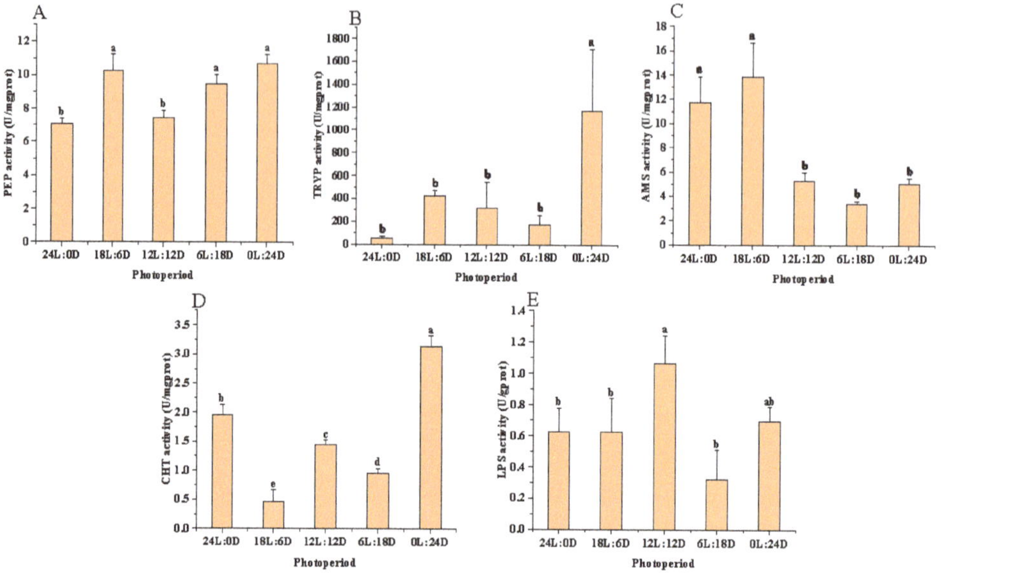

Figure 1. Effect of photoperiod on digestive enzyme activities in scalloped spiny lobsters (n = 18). Pepsin (**A**), trypsin (**B**), amylase (**C**), chymotrypsin (**D**), and lipase (**E**). Different superscripts letters indicate statistically significant differences between treatments ($p < 0.05$).

Table 3. The ANOVA results of PEP, TRYP, AMS, CHT, and LPS.

Item	df	MS	F	*p*
PEP	4	8.12	14.11	0.00
TRYP	4	581,512.53	5.54	0.01
AMS	4	64.03	16.40	0.00
CHT	4	3.19	89.50	0.00
LPS	4	0.21	4.78	0.02

3.3. *Immune Function*

According to the results of variance analysis in Figure 2 and Table 4, ACP activity showed no significant difference between the 0 L:24 D (0.16 ± 0.02 gold unit·gprot^{-1}) and 12 L:12 D (0.12 ± 0.02 gold unit·gprot^{-1}) photoperiods ($p > 0.05$). However, its activity was significantly higher in the 0 L:24 D photoperiods compared to 18 L:6 D (0.10 ± 0.01 gold unit·gprot^{-1}), 24 L:0 D (0.12 ± 0.02 gold unit·gprot^{-1}), and 6 L:18 D (0.11 ± 0.02 gold unit·gprot^{-1}) ($p < 0.05$), as shown in Figure 2A. In terms of AKP activity, the levels of the 18 L:6 D group (0.04 ± 0.00 gold unit·gprot^{-1}) and 12 L:12 D group (0.04 ± 0.00 gold unit·gprot $^{-1}$) were significantly higher than other groups, as shown in Figure 2B. For LZM activity, the 12 L:12 D (76.77 ± 7.01 U·mL^{-1}) and 6 L:18 D (85.30 ± 5.64 U·mL^{-1}) groups had significantly higher activity compared to the 0 L:24 D (58.4 ± 3.71 U·mL^{-1}), 18 L:6 D (67.59 ± 6.50 U·mL^{-1}), and 24 L:0 D (64.96 ± 1.61 U·mL^{-1}) groups ($p < 0.05$). However, there was no significant difference in LZM activity between the 12 L:12 D and 6 L:18 D groups ($p > 0.05$), as indicated in Figure 2C.

3.4. *Antioxidant Capacity*

According to the results of the variance analysis shown in Figure 3 and Table 5, the activities of SOD were significantly higher in the 24 L:0 D (15.69 ± 1.91 U·mgprot^{-1}), 18 L:6 D

(18.85 ± 3.38 U·mgprot^{-1}), and 12 L:12 D (17.87 ± 3.67 U·mgprot^{-1}) groups compared to the 0 L:24 D (8.12 ± 2.17 U·mgprot $^{-1}$) group ($p < 0.05$). However, these SOD levels did not show a significant difference from those in the 6 L:18 D group ($p > 0.05$), as illustrated in Figure 3A. In terms of CAT activity, 18 L:6 D (0.01 ± 0.00 U·gHb^{-1}) was significantly lower than other experimental groups ($p < 0.05$) ($p < 0.05$), as shown in Figure 3B. The GSH-Px activity was significantly greater in the 18 L:6 D (404.20 ± 13.70 activity unit) group compared to all other experimental groups ($p < 0.05$), as depicted in Figure 3C. Additionally, the T-AOC activity and MDA content were notably higher in the 12 L:12 D (0.12 ± 0.00 mM and 2.02 ± 0.10 nmol·mgprot^{-1}) group than in the other groups ($p < 0.05$), as indicated in Figure 3D,E, respectively.

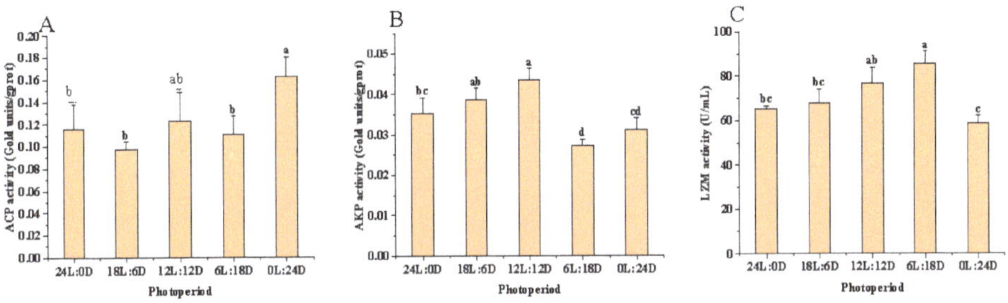

Figure 2. Effect of photoperiod on the immune enzyme activity of scalloped spiny lobsters (n = 18). Acid phosphatase (**A**), alkaline phosphatase (**B**), lysozyme (**C**). Different superscripts letters indicate statistically significant differences between treatments ($p < 0.05$).

Table 4. The results of ANOVA for ACP, AKP, and LZM.

Item	df	MS	F	p
ACP	4	0.00	3.43	0.05
AKP	4	0.00	8.99	0.00
LZM	4	332.992	7.96	0.00

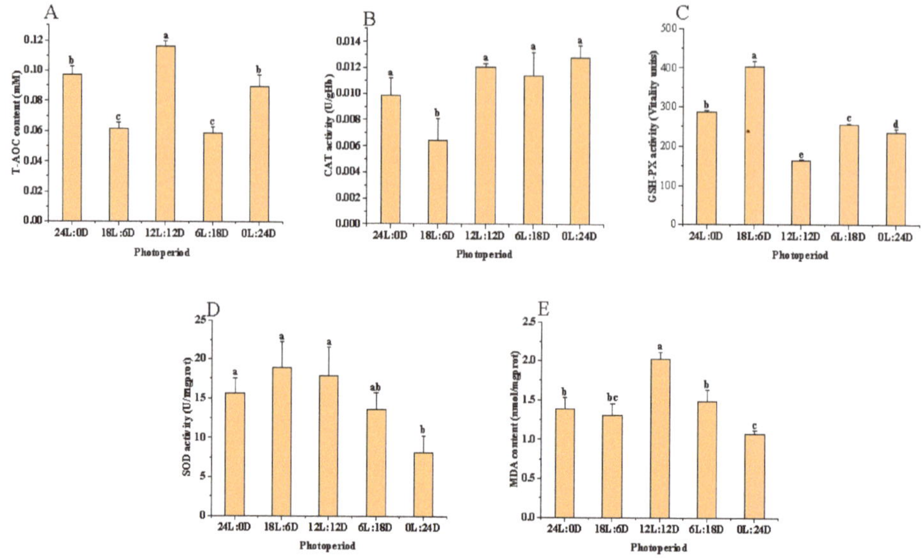

Figure 3. Effect of photoperiod on antioxidant capacity of scalloped spiny lobsters (n = 18). Superoxide dismutase (**A**), peroxidase (**B**), glutathione peroxidase (**C**), total antioxidant capacity (**D**), malondialdehyde (**E**). Different superscripts letters indicate statistically significant differences between treatments ($p < 0.05$).

Table 5. The results of ANOVA for T-AOC, CAT, GSH-Px, SOD, and MDA.

Item	df	MS	F	p
T-AOC	4	0.00	44.84	0.00
CAT	4	0.00	7.06	0.00
GSH-Px	4	23,296.86	260.06	0.00
SOD	4	54.40	4.81	0.02
MDA	4	0.37	16.70	0.00

4. Discussion

4.1. Effects of the Photoperiod on the Growth Performance of the Animals

In this study, the scalloped spiny lobster exhibited optimal WGR and SGR under 12 L:12 D. Although no significant difference was observed in WGR and SGR between long (>12 L) and short (<12 L) photoperiods, lobsters under longer photoperiods showed higher growth rates, suggesting that extended light periods may more effectively promote growth in aquatic animals. This finding aligns with observations in other species, including larval coconut crabs (*Birgus latro*) [57], mud crabs [21], narrow-clawed crayfish (*Astacus leptodactylus*) [46], blue swimming crab (*Portunus pelagicus*) [58], and juvenile winter flounder (*Pseudopleuronectes americanus*) [59]. Contrarily, horsehair crab (*Erimacrus isenbeckii*) larvae showed optimal survival and growth under continuous darkness (0 L:24 D) [60].

Similarly, short photoperiods have been found to enhance growth in other species, such as the African catfish (*Clarias gariepinus*) [61], abalone (*Haliotis discus hannai*) [61,62], and largemouth bass (*Micropterus salmoides*) [36], all of which displayed improved growth rates under limited light conditions. These differences in photoperiod adaptation across species may be attributed to their unique species characteristics and ecological behaviors. Comparative analysis with other aquatic species such as spiny lobster [63], swimming crab (*Portunus trituberculatus*) [64], spanner crab (*Ranina ranina*) [65], neotropical fish (*Hoplias intermedius*) [66], and banded cichlid (*Heros severus*) [67] revealed similar optimal growth under a 12 L:12 D photoperiod. This suggests the possibility of a universal optimal photoperiod for different species, likely influenced by their ecological habits. Consequently, for scalloped spiny lobster aquaculture, 12 L:12 D is recommended to enhance growth performance, shorten production cycles, and improve economic efficiency.

4.2. Effects on the Digestive Performance of Animals Due to the Photoperiod

In this study, the activities of PEP, TRYP, and CHT in scalloped spiny lobsters were highest under continuous darkness (0 L:24 D). This finding suggests that scalloped spiny lobsters are more efficient at protein digestion and absorption in a dark environment. Conversely, LPS and AMS activities peaked in 12 L:12 D and 18 L:6 D light conditions, respectively. This pattern indicates a shift from protein digestion to the utilization of stored nutrients in the hepatopancreas after protein sources are consumed.

The observed differences in enzyme activities may be attributed to the unique ecological and feeding habits of scalloped spiny lobsters. Being nocturnal marine crustaceans, they predominantly feed at night [68] and exhibit a distinct behavior of burrowing and hiding in coral crevices or reefs during the day [69]. Post feeding, the lobsters retreat to their burrows, aiding in the digestion and absorption of proteins. In the absence of daytime feeding, the stored nutrients in the hepatopancreas are utilized [70]. Consequently, photoperiods exceeding 12 h appear to favor the breakdown of fats and starches. Previous research indicates that specific photoperiods can stimulate or inhibit the activity of certain digestive enzymes [71]. Thus, in aquacultural practices, adjusting the photoperiod can be a strategic approach to facilitate the digestion and absorption of specific nutrients in scalloped spiny lobsters.

4.3. Effects of Photoperiod on Animals' Immune Function

In this study, ACP activity in scalloped spiny lobsters was found to be optimal under continuous darkness (0 L:24 D). This suggests that ACP secretion is more favorable in such light conditions, aligning with observations in freshwater shrimp [50]. In contrast, the highest AKP activity in scalloped spiny lobsters was recorded under 12 L:12 D. This difference implies distinct immune response mechanisms to photoperiods in the two types of phosphatases.

The variation in phosphatase activities might be associated with the lobsters' nocturnal feeding behavior [5,68]. Post feeding at night, the lobsters generate energy, which, in turn, could promote the production of these phosphatases, thereby enhancing their immune function [47]. LZM activity reached its peak at 6 L:18 D, and an increase in LZM activity was noted under photoperiods longer than 6 h (24 L:0 D, 18 L:6 D, 12 L:12 D, and 6 L:18 D). This pattern indicates that light exposure positively influences LZM secretion in scalloped spiny lobsters, a finding consistent with studies on tiger puffer [51].

The increased LZM activity under certain photoperiods might be indicative of a photoperiod-induced innate immune response in scalloped spiny lobsters. Therefore, a 12 L:12 D photoperiod is recommended in aquacultural practices to boost their immune function. Such immune enhancement, through the suppression of pathogenic bacteria, could lead to heightened specific immunity in these organisms [72]. The results of this study are in line with findings in white leg shrimp [34], freshwater prawn [50], and red claw crayfish (*Cherax quadricarinatus*) [56], suggesting that an appropriate photoperiod (12 L:12 D) can effectively improve the immune function of scalloped spiny lobsters.

4.4. Effect of Photoperiod on Animals' Antioxidant Capacity

In this study, T-AOC activity and MDA levels in scalloped spiny lobsters were observed to be highest under 12 L:12 D. Similarly, the activities of SOD and CAT were also elevated under these light conditions. These results indicate that the antioxidant capacity of scalloped spiny lobsters is responsive to varying levels of oxidative stress in the external environment. Under the 12 L:12 D photoperiod, the lobsters are likely exposed to external stressors that generate a significant amount of ROS, potentially causing damage to the organisms. MDA is a crucial biomarker for assessing ROS levels in aquatic animals [35], and the activity of T-AOC reflects the overall level of antioxidant enzymes and non-enzymatic antioxidants in an organism, which are vital for combating oxidative stress [53]. Therefore, the activities of T-AOC, SOD, and CAT may fluctuate in response to the changes in MDA levels induced by external stressors. These findings align with research conducted on South American white shrimp [35], bluntmouth bream [39], and abalone [55].

Interestingly, the GSH-Px activity in scalloped spiny lobsters was found to be optimal under a longer photoperiod of 18 L:6 D. This may suggest that the antioxidant system of the lobsters under excessively long or short photoperiods struggles to effectively scavenge excess ROS, resulting in increased GSH-Px activity. In practical aquaculture settings, adjusting the photoperiod could be a strategic measure to mitigate oxidative stress and enhance the antioxidant capacity in scalloped spiny lobsters.

5. Conclusions

This study explored the impact of five different light–dark cycles (0 L:24 D, 6 L:18 D, 12 L:12 D, 18 L:6 D, and 24 L:0 D) on the growth and physiological aspects of scalloped spiny lobsters. Notably, under a 12 L:12 D light–dark cycle, the lobsters demonstrated optimal growth performance, along with enhanced levels of digestive and immune enzymes, and improved antioxidant capacity. These findings suggest that adopting a 12 L:12 D lighting regimen could effectively boost the growth rate and shorten the production cycle in scalloped spiny lobster aquaculture. This research provides valuable insights for creating an ideal lighting environment, contributing significantly to the optimization and enhancement of scalloped spiny lobster cultivation.

Author Contributions: Conceptualization, Z.M. and Z.B.; Methodology, Y.W. and R.Y.; Software, R.Y.; Validation, Z.F.; Formal Analysis, Z.F.; Investigation, R.Y.; Resources, Z.B.; Data Curation, Y.W.; Writing—Original Draft Preparation, Y.W.; Writing—Review and Editing, Z.M. and Z.F.; Visualization, R.Y.; Supervision, Z.M.; Project Administration, R.Y. and Z.B. All authors have read and agreed to the published version of the manuscript.

Funding: This research was funded by Central Public-Interest Scientific Institution Basal Research Fund, South China Sea Fisheries Research Institute, CAFS (NO. 2021SD09), Central Public—interest Scientific Institution Basal Research Fund, CAFS (No. 2023TD58).

Institutional Review Board Statement: The animal study was reviewed and approved by the Animal Care and Use Committee of South China Sea Fisheries Research Institute, Chinese Academy of Fishery Sciences. The ethical code is CAFS (2020TD55).

Informed Consent Statement: Not applicable.

Data Availability Statement: The data presented in this study are available on request from the corresponding author. The data are not publicly available due to the confidentiality of the project.

Conflicts of Interest: Author Zemin Bai was employed by the Yazhou Bay Agriculture and Aquaculture Co., Ltd. The remaining authors declare that the research was conducted in the absence of any commercial or financial relationships that could be construed as a potential conflict of interest.

References

1. Macdonald, A.H.H.; Reddy, M.M.; Groeneveld, J.C.; Schleyer, M.H. Phylogeography of the scalloped spiny-lobster *Panulirus homarus* rubellus in the southwest Indian Ocean. *J. Crustac. Biol.* **2014**, *34*, 773–781. [CrossRef]
2. Singh, S.P.; Groeneveld, J.C.; Willows-Munro, S. Between the current and the coast: Genetic connectivity in the spiny lobster *Panulirus homarus* rubellus, despite potential barriers to gene flow. *Mar. Biol.* **2019**, *166*, 36. [CrossRef]
3. Marzouqi, A.; Al-Nahdi, A.; Jayabalan, N.; Groeneveld, J.C. An Assessment of the Spiny Lobster *Panulirus homarus* Fishery in Oman—Another Decline in the Western Indian Ocean? *West. Indian Ocean J. Mar. Sci.* **2009**, *6*, 1183. [CrossRef]
4. Tirtadanu; Suman, A.; Chodrijah, U.; Kang, B.; Zhang, C. Stock assessment and management implications of three lobster species in Gunungkidul waters, Indonesia. *Ocean. Coast. Manag.* **2021**, *211*, 105780. [CrossRef]
5. Devi, N.R.; Rasheeq, A.A.; Preethi, B.A.; Anand, M.; Titus, C.; Subbiah, S.; Rangesh, K.; Dineshkumar, R.; Arumugam, A. Assessment of Lobster Resources in Coastal Region of Gulf of Mannar, Southeast Coast of India. *Thalass. Int. J. Mar. Sci.* **2023**, *39*, 1169–1186. [CrossRef]
6. Duy Mai, M.; Thi Tran, L. Growth Performance of Scalloped Spiny Lobster *Panulirus homarus* (Linaeus) Fed Formulated Diet in Recirculating System. *Agric. For. Fish.* **2022**, *11*, 1. [CrossRef]
7. Ihsan, M.; Suhirman, S.; Jayadi, E.M.; Sagista, R.; Hardianti, Y.E.; Ilahi, W.B.; Muliasari, H.; Tantilar Wangsajati Sukmaring Kalih, L.A. Analisis makanan alami dalam lambung dan mikrohabitat lobster pasir (*Panulirus homarus*) fase puerulus di teluk awang. *J. Ris. Akuakultur* **2019**, *14*, 183–191. [CrossRef]
8. Suman, A.; Hasanah, A.; Pane, A.R.P.; Panggabean, A.S. Penangkapan, parameter populasi serta tingkat pemanfaatan lobster pasir (*Panulirus homarus*) dan lobster batu (*Panulirus penicillatus*) di perairan gunung kidul, dan sekitarnya. *J. Penelit. Perikan. Indones.* **2020**, *25*, 147–160. [CrossRef]
9. Priyambodo, B.; Jones, C.M.; Sammut, J. Improved collector design for the capture of tropical spiny lobster, *Panulirus homarus* and *P. ornatus* (Decapoda: Palinuridae), pueruli in Lombok, Indonesia. *Aquaculture* **2017**, *479*, 321–332. [CrossRef]
10. Nurfiarini, A.; Wijaya, D. Estimasi potensi DAN tingkat pemanfaatan sumberdaya lobster pasir (*Panulirus homarus*) di perairan prigi kabupanen trenggalek. *J. Penelit. Perikan. Indones.* **2020**, *25*, 169–178. [CrossRef]
11. Priyambodo, B.; Jones, C.M.; Sammut, J. Assessment of the lobster puerulus (*Panulirus homarus* and *Panulirus ornatus*, Decapoda: Palinuridae) resource of Indonesia and its potential for sustainable harvest for aquaculture. *Aquaculture* **2020**, *528*, 735563. [CrossRef]
12. Junaidi, M.; Cokrowati, N.; Diniarti, N.; Setyono, B.D.H.; Mulyani, L.F. Identifying the Environmental Factors Affecting Puerulus Settlement of the Spiny Lobster, *Panulirus homarus* in Awang Bay, Lombok Island. *Asian J. Fish. Aquat. Res.* **2022**, *18*, 1–14. [CrossRef]
13. Chen, H.; Pan, J.; Wang, Y.; Qiao, Y.; Han, F.; Xu, C.; Farhadi, A.; Li, E. Growth, health status and gut microbiota of the scalloped spiny lobster (*Panulirus homarus*) at different salinities. *Aquaculture* **2023**, *562*, 738779. [CrossRef]
14. Adiputra, Y.T.; Zairin Jr, M.; Suprayudi, M.A.; Manalu, W.; Widanarni; Brite, M. The effects of thyroxine hormone on gonadal maturation and growth of male spiny lobster (*Panulirus homarus*). *Malays. J. Sci.* **2020**, *39*, 30–40. [CrossRef]
15. Adiputra, Y.T.; Zairin, M., Jr.; Suprayudi, M.A.; Manalu, W.; Widanarni. Identification of steroid hormones and fatty acids during gonadal maturation of spiny lobster *Panulirus homarus*. *Invertebr. Reprod. Dev.* **2019**, *63*, 77–87. [CrossRef]
16. Sudewi, S.; Widiastuti, Z.; Slamet, B.; Mahardika, K. Experimental inffctions of milky hemolymph disease in spiny lobster *Panulirus homarus*. *Indones. Aquac. J.* **2018**, *13*, 31–40. [CrossRef]

17. Syed Musthaq, S.; Sudhakaran, R.; Balasubramanian, G.; Sahul Hameed, A.S. Experimental transmission and tissue tropism of white spot syndrome virus (WSSV) in two species of lobsters, *Panulirus homarus* and *Panulirus ornatus*. *J. Invertebr. Pathol.* **2006**, *93*, 75–80. [CrossRef]
18. Farhadi, A.; Jeffs, A.G.; Lavery, S.D. Genome-wide SNPs in the spiny lobster *Panulirus homarus* reveal a hybrid origin for its subspecies. *BMC Genom.* **2022**, *23*, 750. [CrossRef]
19. Zhuo, H.B.; Liang, H.F.; Cai, C.X.; Luo, J.J.; Liang, F.S.; Wen, C.Q. Molecular cloning, characterization and expression analysis of the ecdysone receptor from the spiny lobster *Panulirus homarus* (Linnaeus, 1758) (Decapoda, Palinuridae). *Crustaceana* **2020**, *93*, 769–783. [CrossRef]
20. Delghandi, M.; Saif Nasser Al Hinai, M.; Afzal, H.; Khalfan Al-Wahaibi, M. Parentage analysis of tropical spiny lobster (*Panulirus homarus*) by microsatellite markers. *Aquac. Res.* **2017**, *48*, 4718–4724. [CrossRef]
21. Chen, S.; Liu, J.; Shi, C.; Migaud, H.; Ye, Y.; Song, C.; Mu, C.; Ren, Z.; Wang, C. Effect of photoperiod on growth, survival, and lipid metabolism of mud crab *Scylla paramamosain* juveniles. *Aquaculture* **2023**, *567*, 739279. [CrossRef]
22. Gao, X.L.; Pang, G.W.; Luo, X.; You, W.W.; Ke, C.H. Effects of light cycle on circadian feeding activity and digestive physiology in *Haliotis discus hannai*. *Aquaculture* **2021**, *539*, 736642. [CrossRef]
23. Mirko Bögner, C.S.T.G. Effect of ambient light intensity on growth performance and diurnal stress response of juvenile starry flounder (*Platichthys stellatus*) in recirculating aquaculture systems (RAS). *Aquacult Eng.* **2018**, *83*, 20–26. [CrossRef]
24. Hou, Z.; Wen, H.; Li, J.; He, F.; Li, Y.; Qi, X.; Zhao, J.; Zhang, K.; Tao, Y. Effects of photoperiod and light Spectrum on growth performance, digestive enzymes, hepatic biochemistry and peripheral hormones in spotted sea bass (*Lateolabrax maculatus*). *Aquaculture* **2019**, *507*, 419–427. [CrossRef]
25. Wang, K.; Li, K.; Liu, L.; Tanase, C.; Mols, R.; van der Meer, M. Effects of light intensity and photoperiod on the growth and stress response of juvenile Nile tilapia (*Oreochromis niloticus*) in a recirculating aquaculture system. *Aquac. Fish.* **2023**, *8*, 85–90. [CrossRef]
26. Bapary, M.A.J.; Amin, M.N.; Takeuchi, Y.; Takemura, A. The stimulatory effects of long wavelengths of light on the ovarian development in the tropical damselfish, *Chrysiptera cyanea*. *Aquaculture* **2011**, *314*, 188–192. [CrossRef]
27. Kim, B.; Jung, S.J.; Choi, Y.J.; Kim, N.N.; Choi, C.Y.; Kim, J. Effects of different light wavelengths from LEDs on oxidative stress and apoptosis in olive flounder (*Paralichthys olivaceus*) at high water temperatures. *Fish Shellfish. Immun.* **2016**, *55*, 460–468. [CrossRef] [PubMed]
28. Reis, W.G.; Wasielesky, W., Jr.; Abreu, P.C.; Brandão, H.; Krummenauer, D. The influence of different light wavelengths in the culture of the Pacific white shrimp *Litopenaeus vannamei* reared in BFT using LED lights. *Aquaculture* **2023**, *563*, 738924. [CrossRef]
29. Zhang, Y.; Huang, Z.; Zhou, Y.; Ma, H.; Saqib, H.S.A.; Su, Q.; Cui, W.; Ma, H. The effects of different diet, salinity and light condition on growth performance and moulting cycle of juvenile mud crab, *Scylla paramamosain*. *Aquac. Res.* **2022**, *53*, 6333–6342. [CrossRef]
30. Li, H.; Zhao, J.; Ji, B.; Zhang, Y.; Ye, Z.; Zhu, S. Effects of Photoperiod on Growth, Digestive Enzyme Activity, Stress, and Oxidative Status of Juvenile Chinese Soft-Shelled Turtles (*Pelodiscus sinensis*) in a Greenhouse. *Trans. ASABE* **2020**, *63*, 1787–1793. [CrossRef]
31. Boeuf, G.; Le Bail, P. Does light have an influence on fish growth? *Aquaculture* **1999**, *177*, 129–152. [CrossRef]
32. Mel Nikova, Y.B.; Melnikov, A.V. Solar radiation as a synchronizing factor of circadian and ultradian biological rhythms of planktonic communities. *IOP Conf. Ser. Earth Environ. Sci.* **2021**, *853*, 12001. [CrossRef]
33. Wei, H.; Cai, W.; Liu, H.; Han, D.; Zhu, X.; Yang, Y.; Jin, J.; Xie, S. Effects of photoperiod on growth, lipid metabolism and oxidative stress of juvenile gibel carp (*Carassius auratus*). *J. Photochem. Photobiol. B Biol.* **2019**, *198*, 111552. [CrossRef] [PubMed]
34. Wang, X.; Liu, B.; Gao, X.; Wang, X.; Li, H.; Xu, L.; Wang, G.; Zhao, K.; Huang, B. The Effects of Different UVA Photoperiods on the Growth Performance, Immune Responses, Antioxidant Status and Apoptosis-Related Gene Expression of the Pacific White Shrimp (*Penaeus vannamei*). *Antibiotics* **2022**, *11*, 37. [CrossRef]
35. Bembe, S.; Zmora, N.; Williams, E.; Place, A.R.; Liang, D.; Chung, J.S. Effects of temperature and photoperiod on hemolymph vitellogenin levels during spawning events of the blue crab, *Callinectes sapidus*, in captivity. *Aquac. Res.* **2018**, *49*, 2201–2209. [CrossRef]
36. Malinovskyi, O.; Rahimnejad, S.; Stejskal, V.; Boňko, D.; Stará, A.; Velíšek, J.; Policar, T. Effects of different photoperiods on growth performance and health status of largemouth bass (*Micropterus salmoides*) juveniles. *Aquaculture* **2022**, *548*, 737631. [CrossRef]
37. Fitzgibbon, Q.P.; Battaglene, S.C. Effect of photoperiod on the culture of early-stage phyllosoma and metamorphosis of spiny lobster (*Sagmariasus verreauxi*). *Aquaculture* **2012**, *368–369*, 48–54. [CrossRef]
38. Bermudes, M.; Ritar, A.J. Response of early stage spiny lobster *Jasus edwardsii* phyllosoma larvae to changes in temperature and photoperiod. *Aquaculture* **2008**, *281*, 63–69. [CrossRef]
39. Tian, H.; Zhang, D.; Li, X.; Jiang, G.; Liu, W. Photoperiod affects blunt snout bream (*Megalobrama amblycephala*) growth, diel rhythm of cortisol, activities of antioxidant enzymes and mRNA expression of GH/IGF-I. *Comp. Biochem. Physiol. Part B Biochem. Mol. Biol.* **2019**, *233*, 4–10. [CrossRef]
40. Nyuji, M.; Hamada, K.; Kazeto, Y.; Mekuchi, M.; Gen, K.; Soyano, K.; Okuzawa, K. Photoperiodic regulation of plasma gonadotropin levels in previtellogenic greater amberjack (*Seriola dumerili*). *Gen. Comp. Endocr.* **2018**, *269*, 149–155. [CrossRef]
41. Zou, H.; Bai, X.; Feng, Y.; Zhang, Y.; Wang, Y.; Lu, W. Influence of long (16L:8D) and short (8L:16D) photoperiods on blood metabolites and hepatic metabolism in Olive flounder, *Paralichthys olivaceus*. *SpringerPlus* **2016**, *5*, 924. [CrossRef]

42. Farhadi, A.; Harlıoğlu, M.M. Photoperiod affects gamete production, and protein and lipid metabolism in male narrow-clawed Crayfish *Pontastacus leptodactylus* (Eschscholtz, 1823). *Anim. Reprod. Sci.* **2019**, *211*, 106204. [CrossRef] [PubMed]
43. Falahatkar, B.; Poursaeid, S.; Efatpanah, I.; Meknatkhah, B.; Biswas, A. Effect of Photoperiod Manipulation on Growth Performance, Physiological and Hematological Indices in Juvenile Persian Sturgeon, Acipenser persicus. *J. World Aquacult. Soc.* **2012**, *43*, 679–687. [CrossRef]
44. Espinosa-Chaurand, D.; Vega-Villasante, F.; Carrillo-Farnés, O.; Nolasco-Soria, H. Effect of circadian rhythm, photoperiod, and molt cycle on digestive enzymatic activity of *Macrobrachium tenellum* juveniles. *Aquaculture* **2017**, *479*, 225–232. [CrossRef]
45. Farhadi, S.; Atashbar Kangarloei, B.; Imani, A.; Sarvi Moghanlou, K. Biological Impact of Photoperiod on Fairy Shrimp (*Branchinecta orientalis*): Life History and Biochemical Composition. *Biology* **2021**, *10*, 695. [CrossRef]
46. Farhadi, A.; Jensen, M.A. Effects of photoperiod and stocking density on survival, growth and physiological responses of narrow clawed crayfish (*Astacus leptodactylus*). *Aquac. Res.* **2016**, *47*, 2518–2527. [CrossRef]
47. Ray, G.W.; Liang, D.; Yang, Q.; Tan, B.; Dong, X.; Chi, S.; Liu, H.; Zhang, S.; Rimei, L. Effects of replacing fishmeal with dietary soybean protein concentrate (SPC) on growth, serum biochemical indices, and antioxidative functions for juvenile shrimp *Litopenaeus vannamei*. *Aquaculture* **2020**, *516*, 734630. [CrossRef]
48. Xu, Y.; Yin, Y.; Li, T.; Zhao, H.; Li, X.; Li, J.; Sun, T. Effects of lysozyme combined with cinnamaldehyde on storage quality of olive flounder (*Paralichthys olivaceus*) fillets. *J. Food Sci.* **2020**, *85*, 1037–1044. [CrossRef] [PubMed]
49. Wu, J.; Tian, S.; Luo, K.; Zhang, Y.; Pan, H.; Zhang, W.; Mai, K. Dietary recombinant human lysozyme improves the growth, intestinal health, immunity and disease resistance of Pacific white shrimp *Litopenaeus vannamei*. *Fish Shellfish. Immun.* **2022**, *121*, 39–52. [CrossRef]
50. Wei, J.; Tian, L.; Wang, Y.; Yu, L.; Zhu, X. Effects of salinity, photoperiod, and light spectrum on larval survival, growth, and related enzyme activities in the giant freshwater prawn, *Macrobrachium rosenbergii*. *Aquaculture* **2021**, *530*, 735794. [CrossRef]
51. Ma, H.; Wei, P.; Li, X.; Liu, S.; Tian, Y.; Zhang, Q.; Liu, Y. Effects of photoperiod on growth, digestive, metabolic and non-special immunity enzymes of *Takifugu rubripes* larvae. *Aquaculture* **2021**, *542*, 736840. [CrossRef]
52. Hu, N.; Yu, C.; Jin, J.; Zhao, X.; Zhao, Y.; Wei, H.; Li, Y. Impact of photoperiods on the specific activities of immune and antioxidant enzymes in different tissues of Dybowski's frog (*Rana dybowskii*). *Biol. Rhythm. Res.* **2022**, *53*, 1790–1799. [CrossRef]
53. Hao, Z.L.; Tang, X.J.; Ding, J.; Ben, Y.; Chang, Y.Q. Effect of high temperature on survival, oxygen consumption, behavior, ammonia-N excretion, and related immune indicators of the Japanese scallop *Mizuhopecten yessoensis*. *Aquacult. Int.* **2014**, *22*, 1863–1876. [CrossRef]
54. Xu, H.; Shi, C.; Ye, Y.; Mu, C.; Wang, C. Effects of different photoperiods and feeding regimes on immune response, oxidative status, and tissue damage in juvenile rainbow trout (*Oncorhynchus mykiss*). *Front. Mar. Sci.* **2022**, *9*, 1036289. [CrossRef]
55. Gao, X.; Zhang, M.; Li, X.; Song, C.; Liu, Y. Physiological metabolism of *Haliotis discus hannai* Ino under different light qualities and cycles. *Aquac. Res.* **2017**, *48*, 3340–3355. [CrossRef]
56. Cheng, S.; Zheng, J.; Jia, Y.; Chi, M.; Jiang, W.; Liu, S.; Li, F.; Liu, Y.; Gu, Z.; Wang, D. Effects of light color, photoperiod, and growth-related gene interference or overexpression on the survival, growth, or physiological and biochemical indices of red claw crayfish juveniles. *Aquaculture* **2023**, *562*, 738740. [CrossRef]
57. Hamasaki, K.; Ogiso, Y.; Dan, S.; Kitada, S. Survival, development and growth of larvae of the coconut crab, *Birgus latro*, cultured under different photoperiod conditions. *Aquac. Res.* **2016**, *47*, 2506–2517. [CrossRef]
58. Andrés, M.; Rotllant, G.; Zeng, C. Survival, development and growth of larvae of the blue swimmer crab, *Portunus pelagicus*, cultured under different photoperiod conditions. *Aquaculture* **2010**, *300*, 218–222. [CrossRef]
59. Litvak, M.K.; Zadmajid, V.; Butts, I.A.E. Growth and survival of winter flounder (*Pseudopleuronectes americanus*) larvae reared on different photoperiod regimes from hatch to metamorphosis. *Aquac. Res.* **2020**, *51*, 2314–2321. [CrossRef]
60. Ichikawa, T.; Hamasaki, K.; Murakami, K. Larval survival, development and growth in the horsehair crab, *Erimacrus isenbeckii*, cultured under different photoperiod conditions. *Aquac. Res.* **2018**, *49*, 2511–2517. [CrossRef]
61. Indriastuti, C.E.; Junior, M.Z.; Suprayudi, M.A.; Supriyono, E.; Alimuddin, A. Cannibalism, survival and growth performance of juvenile African catfish *Clarias gariepinus* in relation to photoperiod and 17β-Oestradiol treatment. *Aquac. Res.* **2022**, *53*, 4437–4448. [CrossRef]
62. Gao, X.L.; Mo, Z.; Li, X.; Wu, F.C.; Song, C.B.; Liu, Y. Light cycle effects on *Haliotis discus hannai* Ino growth, energy budget, and related gene expression. *Aquaculture* **2018**, *483*, 213–222. [CrossRef]
63. Crear, B.J.; Hart, P.R.; Thomas, C.W. The effect of photoperiod on growth, survival, colour and activity of juvenile southern rock lobster, *Jasus edwardsii*. *Aquac. Res.* **2003**, *34*, 439–444. [CrossRef]
64. Xu, H.; Dou, J.; Wu, Q.; Ye, Y.; Wang, C.; Song, C.; Mu, C.; Ren, Z.; Shi, C. Photoperiod affects the survival rate but not the development of larval swimming crab *Portunus trituberculatus*. *Aquacult. Int.* **2022**, *30*, 1769–1778. [CrossRef]
65. Minagawa, M. Effects of photoperiod on survival, feeding and development of larvae of the red frog crab, *Ranina ranina*. *Aquaculture* **1994**, *120*, 105–114. [CrossRef]
66. Ramos, S.E.; Carvalho, A.F.S.D.; Castro, T.F.D.; Vasconcelos, A.C.N.; Veras, G.C.; Mourão Júnior, C.A.; Murgas, L.D.S. Cannibalism, growth performance, and body composition of giant trahira juveniles under different photoperiods. *Pesqui. Agropecuária Bras.* **2018**, *53*, 664–672. [CrossRef]
67. Veras, G.C.; Paixão, D.J.D.M.; Brabo, M.F.; Soares, L.M.O.; Sales, A.D. Influence of photoperiod on growth, uniformity, and survival of larvae of the Amazonian ornamental *Heros severus* (Heckel, 1840). *Rev. Bras. Zootec.* **2016**, *45*, 422–426. [CrossRef]

68. Lesmana, D.; Supriyono, E.; Junior, M.Z.; Nirmala, K.; Jusadi, D. The colour preference of scalloped spiny lobster, *Panulirus homarus*. *IOP Conf. Ser. Earth Environ. Sci.* **2021**, *744*, 12039. [CrossRef]
69. Lavery, S.D.; Farhadi, A.; Farahmand, H.; Chan, T.; Azhdehakoshpour, A.; Thakur, V.; Jeffs, A.G. Evolutionary Divergence of Geographic Subspecies within the Scalloped Spiny Lobster *Panulirus homarus* (Linnaeus 1758). *PLoS ONE* **2014**, *9*, e97247. [CrossRef] [PubMed]
70. Zhang, B.; Yu, C.; Xu, Y.; Huang, Z.; Cai, Y.; Li, Y. Hepatopancreas immune response during different photoperiods in the Chinese mitten crab, Eriocheir sinensis. *Fish Shellfish. Immun.* **2023**, *132*, 108482. [CrossRef]
71. Zhou, S.J.; Hu, J.; Yu, G.; Yang, Q.B.; Yang, R.; Liu, Y.J.; Ma, Z.H. Effects of photoperiod on digestive enzyme activity in larval and juvenile barramundi Lates calcarifer (Bloch). *Mar. Sci.* **2018**, *42*, 63–69. [CrossRef]
72. Lee, M.; Kovacs-Nolan, J.; Yang, C.; Archbold, T.; Fan, M.Z.; Mine, Y. Hen Egg Lysozyme Attenuates Inflammation and Modulates Local Gene Expression in a Porcine Model of Dextran Sodium Sulfate (DSS)-Induced Colitis. *J. Agr. Food Chem.* **2009**, *57*, 2233–2240. [CrossRef] [PubMed]

Journal of
Marine Science and Engineering

MDPI

Article

Towards Fish Welfare in the Presence of Robots: Zebrafish Case

Andrea Pino [1], Rosario Vidal [1,*], Elisabeth Tormos [2], José Miguel Cerdà-Reverter [2], Raúl Marín Prades [1] and Pedro J. Sanz [1]

[1] Research Centre for Robotics and Underwater Technologies (CIRTESU), Universitat Jaume I, 12006 Castelló de la Plana, Spain; apino@uji.es (A.P.)
[2] Fish NeuroBehaviour Lab, Department of Fish Physiology and Biotechnology, Instituto de Acuicultura de Torre de la Sal, IATS-CSIC, 12595 Castellon, Spain; jm.cerda.reverter@csic.es (J.M.C.-R.)
* Correspondence: vidal@uji.es

Abstract: Zebrafish (Danio rerio) have emerged as a valuable animal model for neurobehavioral research, particularly in the study of anxiety-related states. This article explores the use of conceptual models to investigate stress, fear, and anxiety in zebrafish induced by bio-inspired mini-robotic fish with different components and designs. The objective is to optimize robotic biomimicry and its impact on fish welfare. Previous studies have focused on externally controlled fish models, whereas this study introduces prototypes of freely actuated swimming robots to examine interactions between a bio-inspired robot and individual zebrafish. By means of analysis of behavioral responses, certain robotic components have been identified as potential causes of anxiety in fish, which have provided insights that may be applicable to other species and future aquacultural robot designs.

Keywords: underwater robotics; fish robot; robot biomimicry

Citation: Pino, A.; Vidal, R.; Tormos, E.; Cerdà-Reverter, J.M.; Marín Prades, R.; Sanz, P.J. Towards Fish Welfare in the Presence of Robots: Zebrafish Case. *J. Mar. Sci. Eng.* **2024**, *12*, 932. https://doi.org/10.3390/jmse12060932

Academic Editors: Zhenhua Ma and Jianguang Qin

Received: 30 April 2024
Revised: 24 May 2024
Accepted: 29 May 2024
Published: 31 May 2024

1. Introduction

The field of robotics and underwater design have directed their efforts at the development of unmanned autonomous vehicles (UUVs) [1] to carry out actions that could otherwise pose a risk to humans. This field can be directly applied to the aquaculture environment, where the working atmosphere is demanding and, in many cases, the performance of certain tasks may be challenging to human assets. Manual tasks such as cleaning, inspecting, or repairing net cages can be detrimental to workers [2]; furthermore, there are stricter regulations governing the time divers can spend underwater or the tasks they can perform in order to protect worker conditions. In recent years, the aquaculture industry has witnessed a growing trend towards robotization [3], where robots are employed to carry out various tasks. This development aims to minimize the exposure of workers and enhance the efficiency of farm production processes.

Currently, many projects are working on the development of remote-controlled underwater robots which can be used in this and other environments [4]. One project that has embraced this environment, and on which this study is based, is the ThinkInAzul project [5], which aims, as one of its objectives, to approach sustainable smart precision aquaculture by creating technology which improves inspection, maintenance, and repair operations [6], and for which there is a need to design a mimetic robot that can perform specific monitoring, inspecting, sensing, and sample-collecting tasks within the cages.

However, conventional underwater robots used to date can, in many cases, have an impact on the environment, as well as being intrusive and stressful to marine life, due to their aesthetic and/or mechanical characteristics [7,8].

In the last two decades, the emerging field of biomimetics has sought to explore the design by copying the forms of living beings found in nature, as well as their movements and forms of propulsion [9,10], in order to be friendlier and less invasive to the fauna of the marine environment in which they operate. This has led to a surge of studies in the field of biomimetics focused on the development of bioinspired fish [11]. Most of these studies have primarily concentrated on emulating fish propulsion systems and assessing their efficiency [12–14]. Traditional propellers produce currents, consume more energy, reduce propulsion efficiency, and are noisy and aggressive, while mimicking the swimming propulsion mode of fish is more efficient, less noisy, and provides better robot performance in terms of energy efficiency [15]. Additionally, in the context of aquaculture, keeping fish healthy and stress-free increases production, breeding performance, and profitability [16]. It is therefore desirable that the robot used in fish farms is respectful of the fauna and as non-invasive as possible. To optimize biomimicry and the design of underwater robots operating in fish farms, it is essential to identify and test which robotic elements are the main stressors and causes of disturbance.

To analyze the effects robotic disturbances may elicit on fish, it is imperative to understand the defensive behaviors they exhibit in response to stimuli and situations perceived as stressful threats. To date, controlled studies have been conducted to identify specific responses and consistent behavioral patterns displayed by fish in stressful situations [17]. Such studies have involved the introduction of robotic stimuli in animal behavior research, where bio-inspired fish prototypes were developed and tested to evaluate robot-fish interactions [18,19]. Nevertheless, most of these investigations have utilized robotic platforms and external mechanisms to generate the movement and trajectory of replicas. Although some recent studies have explored the potential for bidirectional interactions between robotic stimuli and live subjects in free-swimming contexts [20], such efforts have not primarily focused on identifying the stress induced by specific robotic components. Several studies have demonstrated the influence of color and/or pattern [21] on conspecific relationships, showing that fish species lack high visual acuity but have a remarkable ability to discriminate contrast. Contrasting patterns may be important cues for social mimicry in discriminating between conspecifics and predators. Animal size may also be an important factor in social interaction with conspecifics and non-conspecifics. Studies focusing on the influence of size have concluded that a larger size may result in the individual being identified as a predator or rejected as a conspecific [22]. Physical parameters like sound and light [23] can also be a source of interference, suggesting that the use of actuators capable of disturbing their acoustic and/or vision channel could be another source of stress. Finally, some studies have also focused on the robot swimming/movements using actuators to control different fins and investigating the efficiency but also the acceptance of these replicas within a group of individuals [24]. Therefore, it is crucial to understand how social and environmental interactions are managed using the senses to achieve the mimicry of bioinspired robots and assess their impact on fish behavior in aquaculture [25].

Therefore, to achieve the mimicry of bioinspired robots, it is crucial to understand how social interactions among fish are dealt with and what specific responses they exhibit in order to evaluate the impact that robots may cause. Such specific responses to challenging situations to which they are exposed are generally classified as anxiety-like behaviors (ALB). However, some researchers have suggested that it may be more appropriate to differentiate between anxiety, fear, and panic based on the perceived immediacy of the threat [26]. When fish perceive the risk as slight, they tend to display exploratory behavior. In situations where risk is perceived as moderate, escape and avoidance behaviors are observed. On the other hand, if fish perceive an imminent threat, they may respond with a defensive attack or freezing [27]. Previous studies [28] have developed models that facilitate the study of these specific behaviors and enable the identification and evaluation of various behavioral variables or endpoints (see Table 1).

In the last decade, zebrafish (Danio rerio) have emerged as an important model organism for behavioral studies [29]. Accordingly, such behavioral model species are used to investigate the emotional effects, such as ALB, which bio-inspired fish robot prototypes can induce on individual zebrafish behavior, and to identify specific robot components as potential causes of ALB in fish.

Table 1. Fish behavior models and evaluative measures for reactions.

Analytical Model	Evaluation Measures for Determining the Presence of Stress
Conditioned Alarm Reaction	Time in Bottom Zone Total Path Length Freeze Time Fast Swimming
Inhibitory Avoidance	Time Until Visiting Aversive Zone
	Time in Aversive Zone
Predator Response	Burst Swimming Freeze Bottom Dwell Time Distance to Predator
Inspection of Novel Objects	Distance Between Fish and Object
	Time Near the Object

All this previous knowledge, therefore, provides insights which enable us to optimize efficiency in the design of robotic prototypes, while also increasing mimicry to reduce robot-induced stress. These concerns are crucial for the development of robots intended to operate in environments where coexistence with fish is required without compromising their well-being. However, while some factors have been extensively studied in the literature, others may warrant further investigation. Therefore, this article is focused on exploring the implications related to stressors associated with the interaction between robots and fish. This article specifically examines various small prototypes to assess how different factors affect fish stress. It investigates the impact of the presence or absence of light, the movement or lack of movement of a body, and the oscillatory or helical movement of a conspecific's tail using different types of actuators. Statistical tests were conducted, and various variables were analyzed to determine the influence of these factors on fish stress so as to use them specifically in bioinspired robotic design for use in real-world aquacultural practices.

2. Materials and Methods

2.1. Fish and Accommodation Conditions

Ten-month-old Tubingen (Tu) zebrafish (length 3–3.5 cm) were reared under standard conditions in the facilities at the Instituto de Acuicultura de Torre la Sal. Naïve animals were acclimated to the behavioral testing room for at least 5 days and maintained at 28 °C with 14 h light/10 h dark. All experiments were performed following the guidelines of the Spanish (Royal Decree 53/2013) and the European Union Directive on the Protection of Animals Used for Scientific Purposes (Directive 2010/63/EU). The protocols applied were approved by IATS Ethics Committee (Register Number 09-0201) under the supervision of the Secretary of State for Research, Development, and Innovation of the Spanish Government.

2.2. Robots Tested

Four different types of bio-inspired robotic replicas, based on the morphology of the zebrafish, were used to classify the different types of actuators and electronic components according to the stress they could generate (Figure 1). The aim was to obtain a size and appearance (colors, line patterns and position of fins) as identical as possible to those of the live individuals to be used in the study and to enable them to swim autonomously and freely around the test area.

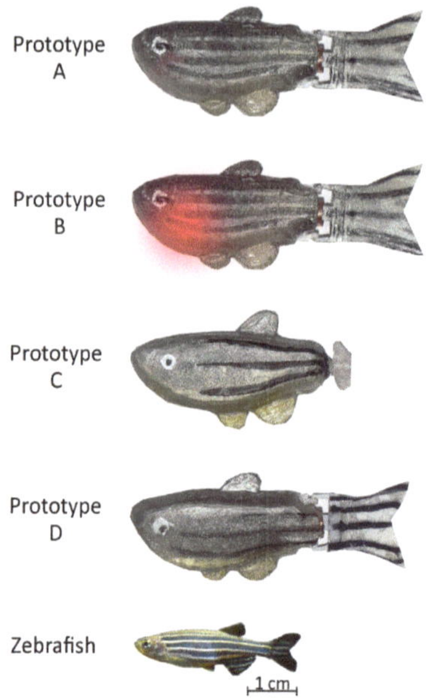

Figure 1. Real images of the four prototypes used and their comparison in size and appearance with a real zebrafish of the same size and aspect as those used in the tests.

The four prototypes were created from scratch and specifically designed for these tests using Solidworks 2020 modeling software. The housing for the prototypes was produced using a high clear ABS-like resin (Anycubic, Shenzhen, China) and an Elegoo Mars 3 (Elegoo, Shenzhen, China) printer through the process of rapid prototyping. This method was selected due to the low water absorption properties of the resin [30], the possibility of achieving smooth, non-porous surfaces, the ability to produce detailed models despite their small size (tolerance of 0.3 mm) [31], and the quality and dimensional precision, which allows for rigorous buoyancy calculations. Each prototype consists of two symmetrical halves which are sealed together using ethyl cyanoacrylate glue, creating an internal cavity where the electronics are placed (see Figure 2). To ensure waterproofing, this cavity was coated with transparent acetic silicone, offering increased resistance to impacts, UV rays, and prolonged exposure to water. The key features of each of the four prototypes are as follows:

- Prototype A is powered by a 0.3 g electromagnetic actuator (provided by Shuaichi, CN), model DIY RC Aircraft, measuring 10 × 10 × 2 mm, with a resistance of 60 ohms, an operating voltage of 3.7–4.2 V, and an operating current of 55 mAh, which is connected to an acetate tail. This propulsion system generates a tail movement characterized by oscillatory beats, the frequency of which can be adjusted in advance. Consequently, changes in the robot's speed and direction are achieved. The electronic system of this prototype includes a rechargeable 2.7 V, 30 mAh lithium battery, model 450909, and a mini-PCB (Figure 3a). The dimensions of this prototype are as follows: length—6.5 cm; height—2 cm; and thickness—1.2 cm (Prototype A in Figure 1). Upon contact with water, the circuit is automatically closed activating the prototype. However, it should be noted that Prototype A only swims on the water's surface and is unable to go deeper than 2 cm below the surface level within the tank environment.
- Prototype B is a replica of the previous model which also includes a red LED light (Prototype B in Figure 1). This red LED is included in the mini-PCB commercially acquired and flashes intermittently at the same frequency as the tail, from within the housing, illuminating the entire body of the prototype. Given that the experiments

are intended to be conducted at a maximum depth of 15 cm and a maximum distance of 35 cm, and that the wavelength of the red LED can be seen by both the cameras and individuals at these distances, this LED is used to simplify the composition of the prototypes.

- Prototype C is actuated by a planetary gear motor (provided by Zhaowei, CN), model ZWPD006006 to 420 rpm, with a weight of 1.6 g, a working torque of 40 g·cm, and a stall torque of 90 g·cm. It measures 6 mm in diameter and 21 mm in length and is linked to a 1.2 cm diameter propeller designed and manufactured following the same process as the outer housings. This propulsion system offers continuous rotation resulting in constant speed and advancement exclusively in the frontal direction. Additionally, the electronic system includes a rechargeable 4.2 V lithium battery and a magnetic switch that allows the system to be actuated by an external magnet which (Figure 3b), in turn, serves as a counterweight to achieve neutral buoyancy. The prototype measures 5.5 cm long, 2 cm tall, and 1.2 cm thick (Prototype C in Figure 1). This prototype can submerge due to the thrust generated by the propeller.
- Prototype D is identical to prototypes A and B, yet all electronic components were removed, resulting in a motionless prototype that can only float or remain stationary at the bottom, depending on its buoyancy (Prototype D in Figure 1). This model allows us to study whether the effects generated by the movement, sounds, and waves of the electronic components of the robots are significant and allows us to analyze whether the presence of a foreign object in the tank, its aesthetics, or size are influential in perceiving the prototypes as stressful.

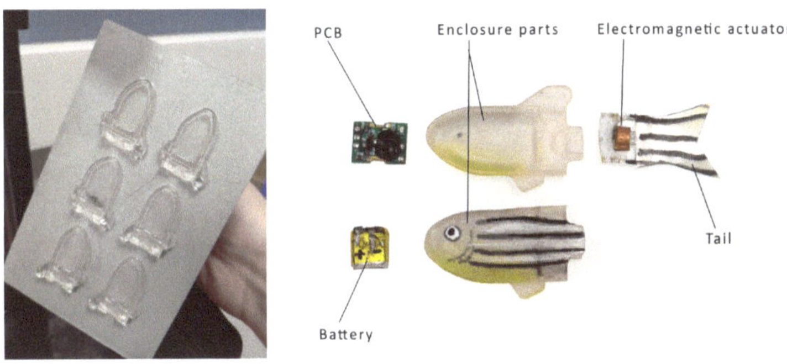

(a) (b)

Figure 2. Manufacturing and construction of robotic prototypes: (**a**) stereolithography printing of watertight housings for the prototypes; (**b**) parts and electronic components of prototypes A and B.

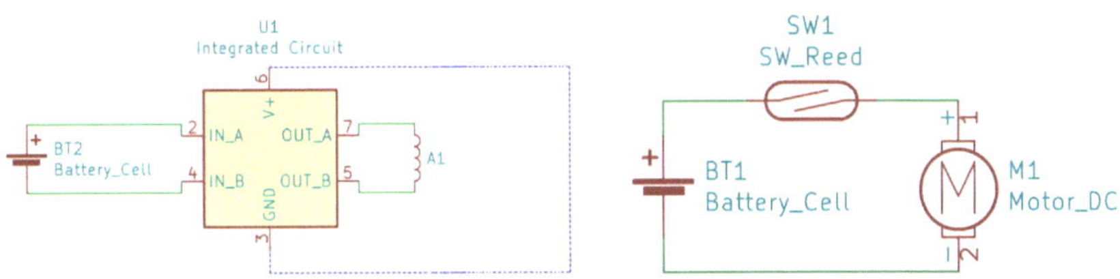

(a) (b)

Figure 3. Electronic schematics of (**a**) Prototypes A and B; (**b**) Prototype C.

2.3. Behavioral Quantification

Individuals were tested in 6 L tanks measuring 27 cm × 22 cm × 15 cm (Aquaneering, San Diego, CA, USA) (Figure 4a,b). The tanks were filled with 5.5 L of chlorine-free water at the same temperature and pH as their home tank. The test lasted 7.5 min following a 60 s period of accommodation. The tests were carried out on four different days, maintaining the same testing schedule from 9:00 a.m. to 1:00 p.m. to ensure similar conditions. The activity of the fish was recorded using industrial digital cameras (IDS (UI-3240CP USB 3.0 uEye CP, IDS Imaging Development Systems GmbH, Obersulm, Germany) and/or Basler (Basler acA1280-60gc GigE camera, Basler AG, Ahrensburg, Germany) equipped with a high-quality monofocal lens (focal length 8 mm) programmed with a resolution of 640 × 426 px and a frame rate of 25 fps. Trajectory tracking was performed using EthoVision®XT v.17.0.1630 software (NoldusInc, Wageningen, The Netherlands).

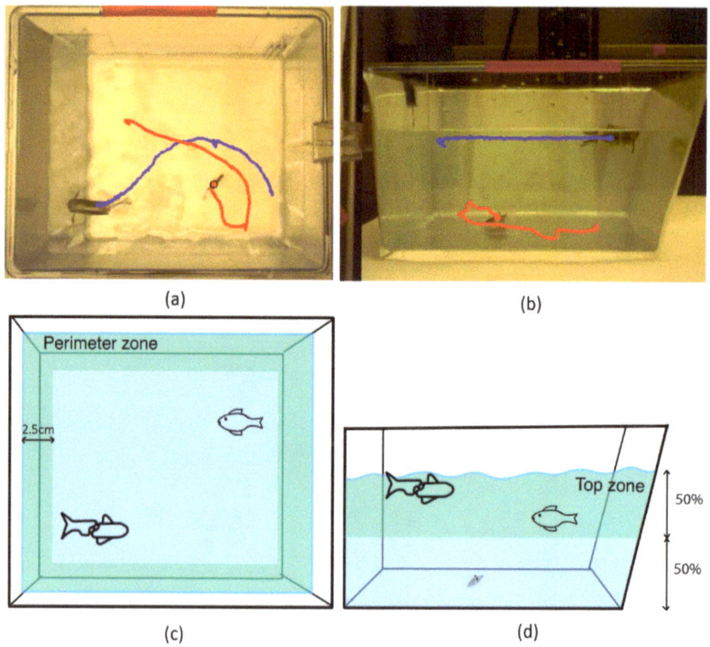

Figure 4. Recording setup of the top (**a**) and frontal (**b**) planes and tracking of the individuals using the AnimalTA software. Overhead (**c**) and frontal (**d**) views of the tank with the aversive zones marked in green: perimeter area from the top view equivalent to the space between the tank edges and its parallel projection at a distance of 2.5 cm; and top zone considered as the top half of the frontal view of the tank.

Fish were recorded simultaneously using frontal and zenithal planes of the tank; therefore, three-dimensional data were obtained. To analyze the natural bottom-dwelling response, each arena was divided into two equal zones: top and bottom (Figure 4d). For the zenithal plane, the arena was divided into two parts corresponding to the center and the perimeter of the tank. The "perimeter" was denoted as the area between the tank edges and its parallel projection at 2.5 cm (Figure 4c). Individual tracking and coordinates were obtained using AnimalTA v.2.3.1 software (http://vchiara.eu/index.php/animalta, accessed on 28 May 2024). Each prototype was tested against 7 naïve fish. The locomotive behaviors of 7 fish without any prototype were used as control.

For analysis, the vision software was programmed to extract the coordinates of both individuals (the real fish and the robot), from which the degree of avoidance of the real fish towards the robot was measured. Based on the behavioral patterns analyzed in the literature [31], the following evaluation measurements calculated and examined included: (i) the proportion of time spent by the fish in the top area of the tank; (ii) the number of visits to top area; (iii) the proportion of time spent in the perimeter area (defined as a distance of 2.5 cm from the tank's edge); (iv) the distance between the robot and the fish; (v) the percentage of time the fish displayed freezing behavior; (vi) average velocity; (vii) tracking distance; (viii) velocity deviation; and (ix) acceleration deviation. These measurements were taken from the real fish in response to its interaction with the robotic fish.

2.4. Statistical Analysis

Data were analyzed parametrically using SPSS software (IBM SPSS Statistics, Version 29.0.1.1 (244)) at a 95% confidence level, with $p < 0.05$ indicating statistical significance. For comparisons between groups, the one-way ANOVA method was used for parameters that follow a normal distribution, considering "type of prototype" as a factor, and the Kruskal–Wallis method was used for samples where the assumption of normality was rejected. Furthermore, post hoc multiple comparisons tests were also used. OriginPro software (Origin (Pro), Version 2022, OriginLab Corporation, Northampton, MA, USA) was used to generate graphs illustrating the results obtained.

Initially, descriptive statistics (shown in the table in Section 3) were calculated for each group and each evaluative measurement. To ensure correct analysis of the variables, a normality test was performed to verify that the samples meet the normality requirement using non-parametric tests. Variables that meet the normality criterion are analyzed using one-way ANOVA to compare means between groups, while those that do not follow a normal distribution are analyzed using the non-parametric Kruskal–Wallis test to compare medians between groups. After conducting a one-way ANOVA on parameters exhibiting a normal distribution, a test for homogeneity of variances is performed to ensure homoscedasticity compliance and to conduct post hoc multiple comparisons, operating under a 95% confidence level. The Tukey method is employed when equality of variances is assumed, whereas the Games–Howell method is utilized when the null hypothesis is rejected.

3. Results

Initially, three parameters were evaluated, which included swimming speed, Euclidean distance between the robot and fish, and latency of the analyzed individual's position over the entire tank area. Swimming speed was relatively low and constant when test animals were exposed to prototypes A and D (Figure 5a,d), a swimming pattern consistent with that of fish swimming alone (Figure 5e). In contrast, animals exposed to prototypes B and C exhibited an erratic swimming pattern, with speeds eventually reaching peaks of up to ten times their own baseline values (Figure 5b,c).

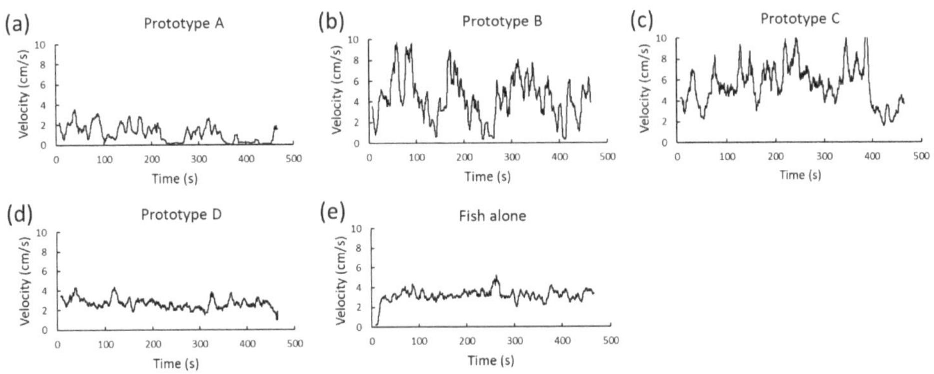

Figure 5. Reference velocities of how fish behaved when interacting with the different prototypes—(**a**) A; (**b**) B; (**c**) C; (**d**) D—and (**e**) fish alone.

The response of fish confronted with prototypes A, B, and C (Figure 6) was somewhat inconsistent as some of the tested animals exhibited freezing bouts either at the beginning of the experiment (Figure 6a/d) or during the experiment (Figure 6b/e,c/f).

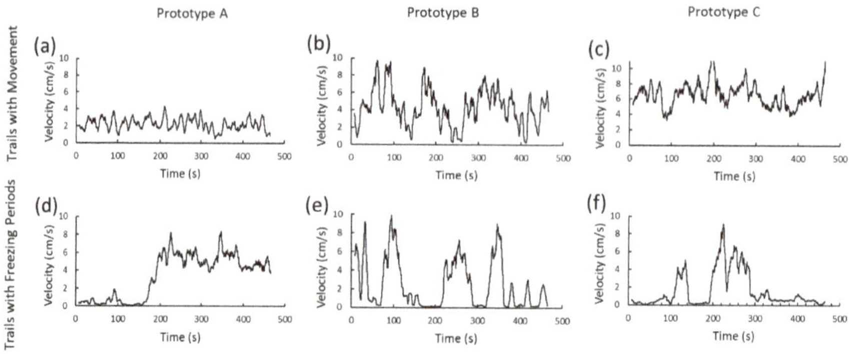

Figure 6. Reference velocities of how fish behaved when interacting with the prototypes: (**a**) Prototype A without freezing periods; (**b**) Prototype B without freezing periods; (**c**) Prototype C without freezing periods; (**d**) Prototype A with freezing periods; (**e**) Prototype B with freezing periods; (**f**) Prototype C with freezing periods.

The velocity was then plotted on the same graph alongside the Euclidean distance between the fish and the robot for each time point. This was carried out in order to analyze whether there was a direct relationship between the velocity of the fish and the distance between the fish and the prototype. This can be observed in Figure 7, where a representative case is shown for each of the evaluated prototypes. When the distance between the robot and the individual remained constant, with fewer abrupt changes over time, the fish also maintained a constant velocity. Conversely, when the distance fluctuates more erratically, the speed also fluctuates, demonstrating that the variation in distance between the animal and the prototype was related to the variation in the speed of the fish. Not only does the speed decrease when the variation in distance is more constant but also when the value of that interindividual distance is greater. Similarly, a decrease in the distance between individuals results in an increase in the velocity of the fish, thus suggesting an avoidance behavior.

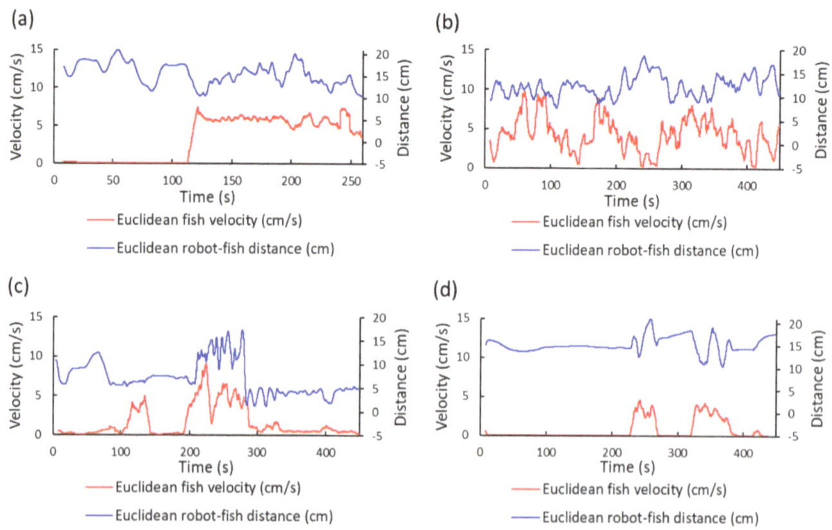

Figure 7. Relationship between velocity and Euclidean distance between the fish and the robot over time for a representative case of each of the tested prototypes: (**a**) Prototype A; (**b**) Prototype B; (**c**) Prototype C; and (**d**) Prototype D.

In order to analyze the positioning of the fish in the tank according to the predefined "top/bottom" and "perimetral" zones, heat maps were generated to visualize the time fish spent in these different areas when confronted with each prototype. This allows a visual comparison of the prototypes and facilitates the identification of freezing behavior, exploratory activity, or areas where individuals spend more time.

Fish swimming alone showed a homogeneous distribution of the position of the individual in the tank and a maximum density of 0.032 (Figure 8e). The distribution is centered in the lower and middle region of the tank, with exploration towards the upper half without reaching the surface of the water. The remaining graphs show a less-uniform distribution compared to the fish alone, with higher densities and longer periods of positioning near the bottom of the tank, indicating potential signs of stress and/or disturbance caused by certain robotic prototypes (Figure 8a–d). However, for prototypes A, C, and D (Figure 8a,c,d), individuals tended to visit the surface by swimming from the sides of the tank rather than from the central area, whereas animals exposed to prototype B exhibited a more evenly distributed pattern across the entire "top" region. Animals exposed to prototype C spent only short periods at the top of the surface, suggesting potential escape behaviors (Figure 8c).

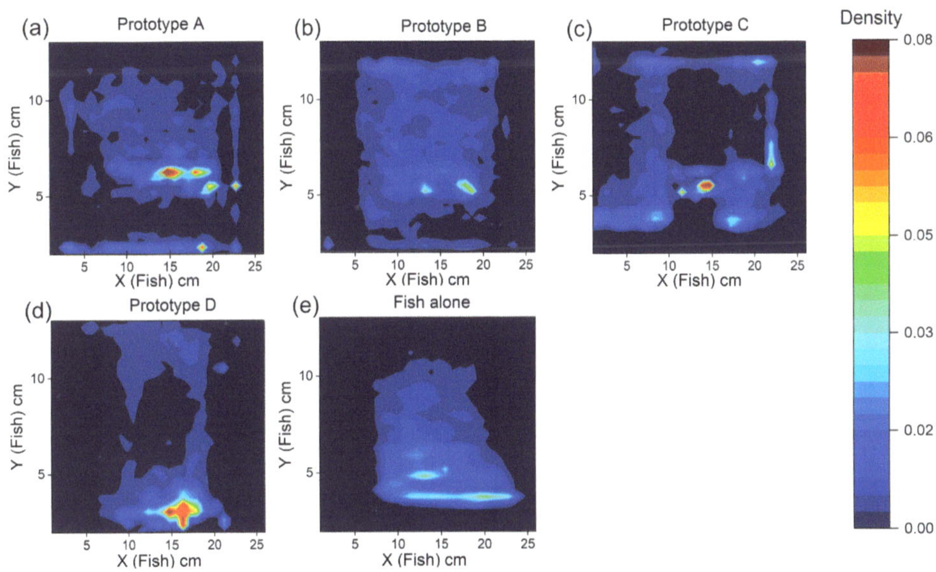

Figure 8. Representative density plots of the position of the fish from the frontal plane when (**a**) exposed to Prototype A; (**b**) exposed to Prototype B; (**c**) exposed to Prototype C; (**d**) exposed to Prototype D; and (**e**) alone. The color scale represents the cumulative time spent in each zone of the tank.

Similarly, fish swimming alone exhibited a homogeneous distribution when observed from the top (superior plane) in Figure 9. Fish swam primarily around the center of the tank, suggesting an absence of anxiety-like behavior and/or stress (Figure 9e). When fish are exposed to the prototypes, the distribution is less uniform; they remain longer in peripheral areas and present freezing episodes (Figure 9a–d).

Following this, Table 2 displays the descriptive statistics for each variable, in addition to normality and variance homoscedasticity studies. The variables that do not meet normality (top time proportion, number of visits to the surface, and velocity absolute deviation) criteria were analyzed using non-parametric methods (Kruskal–Wallis).

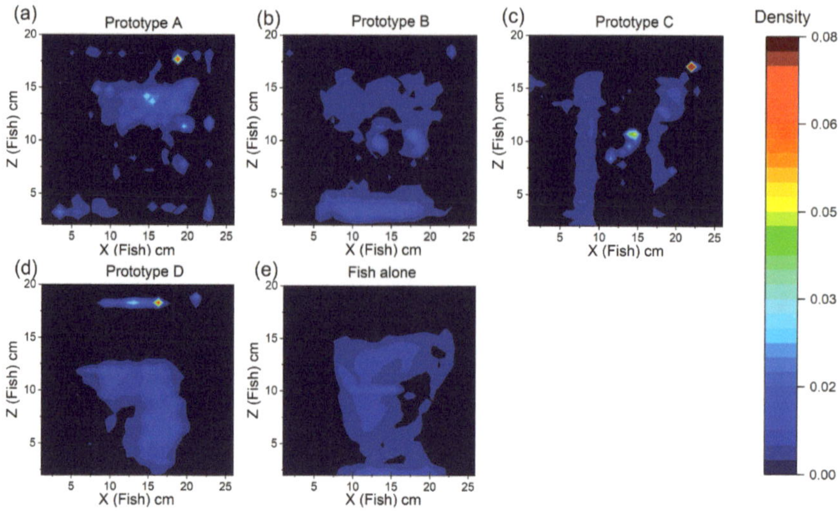

Figure 9. Density plots of the position of the live fish from the superior plane when (**a**) exposed to Prototype A; (**b**) exposed to Prototype B; (**c**) exposed to Prototype C; (**d**) exposed to Prototype D; and (**e**) alone.

Table 2. Descriptive statistics.

Parameter	Model	Mean	Std. Deviation	95% Confidence Interval for Mean: Lower Bound	95% Confidence Interval for Mean: Upper Bound	Normality Sig.	Homogeneity of Variances Sig. *
Top time proportion	A	30.51	18.37	13.52	47.51	0.005	0.23
	B	56.20	20.55	37.20	75.21		
	C	28.92	30.08	3.77	54.06		
	D	22.67	29.93	−5.01	50.35		
	E	15.13	10.65	5.28	24.98		
N° of visits to surface	A	2.81	2.21	0.77	4.86	0.002	0.440
	B	7.25	2.80	4.66	9.84		
	C	7.94	8.45	0.88	15.00		
	D	2.48	2.70	−0.02	4.98		
	E	5.33	4.43	1.23	9.43		
Perimeter time proportion	A	12.83	8.43	5.04	20.63	0.111	0.281
	B	24.48	11.70	13.66	35.30		
	C	43.29	22.38	24.58	62.01		
	D	23.22	15.39	8.98	37.45		
	E	31.40	10.92	21.29	41.50		
Freezing time proportion	A	21.70	18.84	4.28	39.13	<0.001	0.373
	B	6.88	14.45	−6.49	20.25		
	C	10.96	15.70	−2.16	24.07		
	D	12.52	28.67	−13.99	39.04		
	E	7.08	9.30	−1.53	15.69		
Tracking average distance	A	275.34	160.96	126.48	424.2	0.200	0.140
	B	418.24	140.34	288.45	548.03		
	C	470.95	255.39	257.44	684.47		
	D	302.38	127.81	184.17	420.58		
	E	401.09	61.83	343.91	458.27		
Velocity absolute deviation	A	3.39	1.62	1.90	4.88	0.019	0.005
	B	4.15	0.73	3.48	4.82		
	C	4.69	2.84	2.32	7.06		
	D	3.19	0.73	2.52	3.87		
	E	2.23	0.68	1.60	2.86		

Table 2. *Cont.*

Parameter	Model	Mean	Std. Deviation	95% Confidence Interval for Mean		Normality Sig.	Homogeneity of Variances Sig. *
				Lower Bound	Upper Bound		
Acceleration absolute deviation	A	100.41	69.72	35.93	164.90	0.056	0.020
	B	119.52	49.13	74.08	164.96		
	C	69.46	26.26	47.51	91.41		
	D	61.31	34.35	29.54	93.08		
	E	40.99	12.26	29.65	52.33		
Velocity Average	A	2.97	1.74	1.36	4.57	0.200	0.047
	B	3.2	1.55	1.76	4.63		
	C	4.97	2.9	2.55	7.39		
	D	3.55	1.84	1.85	5.26		
	E	4.42	0.93	3.56	5.28		

* Based on mean.

The statistical analyses showed significant differences between the prototypes and the fish alone for three of the parameters analyzed. These differences are indicated by asterisks in the box-and-whisker plots in Figure 10.

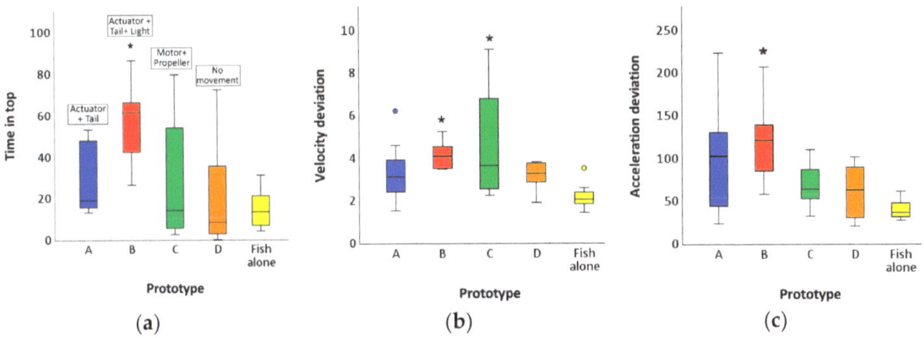

Figure 10. The diagrams display the distribution and central tendency of the numerical values (through quartiles) obtained from each of the prototypes for (**a**) time on the surface; (**b**) velocity deviation; (**c**) acceleration deviation. The asterisks indicate statistical significance between the indicated prototype and the reference group (Fish alone).

Performing non-parametric tests on metrics with non-normal distributions, utilizing the Kruskal–Wallis test with multiple pairwise comparisons, resulted in significant differences between the medians of prototype B and fish alone for the parameter "time in top", with a significance value of 0.028, as shown in Figure 10a.

The Kruskal–Wallis test with multiple pairwise comparisons showed differences between the medians of prototypes B and C compared to fish alone for the parameter "velocity deviation". Such differences were observed to be statistically significant, with significance values of 0.020 and 0.012, respectively, as depicted in Figure 10b.

However, when conducting a one-way ANOVA on parameters that exhibited a normal distribution, significant differences between group means were obtained for the evaluative measurement "acceleration deviation". The F-statistic value was calculated to be 3.803 with a significance value of 0.013, indicating that there are indeed significant differences between populations. Upon performing post hoc comparisons between groups, it was determined that there are differences in absolute acceleration deviation between prototype B and fish alone. The post hoc significance value for this comparison is reported to be 0.028 in Figure 10c.

4. Discussion

Behavioral studies have often been conducted to identify social interactions between conspecifics that promote the hierarchical allocation of individuals [32]. Such studies have often examined interactions with live stimuli and have shown that factors such as size and aesthetics play a crucial role in distinguishing leaders from predators [33,34]. In addition, research suggests that the movement patterns of replicas influence behavior, leading some researchers to use virtual representations or bio-inspired robotic replicas with closed trajectories to control specific parameters [35]. In this study, bio-replicas were meticulously designed to investigate the effects of light, noise, and propulsion mechanisms on fish locomotor behavior. Unlike previous studies, which predominantly analyzed fish behavior in a two-dimensional spatial plane with only two coordinates, this study tracked fish position throughout all three axes, yielding results which are reflective of a three-dimensional reality.

Parameters derived from existing behavioral models were employed to study fish behavioral responses (Table 1). However, some of these models were based on live stimuli without contact or on replicas with closed trajectories. Although this approach enhances experiment reproducibility and controllability, it sacrifices biomimicry and fidelity to reality, as noted by Spinello et al. [36]. Thus, the use of freely swimming robotic replicas in this study brings the experimental conditions closer to real-world scenarios.

The presence of a stationary prototype (D) in the tank was found to be less stressful compared to the presence of moving replicas, with fish perceiving D as a non-threatening entity. Accordingly, fish exposed to D exhibited similar velocity parameters (i.e., low mean velocity and animal deviation) to fish swimming alone.

The influence of the type of motion and the propulsion system of the replica was then investigated by comparing prototypes A, B, and C with the fish swimming alone. Graphical results suggested anxiety-like behaviors related to the interaction with B and C. However, the mean velocity and deviation values were more alike between the fish swimming alone and the electromagnetic-tailed prototype (A) than the propeller-driven prototype (C), which exhibited velocity spikes of up to 10 m/s. Statistical analysis confirmed significant differences in velocity and acceleration deviations between prototype C and fish swimming alone. In addition, two recordings were terminated prematurely due to fish jumping out of the tank in the presence of C, indicating an extreme escape response to a perceived imminent threat. This is consistent with the expected anxiety-like response to this robotic prototype, which can be attributed to the noisier motor and less natural, biomimetic motion associated with the propeller [25].

The effect of light was also investigated by comparing the responses of the fish to prototype A versus B, in comparison to the unstressed fish. Once more, the high velocities, peaking at 10 cm/s and irregular burst swimming patterns, which were evident in the light-presenting prototype (B), were indicative of anxiety/fear-like responses. Although the electromagnetic prototype without light (A) also exhibited differences from the unstressed fish swimming alone, including periods of freezing or less homogeneous tank positioning, these differences were exacerbated when light was added. Statistical analysis revealed significant differences between the light-presenting prototype and the fish alone, which was characterized by increased acceleration and velocity dispersion, indicative of rapid, erratic, burst-like swimming patterns associated with stress or anxiety.

In addition, prototypes B and C showed intermittent periods of freezing in some recordings throughout the trial, thus indicating stress due to a perceived imminent threat, further supporting previous observations. However, statistical support for this parameter was lacking due to non-significant results. The relationship between fish speed changes and robot-fish distance, as well as visual differences in fish tank positioning based on robot exposure, were unable to be statistically supported due to non-significant results for parameters such as "number of surface visits" or "time at perimeter". While significance was obtained for the "time at top" parameter for B, this contradicts interpretations based on velocities and accelerations, where surface latency would indicate non-stress. These limitations hampered the conclusions of the study.

One limitation was the complexity introduced by the free-swimming nature of the robots, which, while enhancing biomimicry and realism, also increased experimental complexity due to random swimming trajectories, hindering exact repeatability and making it difficult to determine whether the robot was approaching the fish or whether the fish were habituating, perceiving the robot as less threatening and reducing inter-individual distance. Similarly, the positioning of fish within defined tank zones, particularly for B, raised concerns, with some authors [23] suggesting that zebrafish may move to the top of the tank as a predator avoidance measure rather than an accommodation response. Another limitation was the sample size, with 37 valid fish tested. Although this is within the typical range for such studies [20] high variability and deviation were observed in each model, in all likelihood due to individual fish personalities and different responses to stimuli, which, in some cases, made it difficult to obtain statistically significant results. Therefore, further expansion would be advisable for future work in order to reduce variability and to improve the accuracy of the statistical analysis.

Finally, it is important to acknowledge a potential confounding factor related to stress, arising from the discrepancy in size between the replicas and the real fish. The latter were approximately twice the size of the replicas. It is known that fish consider size to be an important factor; however, this inconsistency was necessary due to the limitations of the low-cost electronic components available, which made it impossible to further reduce the size of the replicas. To mitigate the impact of this, uniformity in size among all replicas was sought. The fact that all replicas are of the same size and therefore share the same negative implication allows us to exclude size as a differentiating factor upon comparison.

Furthermore, despite some prior demonstrations that fish are attracted to certain tail movements or similar aesthetic patterns [37,38], this study also examined the implication of the type of actuator used to generate such movement. Additionally, this was achieved using the smallest possible replica, similar to the real fish in a context of free-swimming replicas, which allows for a closer approximation of the results to the development of bioinspired robots in a more realistic context.

Such results can be applied to the design of a robot intended for use in a real aquacultural context, tailored to the specific application for which it is designed to be used. The findings are particularly useful for tasks where the robot needs to maintain a close distance to the fish, such as environmental monitoring within cages or behavioral monitoring of the fish inhabiting them. In such cases, the design can be adjusted to avoid or reduce the use of lighting as well as the use of thrusters or cease their activity when the robot is in close proximity to the fish.

As a potential future study, it would be advisable to adapt the size of the prototypes to that of the fish used. This adaptation could be achieved by incorporating larger fish species into the study and creating prototypes at an appropriate scale. This approach would allow for the study of situations and behaviors accurately resembling the real-world applications which prompted this study by offering greater design flexibility for the replicas and reducing the constraints imposed by the size limitations of the components. Additionally, to mitigate data variance and increase statistical robustness, it would be advisable to increase the sample size and re-evaluate the variables for which significant results have not been obtained. Other potential lines of research derived from this study could include exploring the implications of heat, and more specifically, the actuator noise or the magnetic fields generated by electronic components. Finally, it would be interesting to study at what distance the factors involved in stress cease to have a stressful effect and are no longer perceived as a threat by the fish. This would enable us to design recommendations adapted to specific tasks to be performed by the robot.

5. Conclusions

This study highlights discernible differences in the behavioral response of fish to different underwater robots, depending on the propulsion and lighting system. Stressful behaviors were identified via significant differences between the variation in the speed and acceleration of the fish when alone and in the presence of a robot with light or a robot with propeller propulsion. No stressful behaviors were identified via significant differences between the fish when alone or in the presence of an unilluminated, electromagnetically propelled robot. These findings have direct implications for the design of underwater robots to reduce fish stress; thus, for more effective robotic biomimicry, it is advisable to avoid the use of propellers and noisy motors and to carefully consider or limit the use of lights.

Author Contributions: Conceptualization and methodology R.V. and J.M.C.-R.; formal analysis, A.P. and R.V.; investigation, A.P.; resources, A.P. and E.T.; writing—original draft preparation, A.P.; writing—review and editing, A.P., R.V., E.T., J.M.C.-R. and R.M.P.; visualization, A.P.; supervision, R.V., R.M.P. and P.J.S.; funding acquisition, P.J.S. and J.M.C.-R. All authors have read and agreed to the published version of the manuscript.

Funding: This study forms part of the projects JELLYFISH and included in the Aquaculture programme supported by MICINN with funding from European Union NextGenerationEU (PRTR-C17.I1) and by Generalitat Valenciana (ThinkInAzul/2021/037, 2021-25) and GEIROV, funded by Universitat Jaume I (UJI-B2021-61) and by MCIN/AEI/10.13039/501100011033 (PID2022-136288OB-C33) to J.M.C.-R.) and, as appropriate, by "ERDF A way of making Europe", by the "European Union" or by the "European Union Next Generation EU/PRTR".

Institutional Review Board Statement: All experiments were performed following guidelines of the Spanish (Royal Decree 53/2013) and the European Union Directive on the Protection of Animals Used for Scientific Purposes (Directive 2010/63/EU). The protocols applied were approved by IATS Ethics Committee (Register Number 09-0201) under the supervision of the Secretary of State for Research, Development and Innovation of the Spanish Government.

Informed Consent Statement: Not applicable.

Data Availability Statement: The videos and datasets generated during and/or analyzed during the current study are available from the first author on reasonable request.

Conflicts of Interest: The authors declare no conflicts of interest.

References

1. Huang, H.; Tang, Q.; Li, J.; Zhang, W.; Bao, X.; Zhu, H.; Wang, G. A Review on Underwater Autonomous Environmental Perception and Target Grasp, the Challenge of Robotic Organism Capture. *Ocean. Eng.* **2020**, *195*, 106644. [CrossRef]
2. Ranapurwala, S.I.; Kucera, K.L.; Denoble, P.J. The Healthy Diver: A Cross-Sectional Survey to Evaluate the Health Status of Recreational Scuba Diver Members of Divers Alert Network (DAN). *PLoS ONE* **2018**, *13*, e0194380. [CrossRef]
3. Bannister, J.; Sievers, M.; Bush, F.; Bloecher, N. Biofouling in Marine Aquaculture: A Review of Recent Research and Developments. *Biofouling* **2019**, *35*, 631–648. [CrossRef]
4. Amundsen, H.B.; Caharija, W.; Pettersen, K.Y. Autonomous ROV Inspections of Aquaculture Net Pens Using DVL. *IEEE J. Ocean. Eng.* **2022**, *47*, 1–19. [CrossRef]
5. Thinkinazul. Available online: https://thinkinazul.es/en/ (accessed on 25 May 2024).
6. López-Barajas, S.; Sanz, P.J.; Marín-Prades, R.; Gómez-Espinosa, A.; González-García, J.; Echagüe, J. Inspection Operations and Hole Detection in Fish Net Cages through a Hybrid Underwater Intervention System Using Deep Learning Techniques. *J. Mar. Sci. Eng.* **2024**, *12*, 80. [CrossRef]
7. Kruusmaa, M.; Gkliva, R.; Tuhtan, J.A.; Tuvikene, A.; Alfredsen, J.A. Salmon Behavioural Response to Robots in an Aquaculture Sea Cage. *R. Soc. Open Sci.* **2020**, *7*, 191220. [CrossRef]
8. Vidyarthi, P.K.; Mukherjee, K.; Roy, B.K. Fish-Like Robots and Applications of Sensor-A Review. In Proceedings of the 4th International Conference on Energy, Power, and Environment, ICEPE 2022, Shillong, India, 29 April–1 May 2022.
9. Free, B.A.; Lee, J.; Paley, D.A. Bioinspired Pursuit with a Swimming Robot Using Feedback Control of an Internal Rotor. *Bioinspir. Biomim.* **2020**, *15*, 035005. [CrossRef]

10. Li, L.; Ravi, S.; Xie, G.; Couzin, I.D. Using a Robotic Platform to Study the Influence of Relative Tailbeat Phase on the Energetic Costs of Side-by-Side Swimming in Fish. *Proc. R. Soc. A Math. Phys. Eng. Sci.* **2021**, *477*, 20200810. [CrossRef]
11. Van Den Berg, S.C.; Scharff, R.B.N.; Rusák, Z.; Wu, J. OpenFish: Biomimetic Design of a Soft Robotic Fish for High Speed Locomotion. *HardwareX* **2022**, *12*, e00320. [CrossRef]
12. Katzschmann, R.K.; Delpreto, J.; Maccurdy, R.; Rus, D. Exploration of Underwater Life with an Acoustically Controlled Soft Robotic Fish. *Sci. Robot.* **2018**, *3*, eaar3449. [CrossRef]
13. Jarque, A.P.; Solis, A.; Vidal, R.; Marín, R.; Cerdá-Reverter, J.M.; Sanz, P.J. Towards Precision Aquaculture: "The UJI Micro Robot Fish Prototype". In Proceedings of the OCEANS 2023-Limerick, OCEANS Limerick 2023, Limerick, Ireland, 5–8 June 2023.
14. Chiara, V.; Kim, S.Y. AnimalTA: A Highly Flexible and Easy-to-Use Program for Tracking and Analysing Animal Movement in Different Environments. *Methods Ecol. Evol.* **2023**, *14*, 1699–1707. [CrossRef]
15. Maertens, A.P.; Triantafyllou, M.S.; Yue, D.K.P. Efficiency of Fish Propulsion. *Bioinspir. Biomim.* **2015**, *10*, 046013. [CrossRef]
16. Tschirren, L.; Bachmann, D.; Güler, A.C.; Blaser, O.; Rhyner, N.; Seitz, A.; Zbinden, E.; Wahli, T.; Segner, H.; Refardt, D. Myfishcheck: A Model to Assess Fish Welfare in Aquaculture. *Animals* **2021**, *11*, 145. [CrossRef]
17. Blanchard, D.C.; Blanchard, R.J. Chapter 2.4 Defensive Behaviors, Fear, and Anxiety. In *Handbook of Anxiety and Fear*; Academic Press: Boston, MA, USA; Amsterdam, The Netherlands, 2008; Volume 17, ISBN 0-444-53065-7.
18. Kopman, V.; Laut, J.; Polverino, G.; Porfiri, M. Closed-Loop Control of Zebrafish Response Using a Bioinspired Robotic-Fish in a Preference Test. *J. R. Soc. Interface* **2013**, *10*, 20120540. [CrossRef]
19. Bierbach, D.; Landgraf, T.; Romanczuk, P.; Lukas, J.; Nguyen, H.; Wolf, M.; Krause, J. Using a Robotic Fish to Investigate Individual Differences in Social Responsiveness in the Guppy. *R. Soc. Open Sci.* **2018**, *5*, 181026. [CrossRef]
20. Kruusmaa, M.; Rieucau, G.; Montoya, J.C.C.; Markna, R.; Handegard, N.O. Collective Responses of a Large Mackerel School Depend on the Size and Speed of a Robotic Fish but Not on Tail Motion. *Bioinspir. Biomim.* **2016**, *11*, 056020. [CrossRef]
21. Phamduy, P.; Polverino, G.; Fuller, R.C.; Porfiri, M. Fish and Robot Dancing Together: Bluefin Killifish Females Respond Differently to the Courtship of a Robot with Varying Color Morphs. *Bioinspir. Biomim.* **2014**, *9*, 036021. [CrossRef]
22. Bierbach, D.; Mönck, H.J.; Lukas, J.; Habedank, M.; Romanczuk, P.; Landgraf, T.; Krause, J. Guppies Prefer to Follow Large (Robot) Leaders Irrespective of Own Size. *Front. Bioeng. Biotechnol.* **2020**, *8*, 441. [CrossRef]
23. Ruchin, A.B. Effect of Illumination on Fish and Amphibian: Development, Growth, Physiological and Biochemical Processes. *Rev. Aquac.* **2021**, *13*, 567–600. [CrossRef]
24. Glaze, J.; Salazar, R.; Vasconcellos, R.; Abdelkefi, A. Comparative Design, Hydrodynamic Analysis, and Physical Performance of Fish-like Robots. *Appl. Ocean. Res.* **2021**, *106*, 102443. [CrossRef]
25. Dadswell, M.J.; Spares, A.D.; Mclean, M.F.; Harris, P.J.; Rulifson, R.A. Long-Term Effect of a Tidal, Hydroelectric Propeller Turbine on the Populations of Three Anadromous Fish Species. *J. Fish Biol.* **2018**, *93*, 192–206. [CrossRef]
26. Maximino, C.; de Brito, T.M.; da Silva Batista, A.W.; Herculano, A.M.; Morato, S.; Gouveia, A. Measuring Anxiety in Zebrafish: A Critical Review. *Behav. Brain Res.* **2010**, *214*, 157–171. [CrossRef]
27. Stewart, A.; Gaikwad, S.; Kyzar, E.; Green, J.; Roth, A.; Kalueff, A.V. Modeling Anxiety Using Adult Zebrafish: A Conceptual Review. *Neuropharmacology* **2012**, *62*, 135–143. [CrossRef]
28. Johnson, A.; Loh, E.; Verbitsky, R.; Slessor, J.; Franczak, B.C.; Schalomon, M.; Hamilton, T.J. Examining Behavioural Test Sensitivity and Locomotor Proxies of Anxiety-like Behaviour in Zebrafish. *Sci. Rep.* **2023**, *13*, 3768. [CrossRef]
29. Phamduy, P. Robotic Fish to Aid Animal Behavior Studies and Informal Science Learning. Ph.D. Thesis, New York University, New York, NY, USA, 2017.
30. Blackstock, J.M.; Covington, M.D.; Perne, M.; Myre, J.M. Monitoring Atmospheric, Soil, and Dissolved CO_2 Using a Low-Cost, Arduino Monitoring Platform (CO2-LAMP): Theory, Fabrication, and Operation. *Front. Earth Sci.* **2019**, *7*, 313. [CrossRef]
31. Kalueff, A.V.; Gebhardt, M.; Stewart, A.M.; Cachat, J.M.; Brimmer, M.; Chawla, J.S.; Craddock, C.; Kyzar, E.J.; Roth, A.; Landsman, S.; et al. Towards a Comprehensive Catalog of Zebrafish Behavior 1.0 and Beyond. *Zebrafish* **2013**, *10*, 70–86. [CrossRef]
32. Romenskyy, M.; Herbert-Read, J.E.; Ward, A.J.W.; Sumpter, D.J.T. Body Size Affects the Strength of Social Interactions and Spatial Organization of a Schooling Fish (*Pseudomugil signifer*). *R. Soc. Open Sci.* **2017**, *4*, 161056. [CrossRef]
33. Polverino, G.; Porfiri, M. Mosquitofish (*Gambusia affinis*) Responds Differentially to a Robotic Fish of Varying Swimming Depth and Aspect Ratio. *Behav. Brain Res.* **2013**, *250*, 133–138. [CrossRef]
34. Landgraf, T.; Bierbach, D.; Nguyen, H.; Muggelberg, N.; Romanczuk, P.; Krause, J. RoboFish: Increased Acceptance of Interactive Robotic Fish with Realistic Eyes and Natural Motion Patterns by Live Trinidadian Guppies. *Bioinspir. Biomim.* **2016**, *11*, 15001. [CrossRef]
35. Rocha, A.; Godino-Gimeno, A.; Rotllant, J.; Cerdá-Reverter, J.M. Agouti-Signalling Protein Overexpression Reduces Aggressiveness in Zebrafish. *Biology* **2023**, *12*, 712. [CrossRef]
36. Spinello, C.; Yang, Y.; Macrì, S.; Porfiri, M. Zebrafish Adjust Their Behavior in Response to an Interactive Robotic Predator. *Front. Robot. AI* **2019**, *1*, 38. [CrossRef]

37. Polverino, G.; Phamduy, P.; Porfiri, M. Fish and Robots Swimming Together in a Water Tunnel: Robot Color and Tail-Beat Frequency Influence Fish Behavior. *PLoS ONE* **2013**, *8*, e77589. [CrossRef]
38. Butail, S.; Polverino, G.; Phamduy, P.; Del Sette, F.; Porfiri, M. Influence of Robotic Shoal Size, Configuration, and Activity on Zebrafish Behavior in a Free-Swimming Environment. *Behav. Brain Res.* **2014**, *275*, 269–280. [CrossRef]

Journal of
Marine Science and Engineering

MDPI

Article

Antioxidant and Metabolic Response to Acute Acidification Stress of Juvenile Yellowfin Tuna (*Thunnus albacares*)

Xiaoyan Wang [1,2,3,4,5,†], Rui Yang [2,3,4,5,†], Zhengyi Fu [2,3,4,5,6], Lei Zhao [7,*] and Zhenhua Ma [2,3,4,5,6,*]

1 College of Fisheries and Life Sciences, Dalian Ocean University, Dalian 116023, China
2 Key Laboratory of Efficient Utilization and Processing of Marine Fishery Resources of Hainan Province, Sanya Tropical Fisheries Research Institute, Sanya 572018, China
3 South China Sea Fisheries Research Institute, Chinese Academy of Fishery Sciences, Guangzhou 510300, China
4 Hainan Engineering Research Center for Deep-Sea Aquaculture and Processing, Sanya 572018, China
5 International Joint Research Center for Conservation and Application of Fishery Resources in the South China Sea, Sanya 572018, China
6 College of Science and Engineering, Flinders University, Adelaide 5001, Australia
7 Yazhou Bay Agriculture and Aquaculture Co., Ltd., Sanya 572025, China
* Correspondence: leiforever@hotmail.com (L.Z.); zhenhua.ma@scsfri.ac.cn (Z.M.)
† These authors contributed equally to this work.

Abstract: This study aimed to explore the impact of acute acidification on the antioxidant, metabolic performance, and liver histology of juvenile yellowfin tuna. The experiment subjected juvenile yellowfin tuna to a pH gradient environment of 8.1, 7.6, 7.1, and 6.6 for 48 h. The findings indicate that a seawater pH of 7.1 significantly impacts the antioxidant and metabolic systems of the juvenile yellowfin tuna in comparison to the control group. At pH 7.1, there were observed increases in glutathione reductase (GR), total antioxidant capacity (T-AOC), lactate dehydrogenase (LDH), hexokinase (HK), pyruvate kinase (PK), sodium-potassium ATPase ($Na^{+}K^{+}$-ATP), and calcium-magnesium ATPase ($Ca^{2+}Mg^{2+}$-ATP). Conversely, low-density lipoprotein cholesterol (LDL-C), high-density lipoprotein cholesterol (HDL-C), and triglycerides (TGs) were not significantly different across the treatment groups. However, an increase in transaminases at pH 7.1 suggested potential liver damage, which was further supported by observed structural liver tissue degeneration and hepatocyte vacuolation. In conclusion, under conditions of acute acidification stress, there is a decrease in antioxidant capacity and a suppression of metabolic levels in juvenile yellowfin tuna, leading to oxidative damage. This study lays the foundation for an in-depth understanding of the response mechanisms of juvenile yellowfin tuna in response to seawater acidification as well as healthy tuna farming in the broader context of seawater acidification.

Keywords: environmental stress; antioxidant enzymes; metabolic enzymes; serum biochemistry index; liver histology

Citation: Wang, X.; Yang, R.; Fu, Z.; Zhao, L.; Ma, Z. Antioxidant and Metabolic Response to Acute Acidification Stress of Juvenile Yellowfin Tuna (*Thunnus albacares*). *J. Mar. Sci. Eng.* **2024**, *12*, 970. https://doi.org/10.3390/jmse12060970

Academic Editors: Alberta Mandich and Jose Pedro Andrade

Received: 5 April 2024
Revised: 23 May 2024
Accepted: 6 June 2024
Published: 8 June 2024

1. Introduction

The pH in aquatic environments is one of the factors affecting the physiological and biochemical functions of fish. Deviations from the optimal pH range directly impact fish's acid–base balance, respiratory function, circulation, and lipid accumulation [1–3]. In recent years, there has been a growing concern for the aquatic environment, highlighting the necessity to focus on the resilience and adaptability of marine species in low pH conditions. However, several factors contribute to the instability of seawater pH. For example, an increase in CO_2 concentration leads to a reduction in seawater pH, with predictions suggesting a decrease of 0.7 pH units by the year 2300 [4]. Additionally, the coastal regions of China, being low-lying, are more susceptible to changes in water body pH due to extreme weather conditions such as typhoons and hurricanes. Human activities also have a more severe impact on the water environment, sometimes even causing extreme disturbances [5,6].

Studies have shown that a decrease in environmental pH leads to an increase in levels of reactive oxygen species (ROS). Antioxidant defenses play a crucial role in neutralizing ROS and protecting fish from oxidative damage [7]. Long-term exposure to low pH environments can damage these antioxidant mechanisms, potentially weakening the fish's ability to effectively manage oxidative stress [8]. Additionally, acidic waters can lead to a decrease in the pH of fish blood, thereby affecting the binding of hemoglobin to oxygen [9,10]. This impairment restricts oxygen transport, leading to hypoxia in fish. Symptoms include reduced activity, significant declines in metabolic function, decreased appetite, digestive issues, and stunted growth [11]. Ding Zhaokun et al. found that a reduction in water body pH not only affects gene expression but also adversely impacts the energy, protein, lipid, and nucleic acid metabolism of marine organisms [12]. Changes in external environmental conditions can also cause shifts in fish osmotic pressure and serum ion levels, making blood parameters a reliable indicator of fish health under environmental stress [13]. Petochi et al. exposed European sea bass (*Dicentrarchus labrax*) to low pH environments and discovered that their metabolic rates increased as the water's pH decreased [14]. Md and colleagues exposed juvenile catfish (*Hexanematichthys sagor*) to a low pH environment and analyzed their blood for aspartate aminotransferase, alanine aminotransferase, alkaline phosphatase, glucose, and cortisol. The results showed that exposure to a high-acidity environment can damage the health of the fish, potentially leading to poor growth and reduced survival rates [15]. Although adult fish exhibit a strong adaptability to changes in environmental pH, their juveniles lack effective mechanisms for intracellular pH regulation, making them particularly susceptible to early-life damage caused by acidification [16].

A decrease in water pH poses significant risks to marine ecosystems and threatens the survival and reproductive capabilities of marine fish species [17,18]. The yellowfin tuna, known for its high economic value, is an essential global fishery resource. By 2022, the annual global catch of yellowfin tuna surpassed 1.4 million tons, making it the second most extensively fished tuna species after the skipjack tuna [19]. Understanding how yellowfin tuna manages oxidative stress through its antioxidant system becomes particularly important. Studying the antioxidant and metabolic responses helps to better comprehend the adaptation mechanisms of yellowfin tuna to acidic environments, which is crucial for predicting their survival and reproduction in future marine conditions. Despite the significance of these issues, research on how seawater acidification affects physiological changes in yellowfin tuna juveniles is still limited. This study focuses on juvenile yellowfin tuna as a model organism. Based on predictions from ocean models and experimental results from Bromhead et al., it evaluates the impact of different levels of seawater acidity on the antioxidant and metabolic performance, serum indicators, and liver histology of yellowfin tuna [20,21]. The aim is to provide reference data for the stress response of marine fish to seawater acidification.

2. Materials and Methods

2.1. Experimental Materials

The experimental fish were caught in the sea area near the Xincun port (18°22′31.9″ N, 109°58′28.9″ E). These fish were transported to the Lingshui Experimental Center of Sanya Tropical Fisheries Research Institute (Sanya, China) for a week to acclimate and subsequent acute acidification experiment. A total of 72 healthy juvenile yellowfin tuna of similar size (mean length: 18.21 ± 1.09 cm, mean weight: 307.49 ± 49.38 g) were selected and randomly placed into 12 circular fiberglass tanks (5000 L). To manipulate seawater pH levels, solutions of 1.0 mmol·L^{-1} NaOH and 1.0 mmol·L^{-1} HCl [22,23] were used, setting a control pH at 8.1 [24] and treatment pH levels at 6.6, 7.1, and 7.6 [25–28], with three replications of each condition, and six fish were set up for each replication. The acidification stress test lasted 48 h [29,30], during which time feeding was stopped and seawater acidity was measured using a pH analyzer (Shenzhen Jingxin Microelectronics Co., Shenzhen, China), which was used to monitor seawater pH every two hours to maintain fluctuations within ±0.1 pH units. During the temporary rearing and experimental period, 12 h of light and 12 h of darkness were maintained, the water temperature was kept at 22.5 ± 1 °C, dissolved oxygen levels were maintained above 7.5 mg·L^{-1}, nitrite nitrogen concentrations were kept below 0.05 mg·L^{-1}, and ammonia nitrogen concentrations were also maintained below <0.05 mg·L^{-1}.

2.2. Sample Collection and Processing

After 48 h of acidification stress, the fish were anesthetized with eugenol (Shangchi Dental Material Co., Ltd., Changshu, China), six fish were removed from each bucket, weighed on a scale (to the nearest 0.01 cm), measured for body length with calipers (to the nearest 0.01 cm), and dissected according to the method of Rosseland et al. [31]. Blood samples were drawn from the caudal vein of the fish and collected using a 2 mL syringe precoated with 1% sodium heparin. Blood samples were centrifuged in a cryo-centrifuge (XinAoYi Instrument Co., Ltd., Changsha, China) at 3000 r·min^{-1} for 10 min, and the serum was collected. The liver was quickly removed, rinsed with pre-cooled saline, blotted dry on filter paper, and collected into a 2 mL sterile cryo-centrifuge tube. The serum and the liver samples for enzyme activity determination were then stored at −80 degrees Celsius.

A total of 0.1~0.2 g of liver tissue was accurately weighed, 9 times volume of 0.9% saline was added, it was mechanically homogenized under ice-water bath condition and centrifuged at 3000 r·min^{-1} for 10 min, and the supernatant was taken to determine the contents of total protein (TP) content, total antioxidant capacity (T-AOC), glutathione reductase (GR), lipid peroxidation (LPO), lactate dehydrogenase (LDH), hexokinase (HK), pyruvate kinase (PK), sodium-potassium ATPase (Na^+K^+-ATP), and calcium-magnesium ATPase ($Ca^{2+}Mg^{2+}$-ATP). The serum samples were used to determine alkaline phosphatase (AKP), glutamine transaminase (GPT), glutamic oxaloacetic transaminase (GOT), high-density lipoprotein cholesterol (HDL-C), low-density lipoprotein cholesterol (LDL-C), triglycerides (TGs), total cholesterol (TCH), and glucose (GLU). All analyses were performed using kits (Nanjing Jiancheng Bioengineering Institute, Nanjing, China). TP was detected using the bicinchoninic acid assay (minimum detection limit: 20 μg·mL^{-1}). T-AOC was detected using the ABTS (2,2′-azino-bis (3-ethylbenzothiazoline-6-sulfonic acid)) radical cation decolorization assay (minimum detection limit: 0.5 mmol). GR was detected using the glutathione reductase activity assay (minimum detection limit: 0.1 μ·L^{-1}). LPO was detected using the thiobarbituric acid reactive substances assay (minimum detection limit: 0μ·L^{-1}). LDH was detected using the lactate dehydrogenase activity assay (minimum detection limit: 1 μ·L^{-1}). PK was detected using the pyruvate kinase–lactate dehydrogenase-coupled assay (minimum detection limit: 1.3 μ·L^{-1}). HK was detected using the hexokinase-glucose-6-phosphate dehydrogenase-coupled assay (minimum detection limit: 2.3 μ·L^{-1}). Na^+K^+-ATP and $Ca^{2+}Mg^{2+}$-ATP were detected using the colorimetric method based on inorganic phosphate detection (minimum detection limit: 0.0026 μmol·mL^{-1}). GLU was detected using the glucose oxidase method (minimum detection limit: 2.2 mmol·L^{-1}). LDL-C and HDL-C were detected using the double reagent direct method (minimum detection limit: 0 mmol·L^{-1}). TG was detected using the glycerol-3-phosphate oxidase-peroxidase method (minimum detection limit: 0.2 mmol·L^{-1}). TCH was detected using the cholesterol oxidase-peroxidase method (minimum detection limit: 0 mmol·L^{-1}). GOT and GPT were detected using the Reitman–Frankel method (minimum detection limit: 0 μ·L^{-1}). AKP was detected using the p-nitrophenyl phosphate method (minimum detection limit: 45 μ·L^{-1}). Determination was carried out using a hybrid microplate reader (BioTek Instruments, Winooski, VT, USA) and a UV–visible spectrophotometer (Shanghai Meppan Instruments Co., Ltd., Shanghai, China) in strict accordance with the operating instructions. Each sample was replicated three times, with negative control and positive control set up for each determination.

Tissue sections were prepared with reference to Yancheva et al. [32]. Juvenile fish livers were collected, preserved in 4% paraformaldehyde (BL 539A, Biosharp, Hefei, China), and then embedded in paraffin. These prepared tissues were then sectioned into 4 μm thick cross sections utilizing a Leica rotary microtome (Leica Instruments Shanghai Co.,

Ltd., Shanghai, China). The sections were stained with hematoxylin and eosin (H&E) and permanently mounted with a neutral resin. Observations were made with a DMI8 fluorescence inverted microscope (Leica, Wetzlar, Germany), photographs were taken at 400× magnification, and pictures were saved.

2.3. Statistical Analyses

Shapiro–Wilk test [33] was used to examine the normal distribution of the data. The homogeneity of variance was assessed using the Levene test. One-way analysis of variance (ANOVA) and Duncan's test were employed to analyze the significance of differences between the control group and the treatment group in liver and serum parameters. A p-value of less than 0.05 was considered significant, while equal to or above this threshold was considered insignificant. Data were analyzed using SPSS 25.0 software, expressed as mean ± standard deviation (Mean ± SD), and displayed as bar charts using Origin 2022.

3. Results

3.1. Effect of Acute Acidification Stress on Antioxidant Parameters of Yellowfin Tuna Liver

The change in GR activity exhibited a general downward trend with the decrease in pH, reaching its lowest at pH 7.6 (5.05 ± 1.61 μ·gprot^{-1}), which is significantly different from the control group ($p < 0.05$, Figure 1a). Conversely, T-AOC displayed an upward trend with the decrease in pH, peaking at pH 6.6 (0.84 ± 0.19 mmol·gprot^{-1}), though this did not significantly differ from the control ($p > 0.05$, Figure 1b). The LPO concentration demonstrated an upward trend with the decrease in pH, achieving its highest level at pH 7.6 (0.04 ± 0.01 μ·gprot^{-1}), having a significant departure from the control ($p < 0.05$, Figure 1c).

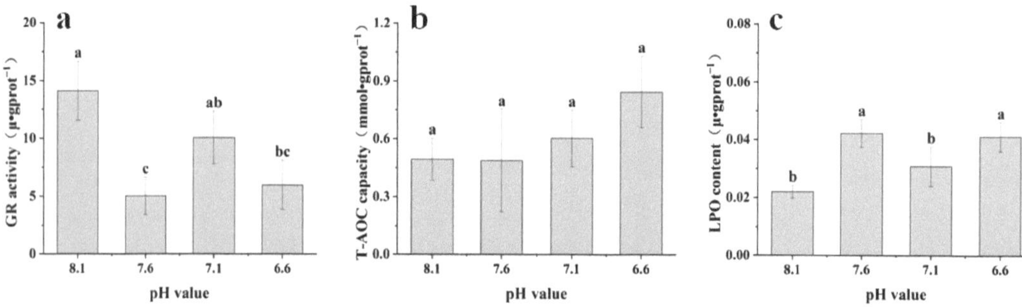

Figure 1. Effect of acute acidification stress on antioxidant parameters in juvenile yellowfin tuna liver. The bar graph represents the mean ± SD of the measurements taken at 48 h. (**a**) Glutathione reductase (GR) activity (ANOVA, GR: $F = 10.968$, $p < 0.05$), (**b**) total antioxidant (T-AOC) capacity (ANOVA, T-AOC: $F = 2.396$, $p > 0.05$), and (**c**) lipid peroxidation (LPO) (ANOVA, LPO: $F = 10.714$, $p < 0.05$). Different letters on the columns indicate significant differences between groups ($p < 0.05$), and the same letters indicate non-significant differences between groups ($p > 0.05$).

3.2. Effects of Acute Acidification Stress on Metabolic Parameters of Yellowfin Tuna in Liver and Trunk Kidney

The activities of LDH, HK, and PK all peaked at pH 7.1 (LDH: 6.12 ± 0.24 μ·gprot^{-1}, HK: 2.26 ± 0.24 μ·gprot^{-1}, and PK: 14.18 ± 0.97 μ·gprot^{-1}, Figure 2a–c). The LDH and PK activities demonstrated an increasing trend, and a significant difference was observed between the treatment groups and control groups at pH 7.1 ($p < 0.05$, Figure 2a,c). Na^+K^+-ATPase activity showed irregular changes with decreasing seawater pH and was highest at pH 7.1 (0.15 ± 0.01 μ·mgprot^{-1}, Figure 2d), and $Ca^{2+}Mg^{2+}$-ATP activity was significantly lower than that of the control group in all treatment groups and reached the lowest level at pH 7.6 (0.06 ± 0.00 μ·mgprot^{-1}, Figure 2e), all of which were significantly different from the control group ($p < 0.05$).

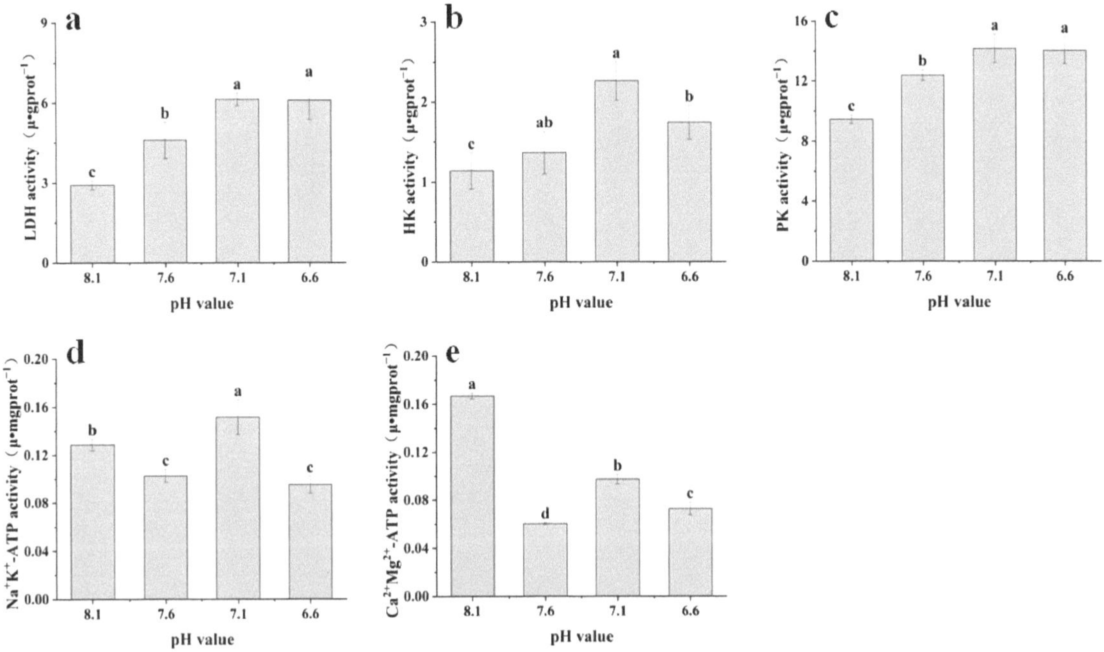

Figure 2. Effect of acute acidification stress on metabolic enzymes in juvenile yellowfin tuna liver. The bar graph represents the mean ± SD of the measurements taken at 48 h. (**a**) Lactate dehydrogenase (LDH) activity (ANOVA, LDH: F = 24.458, $p < 0.05$), (**b**) hexokinase (HK) activity (ANOVA, HK: F = 12.447, $p < 0.05$), (**c**) pyruvate kinase (PK) activity (ANOVA, PK: F = 28.622, $p < 0.05$), (**d**) sodium-potassium ATPase ($Na^{+}K^{+}$-ATP) activity (ANOVA, $Na^{+}K^{+}$-ATP: F = 25.508, $p < 0.05$), and (**e**) calcium-magnesium ATPase ($Ca^{2+}Mg^{2+}$-ATP) activity (ANOVA, $Ca^{2+}Mg^{2+}$-ATP: F = 347.865, $p < 0.05$). Different letters on the columns indicate significant differences between groups ($p < 0.05$), and the same letters indicate non-significant differences between groups ($p > 0.05$).

3.3. Effect of Acute Acidification Stress of Yellowfin Tuna on Serum Indices

The GLU levels initially decreased, then increased, peaking at pH 6.6 (11.47 ± 0.6 mmol·L^{-1}), marking a significant difference from the control group ($p < 0.05$, Figure 3a). The LDL-C content showed an increasing trend and was highest at pH 6.6 (3.95 ± 1.04 mmol·L^{-1}, Figure 3b), and the HDL-C content was highest at pH 7.6 (15.64 ± 3.00 mmol·L^{-1}, Figure 3c) with no significant difference ($p > 0.05$) compared with the control group. TG levels were highest at pH 7.6 (0.97 ± 0.10 mmol·L^{-1}, Figure 3d) and TCH levels were highest at pH 6.6 (6.30 ± 0.50 nmol·L^{-1}, Figure 3e) with no significant difference ($p > 0.05$) compared with the control group. The GOT and GPT levels showed an increasing then decreasing pattern, peaking at pH 7.1 (GOT: 63.83 ± 2.30 μ·L^{-1}, GPT: 19.61 ± 4.69 μ·L^{-1}), with only the GOT level showing a significant difference from the control group ($p < 0.05$, Figure 3f), whereas the GPT level did not ($p > 0.05$, Figure 3g). The AKP activity displayed a tendency to increase, reaching the maximum at pH 6.6 (1.37 ± 0.08 Kim's unit·mL^{-1}), significantly differing from the control group ($p < 0.05$, Figure 3h).

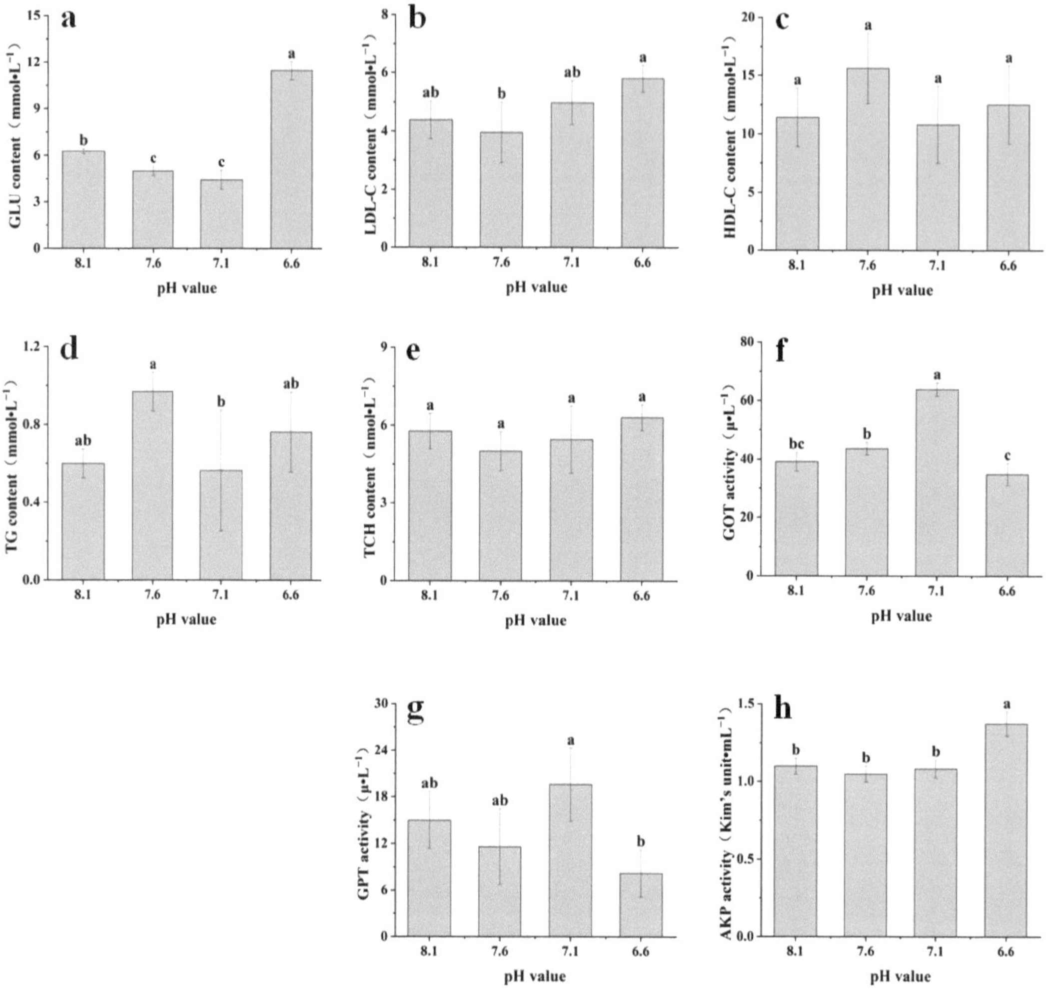

Figure 3. Effect of acidification stress on serum indices of juvenile yellowfin tuna. The bar graph represents the mean ± SD of the measurements taken at 48 h. (**a**) Glucose (GLU) content (ANOVA, GLU: $F = 148.637$, $p < 0.05$), (**b**) low-density lipoprotein cholesterol (LDL-C) content (ANOVA, LDH-C: $F = 3.385$, $p > 0.05$), (**c**) high-density lipoprotein cholesterol (HDL-C) content (ANOVA, HDH-C: $F = 1.148$, $p > 0.05$), (**d**) triglycerides (TGs) content (ANOVA, TG: $F = 2.688$, $p > 0.05$), (**e**) total cholesterol (TCH) content (ANOVA, TCH: $F = 1.209$, $p > 0.05$), (**f**) glutamic oxaloacetic transaminase (GOT) activity (ANOVA, GOT: $F = 57.086$, $p < 0.05$), (**g**) glutamine pyruvate transaminase (GPT) activity (ANOVA, GPT: $F = 4.197$, $p < 0.05$), and (**h**) alkaline phosphatase (AKP) activity (ANOVA, AKP: $F = 18.039$, $p < 0.05$). Different letters on the columns indicate significant differences between groups ($p < 0.05$), and the same letters indicate non-significant differences between groups ($p > 0.05$).

3.4. Effect of Acute Acidification Stress of Yellowfin Tuna on Liver Histology

In the control group and the pH 7.6 treatment group, hepatocytes displayed a homogeneous cytoplasm, distinct cell boundaries, and a compact arrangement, with normal morphology. The distribution of sinusoidal spaces between hepatocytes was normal (Figure 4a,b). In contrast, hepatocytes in the pH 7.1 treatment group were loosely arranged (Figure 4c). pH 6.6-treated hepatocytes had blurred edges, reduced and displaced nuclei, enlarged sinusoidal spaces, and irregular rounded vacuoles, indicating the presence of fat vacuoles (Figure 4d).

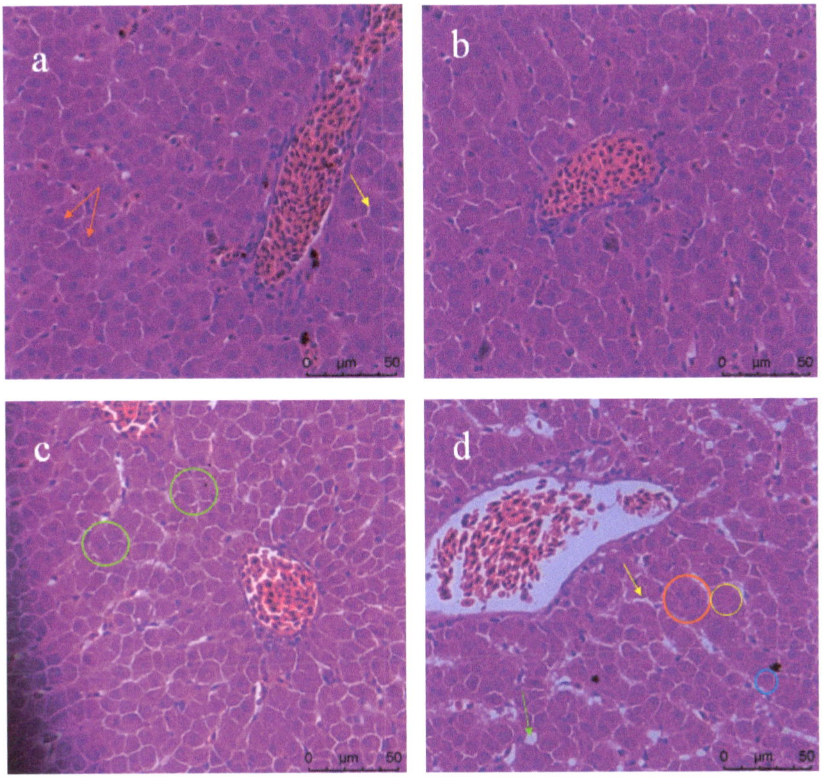

Figure 4. Effect of acute acidification stress of yellowfin tuna on liver histology (400×). (**a**) The pH 8.1 treatment group (control), (**b**) the pH 7.6 treatment group, (**c**) the pH 7.1 treatment group, and (**d**) the pH 6.6 treatment group. Red arrows indicate hepatocyte nuclei; red circles indicate blurred hepatocyte structures; yellow arrows indicate sinusoidal gaps; yellow circles indicate loss of nuclei; green arrows indicate rounded vacuoles; green circles indicate abnormal cellular morphology; and blue circles indicate nuclear excursions.

4. Discussion

4.1. Effects of Acute Acidification Stress on the Antioxidant Defense System of Yellowfin Tuna

When organisms encounter external stimuli, they often produce a substantial amount of reactive oxygen species (ROS), leading to various oxidative damages like lipid peroxidation and enzyme deactivation. This effect, known as oxidative stress, occurs if these ROS are not efficiently neutralized [34]. Antioxidant enzymes, such as glutathione reductase (GR) and total antioxidant capacity (T-AOC), play a pivotal role in combating oxidative damage by neutralizing excess ROS and superoxide anion radicals (O_2^-) [34]. In our research, we observed a reduction in GR activity across all experimental groups compared to controls, aligning with findings in Antarctic fishes (*Notothenia rossii* and *Notothenia coriiceps*) exposed to sewage [35]. The decline in GR activity in Antarctic fish was not explicitly explained due to the unpredictable nature of antioxidant enzyme responses to complex contaminant mixtures. Nevertheless, our study utilized natural seawater purified through sand filtration, suggesting the reduced GR activity might stem from the antioxidant system's failure to efficiently clear accumulated ROS, leading to the impairment or breakdown of crucial antioxidant enzymes. T-AOC represents a comprehensive measure of an organism's overall antioxidant defense, covering both enzymatic and non-enzymatic systems [36]. Studies have shown that fingerling black sea bream (*Acanthopagrus schlegelii*) exhibit an increase in serum T-AOC levels under ammonia stress [37]. Similarly, our study detected elevated T-AOC levels in the liver, indicating that organisms bolster their antioxidant defenses in response to stressors like oxidative damage. Our findings elucidate the intricate relationship between oxidative stress and antioxidant defense mechanisms in organisms, emphasizing the critical role of a balanced antioxidant system in minimizing oxidative damage and maintaining health under stressful conditions.

The fish oxidative stress system is not sufficient to scavenge all the oxygen radicals, and the excess free radicals attack the polyunsaturated fatty acids in the biofilm, causing lipid peroxidation and the production of lipid peroxidation [38]. Our study observed an uptick in LPO activity across all experimental groups in comparison to the control group, mirroring findings of rainbow trout [39] (*Oncorhynchus mykiss*) after heat stress in a similar situation. This increase might be due to the accelerated respiratory rates in fish under acidic stress, leading to heightened oxygen consumption and, consequently, an increase in oxidative stress. This results in elevated levels of reactive oxygen radicals and, thereby, an increase in LPO activity. The observed shifts in the rise in lipid peroxidation activity in our study reflect the stress response and oxidative damage in fish subjected to acidic conditions. Further research is essential to shed light on the mechanisms behind these responses and their potential impact on fish health and welfare.

Blood biochemical markers are indicative of an organism's metabolic status and the health of its organs and tissues under environmental stress [40]. In our investigation, we focused on serum alkaline phosphatase (AKP), acting as a non-specific phosphohydrolase pivotal for protecting against the intrusion of external pathogens [13]. AKP is intricately involved in the transfer and metabolism of phosphate groups, highlighting its critical role among enzymes that regulate metabolism [41]. We observed an increase in AKP activity at a pH of 6.6, mirroring the trend seen in Sagor catfish [14], suggesting that AKP activity may increase with a sudden drop in seawater pH. This observation underscores the potential impact of heightened seawater CO_2 concentrations on the metabolic frameworks of marine fish. Such species appear to navigate environmental shifts by modulating the activity of key metabolic enzymes within their systems, indicating a sophisticated adaptive mechanism to environmental stressors. This adjustment in enzyme activity reflects a broader biological strategy employed by marine organisms to maintain metabolic equilibrium in the face of changing environmental conditions.

GOT (glutamate oxaloacetate transaminase) and GPT (glutamate pyruvate transaminase) are critical serum enzymes for assessing animal health, primarily found in cellular mitochondria. Their activity levels serve as crucial indicators of hepatocyte integrity [42]. In our study, we monitored the serum activities of GOT and GPT, which are emblematic of liver damage and have their main residency within the mitochondria of cells. We noted a pattern where these enzyme levels initially increased before decreasing, reaching a zenith at pH 7.1. Significantly, GOT levels were observed to be notably higher than those of GPT, suggesting possible mitochondrial impairment within hepatocytes—a phenomenon mirrored in fingerling black sea bream [37], yellowfin sea bream (*Acanthopagrus latus*), and Asian sea bass (*Lates calcarifer*) [43] under ammonia nitrogen stress. This pattern suggests that heightened acidity precipitates liver damage in fish, triggering an acute release of these transaminases into the serum. Tuna, when exposed to acidic conditions, appear to mitigate this effect by downregulating transaminase synthesis, thus diminishing the levels of serum transaminases. This dynamic reflects a complex biological response to environmental stressors, underscoring the delicate balance organisms maintain to preserve physiological homeostasis under adverse conditions.

The liver, an indispensable organ for metabolic processes in fish, is charged with the crucial functions of nutrient metabolism, detoxification of substances, and the breakdown of medications and toxins. However, when there is a discord between the rate of hepatocyte synthesis and their entry into systemic circulation, it can lead to hepatocyte vacuolar degeneration [44]. This condition was notably observed in juvenile blunt snout bream [45] (*Megalobrama amblycephala*) under acute heat stress as well as in tilapia [46] (*Oreochromis niloticus*) subjected to cyfluthrin exposure, highlighting a pronounced pathological response of fish livers to acute stress from environmental factors.

In our study, the worsening of liver tissue structure with the increase in seawater pH, characterized by hepatocyte reduction, vacuolation, and nuclear displacement, underscores the escalating adverse effects of seawater acidification on juvenile yellowfin tuna. This is corroborated by alterations in enzyme activities such as GOT and GPT. We propose that the osmotic pressure imbalance in fish within a highly acidic environment (pH 7.1), coupled with the heightened demand for nutrient uptake to sustain osmotic equilibrium and an increased liver burden, contributes to these liver tissue changes. Nonetheless, vacuolar degeneration represents a reversible form of injury, prompting a need for additional research to determine the liver's capacity to recuperate post-stress alleviation. This insight into the liver's adaptive and pathological responses to environmental stresses provides a crucial understanding of fish physiology and health, offering a foundation for further investigations into recovery mechanisms and resilience strategies in marine life.

4.2. Effects of Acute Acidification Stress on Metabolic Function of Yellowfin Tuna

The observed decrease in lactate dehydrogenase (LDH) activity in our study might signal a diminished anaerobic capacity and a reduction in tissue glycolysis [47]. Notably, juvenile yellowfin tuna showed an increasing trend in LDH activity with decreasing environmental pH. This pattern is consistent with findings in Mediterranean sea bream [3] (*Sparus aurata*) and Sagor catfish [15] juveniles and large yellow croaker [48] (*Larimichthys crocea*), pointing to the possibility that the environmental pH may drop below the optimal range for juvenile tuna, leading to less movement and, consequently, lower rates of oxygen consumption. This reduction in oxygen could prompt a shift from aerobic to anaerobic respiration, as evidenced by the increased LDH activity. LDH plays a crucial role in anaerobic respiration, catalyzing the conversion of lactate to pyruvate and vice versa, depending on oxygen availability. Thus, our findings imply that changes in environmental pH may interfere with the aerobic metabolism of juvenile tuna, triggering an adaptive increase in anaerobic respiration as reflected by the LDH activity. Further investigations are essential to comprehend the full impact of these metabolic adjustments on the health and viability of juvenile tuna under evolving environmental conditions.

Hexokinase (HK) and pyruvate kinase (PK) are critical enzymes in the glycolytic pathway, essential for regulating blood glucose levels [49]. In our study, we found that HK activity tended to increase and then decrease. However, similar to Mediterranean sea bream [3], which exhibited heightened PK activity under environmental stress, our study observed comparable changes. Furthermore, in the Pacific oyster [50] (*Crassostrea gigas*), both HK and PK activities were found to increase in response to environmental stress, suggesting that variations between species could account for these differences. These findings imply that juvenile tuna may experience shifts in energy distribution patterns, necessitating additional energy to cope with environmental alterations. Yet, the specific mechanisms that regulate these changes remain to be thoroughly investigated and clarified. While HK and PK play fundamental roles in glycolysis and energy metabolism, the distinct responses seen in juvenile tuna under varying environmental conditions highlight the intricacies of metabolic regulation. Further research is crucial to decipher the mechanisms behind these adaptations and their potential effects on the health and sustainability of juvenile tuna.

$Na^{+}K^{+}$-ATP and $Ca^{2+}Mg^{2+}$-ATP play vital roles as key transmembrane proteins in osmoregulation [51]. The functioning of these proteins often serves as a gauge for assessing the health, metabolic efficiency, and osmoregulatory capabilities of the organism. Notably, the ionic activity in Atlantic salmon smolts [52] (*Salmo salar*) declines with a decrease in seawater pH. Similarly, juvenile European sea bass [53] exhibit an increase in $Na^{+}K^{+}$-ATP activity in low-salt conditions, possibly as an adaptive response to mitigate ionic losses due to reduced salinity. In our research, a reduction in the activities of $Na^{+}K^{+}$-ATP and $Ca^{2+}Mg^{2+}$-ATP was observed, potentially due to acidification stress affecting the cell membrane's permeability and, consequently, impairing osmoregulation in fish. These observations indicate that alterations in the activities of $Na^{+}K^{+}$-ATP and $Ca^{2+}Mg^{2+}$-ATP under acidic conditions could signify disruptions in osmoregulatory processes, underscoring the impact of environmental stressors on fish physiology. Further investigation is necessary to fully understand the mechanisms behind these changes and their implications for fish health and their ability to adapt to shifting aquatic environments.

Sugars, serving as the primary energy substrates, are metabolized into proteins and fats, with triglycerides (TGs) playing a pivotal role in cellular structures as the main form of energy storage [40]. Cholesterol, a precursor for steroid hormones and cellular membranes, is essential for maintaining cell membrane integrity. Its functionality hinges on the balance between low-density lipoprotein cholesterol (LDL-C) and high-density lipoprotein cholesterol (HDL-C), where HDL-C facilitates the removal of cholesterol from tissues to the liver for degradation, thus preserving cholesterol equilibrium [54]; conversely, LDL-C distributes cholesterol from the liver to the body's tissues [55]. Mediterranean gilthead sea bream exposed to temperature stress exhibited a marked increase in blood glucose (GLU) on the fifth day, with TG levels remaining unchanged from the control [56]. In our study, a trend of decreasing GLU levels was noted, with a significant rise at pH 6.6. However, total cholesterol (TCH) and HDL-C remained unaffected. The heightened acidity in seawater may trigger a stress response in fish, leading to the mobilization of hepatic glucose into GLU to fulfill increased metabolic demands and maintain fat equilibrium. The maintenance of TCH levels in tuna, despite acidification stress, could reflect a timely adaptive modulation of LDL-C and HDL-C concentrations. The changes in this column illustrate that under stressful conditions, the body prioritizes rapidly accessible energy sources over those that are energy dense, such as lipids. Initially, the body uses glucose, which mobilizes and utilizes energy more quickly, thus preserving lipid reserves in more severe stressful situations [57]. Nonetheless, significant differences in LDL-C between pH 7.6 and 6.6 and TG levels between pH 7.6 and 7.1 suggest that oxidative stress leads to liver metabolic disturbances under acidic conditions. Our observations underscore the intricate metabolic adjustments in tuna when faced with acidic environments, highlighting shifts in glucose metabolism and lipid regulation. Further studies are required to uncover the precise mechanisms behind these adaptations and their consequences for the health and environmental adaptability of fish.

5. Conclusions

In conclusion, juvenile yellowfin tuna exhibited stress responses to acute acidification, showcasing significant differences in various biomarkers. The key impacts observed were a reduction in antioxidant enzyme activity and a downturn in metabolic function. When faced with highly acidic seawater conditions (pH 6.6), the juvenile yellowfin tuna adapted by utilizing glucose to sustain their metabolism and ensure lipid stability, showing that lipids were the least affected. Moreover, liver analyses revealed hepatocytopenia and vacuolar degeneration, conditions that appeared to intensify with an increase in pH levels. These findings underscore the importance of studying the physiological responses of marine organisms to acidification, providing valuable insights into their adaptive mechanisms and highlighting the potential ecological consequences of seawater acidification on tuna populations. This study provides an experimental model to understand the effects of reduced seawater pH on antioxidant, metabolic function, and organizational structure in tuna. This is critical for the development of predictive models of marine biodiversity as changes in the aquatic environment intensify.

Author Contributions: Conceptualization, X.W. and R.Y.; methodology, X.W. and L.Z.; software, Z.F. and L.Z.; validation, Z.F.; formal analysis, X.W. and R.Y.; investigation, X.W. and R.Y.; resources, L.Z. and Z.M.; data curation, X.W.; writing—original draft preparation, X.W. and R.Y.; writing—review and editing, Z.F., L.Z., and Z.M.; visualization, X.W. and R.Y.; supervision, Z.M.; project administration, Z.M. and L.Z.; funding acquisition, Z.M. All authors have read and agreed to the published version of the manuscript.

Funding: This work was supported by the Hainan Major Science and Technology Project (grant number ZDKJ2021011); the Project of Sanya Yazhou Bay Science and Technology City (grant number SKJC-2022-PTDX-015); the Central Public-Interest Scientific Institution Basal Research Fund, CAFS (grant number 2023TD58, 2024XT04); the Central Public-Interest Scientific Institution Basal Research Fund South China Sea Fisheries Research Institute, CAFS (grant number 2021SD09); the Hainan Provincial Natural Science Foundation of China (grant number 323QN331); and the earmarked fund for Agriculture Research System in Hainan Province (grant number HNARS2022-03-Z2).

Institutional Review Board Statement: The animal study was reviewed and approved by the Animal Care and Use Committee of South China Sea Fisheries Research Institute, Chinese Academy of Fishery Sciences. The ethical code is 2020TD55, which was approved on 5 January 2020.

Informed Consent Statement: Not applicable.

Data Availability Statement: The data that support the findings of this study are available from the corresponding author upon reasonable request.

Conflicts of Interest: Author Zhao Lei was employed by the company Yazhou Bay Agriculture and Aquaculture Co., Ltd. The remaining authors declare that the research was conducted in the absence of any commercial or financial relationships that could be construed as a potential conflict of interest..

References

1. Frommel, A.Y.; Maneja, R.; Lowe, D.; Malzahn, A.M.; Geffen, A.J.; Folkvord, A.; Piatkowski, U.; Reusch, T.B.H.; Clemmesen, C. Severe tissue damage in Atlantic cod larvae under increasing ocean acidification. *Nat. Clim. Chang.* **2012**, *2*, 42–46. [CrossRef]
2. Kurihara, H.; Ishimatsu, A. Effects of high CO_2 seawater on the copepod (*Acartia tsuensis*) through all life stages and subsequent generations. *Mar. Pollut. Bull.* **2008**, *56*, 1086–1090. [CrossRef] [PubMed]
3. Michaelidis, B.; Spring, A.; Pörtner, O.H. Effects of long-term acclimation to environmental hypercapnia on extracellular acid–base status and metabolic capacity in Mediterranean fish *Sparus aurata*. *Mar. Biol.* **2007**, *150*, 1417–1429. [CrossRef]
4. Denman, K.; Christian, J.R.; Steiner, N.; Pörtner, H.-O.; Nojiri, Y. Potential impacts of future ocean acidification on marine ecosystems and fisheries: Current knowledge and recommendations for future research. *ICES J. Mar. Sci.* **2011**, *68*, 1019–1029. [CrossRef]
5. Glaspie, N.C.; Longmire, K.; Seitz, D.R. Acidification alters predator-prey interactions of blue crab *Callinectes sapidus* and soft-shell clam *Mya arenaria*. *J. Exp. Mar. Biol. Ecol.* **2017**, *489*, 58–65. [CrossRef]
6. Miller, S.; Breitburg, D.; Burrell, R.; Keppel, A. Acidification increases sensitivity to hypoxia in important forage fishes. *Mar. Ecol. Prog. Ser.* **2016**, *549*, 1–8. [CrossRef]
7. Regaudie-de-Gioux, A.; Duarte, C.M. Temperature dependence of planktonic metabolism in the ocean. *Glob. Biogeochem. Cycles* **2012**, *26*, 1015. [CrossRef]
8. Lushchak, I.V. Environmentally induced oxidative stress in aquatic animals. *Aquat. Toxicol.* **2010**, *101*, 13–30. [CrossRef] [PubMed]
9. Berenbrink, M. Evolution of vertebrate haemoglobins: Histidine side chains, specific buffer value and Bohr effect. *Respir. Physiol. Neurobiol.* **2006**, *154*, 165–184. [CrossRef]
10. Pörtner, O.H.; Knust, R. Climate Change Affects Marine Fishes Through the Oxygen Limitation of Thermal Tolerance. *Science* **2007**, *315*, 95–97. [CrossRef]
11. Yamada, N.; Suzumura, M. Effects of Seawater Acidification on Hydrolytic Enzyme Activities. *J. Oceanogr.* **2010**, *66*, 233–241. [CrossRef]
12. Ding, Z.K.; Wang, F.P.; Xu, Y.Q. Effect of Ocean Acidification on Metabolism of Marine Organisms. *Fish. Sci.* **2015**, *34*, 331–334. [CrossRef]
13. Tao, L.; Xiaobo, Y.; Xiaohui, D.; Pan, S.; Tan, B.; Zhang, S.; Suo, X.; Huang, W.; Zhou, M.; Yang, Y. Effects of choline supplementation on growth performance, liver histology, nonspecific immunity and related genes expression of hybrid grouper (♀*Epinephelus fuscoguttatus* × ♂*E. lanceolatu*) fed with high-lipid diets. *Fish Shellfish. Immunol.* **2023**, *138*, 108815.
14. Petochi, T.; Di Marco, P.; Priori, A.; Finoia, M.; Mercatali, I.; Marino, G. Coping strategy and stress response of European sea bass *Dicentrarchus labrax* to acute and chronic environmental hypercapnia under hyperoxic conditions. *Aquaculture* **2011**, *315*, 312–320. [CrossRef]
15. Noor, N.M.; De, M.; Cob, Z.C.; Das, S.K. Welfare of scaleless fish, Sagor catfish (*Hexanematichthys sagor*) juveniles under different carbon dioxide concentrations. *Aquac. Res.* **2021**, *52*, 2980–2987. [CrossRef]

16. Hoyle, S.D.; Williams, A.J.; Minte-Vera, C.V.; Maunder, M.N. Approaches for estimating natural mortality in tuna stock assessments: Application to global yellowfin tuna stocks. *Fish. Res.* **2023**, *257*, 106498. [CrossRef]
17. Harley, C.D.G.; Randall Hughes, A.; Hultgren, K.M.; Miner, B.G.; Sorte, C.J.B.; Thornber, C.S.; Rodriguez, L.F.; Tomanek, L.; Williams, S.L. The impacts of climate change in coastal marine systems. *Ecol. Lett.* **2006**, *9*, 228–241. [CrossRef]
18. Lingbin, S.; Jinpeng, R.; Mengchao, L.; Chen, M.; Dai, Z.; Zuo, Z. Combined effects of ocean acidification and crude oil pollution on tissue damage and lipid metabolism in embryo-larval development of marine medaka (*Oryzias melastigma*). *Environ. Geochem. Health* **2019**, *41*, 1847–1860.
19. Kurihara, H. Effects of CO_2-driven ocean acidification on the early developmental stages of invertebrates. *Mar. Ecol. Prog. Ser.* **2008**, *373*, 275–284. [CrossRef]
20. Caldeira, K.; Wickett, E.M. Ocean model predictions of chemistry changes from carbon dioxide emissions to the atmosphere and ocean. *J. Geophys. Res. Ocean.* **2005**, *110*, C09S04. [CrossRef]
21. Bromhead, D.; Scholey, V.; Nicol, S.; Margulies, D.; Wexler, J.; Stein, M.; Hoyle, S.; Lennert-Cody, C.; Williamson, J.; Havenhand, J.; et al. The potential impact of ocean acidification upon eggs and larvae of yellowfin tuna (*Thunnus albacares*). *Deep-Sea Res. Part II* **2015**, *113*, 268–279. [CrossRef]
22. Xu, M.; Sun, T.; Tang, X.; Lu, K.; Jiang, Y.; Cao, S.; Wang, Y. Title: CO_2 and HCl-induced seawater acidification impair the ingestion and digestion of blue mussel *Mytilus edulis*. *Chemosphere* **2020**, *240*, 124821. [CrossRef]
23. Sun, T.; Tang, X.; Zhou, B.; Wang, Y. Comparative studies on the effects of seawater acidification caused by CO_2 and HCl enrichment on physiological changes in Mytilus edulis. *Chemosphere* **2016**, *144*, 2368–2376. [CrossRef]
24. Fu, Z.; Qin, J.G.; Ma, Z.; Yu, G. Acute acidification stress weakens the head kidney immune function of juvenile Lates calcarifer. *Ecotoxicol. Environ. Saf.* **2021**, *225*, 112712. [CrossRef] [PubMed]
25. Navarro, J.M.; Torres, R.; Acuña, K.; Duarte, C.; Manriquez, P.H.; Lardies, M.; Lagos, N.A.; Vargas, C.; Aguilera, V. Impact of medium-term exposure to elevated p CO_2 levels on the physiological energetics of the mussel *Mytilus chilensis*. *Chemosphere* **2013**, *90*, 1242–1248. [CrossRef] [PubMed]
26. Jonathan, M. IPCC Special Report on the Ocean and Cryosphere in a Changing Climate. *New York Rev. Books* **2020**, *67*, 49–51.
27. Arthur, J.B.; Bjerkeng, B.; Pettersen, O.; Schaanning, M.T.; Øxnevad, S. Effects of increased sea water concentrations of CO_2 on growth of the bivalve *Mytilus edulis* L. *Chemosphere* **2006**, *62*, 681–687.
28. Orr, J.C.; Fabry, V.J.; Aumont, O.; Bopp, L.; Doney, S.C.; Feely, R.A.; Gnanadesikan, A.; Gruber, N.; Ishida, A.; Joos, F.; et al. Anthropogenic ocean acidification over the twenty-first century and its impact on calcifying organisms. *Nature* **2005**, *437*, 681–686. [CrossRef] [PubMed]
29. Zhang, N.; Yang, R.; Fu, Z.; Yu, G.; Ma, Z. Mechanisms of Digestive Enzyme Response to Acute Salinity Stress in Juvenile Yellowfin Tuna (*Thunnus albacares*). *Animals* **2023**, *13*, 3454. [CrossRef]
30. Liu, H.; Fu, Z.; Yu, G.; Ma, Z.; Zong, H. Effects of Acute High-Temperature Stress on Physical Responses of Yellowfin Tuna (*Thunnus albacares*). *J. Mar. Sci. Eng.* **2022**, *10*, 1857. [CrossRef]
31. Rosseland, B.O.; Massabuau, J.C.; Grimalt, J.; Rosseland, B.O.; Massabuau, J.C.; Grimalt, J.; Hofer, R.; Lackner, R.; Raddum, G.; Rognerud, S.; et al. *Fish Ecotoxicology: European Mountain Lake Ecosystems Regionalisation, Diagnostic and Socio-Economic Evaluation (EMERGE)*; Norwegian Institute for Water Research (NIVA): Oslo, Norway, 2003; p. 23.
32. Yancheva, V.; Georgieva, E.; Velcheva, I.; Iliev, I.; Stoyanova, S.; Vasileva, T.; Bivolarski, V.; Todorova-Bambaldokova, D.; Zulkipli, N.; Antal, L.; et al. Assessment of the exposure of two pesticides on common carp (Cyprinus carpio Linnaeus, 1758): Are the prolonged biomarker responses adaptive or destructive? *Comp. Biochem. Physiol. Part C: Toxicol. Pharmacol.* **2022**, *261*, 109446.
33. Georgieva, E.; Yancheva, V.; Stoyanova, S.; Velcheva, I.; Iliev, I.; Vasileva, T.; Bivolarski, V.; Petkova, E.; László, B.; Nyeste, K.; et al. Which Is More Toxic? Evaluation of the Short-Term Toxic Effects of Chlorpyrifos and Cypermethrin on Selected Biomarkers in Common Carp (*Cyprinus carpio*, Linnaeus 1758). *Toxics* **2021**, *9*, 125. [CrossRef] [PubMed]
34. Wang, Y.; Yang, R.; Fu, Z.; Ma, Z.; Bai, Z. The Photoperiod Significantly Influences the Growth Rate, Digestive Efficiency, Immune Response, and Antioxidant Activities in the Juvenile Scalloped Spiny Lobster (*Panulirus homarus*). *J. Mar. Sci. Eng.* **2024**, *12*, 389. [CrossRef]
35. Edson, R.; Mariana, F.; Suda, C.N.K.; Vani, G.S.; Donatti, L.; Rodrigues, E.; Lavrado, H.P. Metabolic responses of the Antarctic fishes *Notothenia rossii* and *Notothenia coriiceps* to sewage pollution. *Fish Physiol. Biochem.* **2015**, *41*, 1205–1220.
36. Chen, S.; Guo, Y.; Espe, M.; Yang, F.; Fang, W.P.; Wan, M.G.; Niu, J.; Liu, Y.J.; Tian, L.X. Growth performance, haematological parameters, antioxidant status and salinity stress tolerance of juvenile Pacific white shrimp (*Litopenaeus vannamei*) fed different levels of dietary myo-inositol. *Aquac. Nutr.* **2018**, *24*, 1527–1539. [CrossRef]
37. Wang, L.; Sun, Y.; Xu, B.; Sagada, G.; Chen, K.; Xiao, J.; Zhang, J.; Shao, Q. Effects of berberine supplementation in high starch diet on growth performance, antioxidative status, immune parameters and ammonia stress response of fingerling black sea bream (*Acanthopagrus schlegelii*). *Aquaculture* **2020**, *527*, 735473. [CrossRef]
38. Radi, R.; Beckman, J.S.; Bush, K.M.; Freeman, B.A. Peroxynitrite-induced membrane lipid peroxidation: The cytotoxic potential of superoxide and nitric oxide. *Arch. Biochem. Biophys.* **1991**, *288*, 481–487. [CrossRef] [PubMed]
39. Shi, K.P.; Dong, S.L.; Zhou, Y.G.; Gao, Q.F.; Sun, D.J. Antioxidant responses of rainbow trout with different ploidies to acute temperature stress. *J. Appl. Ecol.* **2018**, *29*, 3102–3110.

40. Zhenkun, X.; Hongzhi, Z.; Meijie, G.; Fang, D.; Mei, J.; Xie, J. Analysis of Acute Nitrite Exposure on Physiological Stress Response, Oxidative Stress, Gill Tissue Morphology and Immune Response of Large Yellow Croaker (*Larimichthys crocea*). *Animals* **2022**, *12*, 1791.
41. Wu, Y.C.; Li, R.M.; Shen, G.R.; .Shen, G.R.; Huang, F.; Yang, Q.H.; Tan, B.P.; Chi, S.Y. Effects of dietary small peptides on growth, antioxidant capacity, nonspecific immunity and ingut microflora structure of *Litopenaeus vannamei*. *Guangdong Ocean. Univ.* **2021**, *41*, 1–9.
42. Wang, X.; Li, Y.; Hou, C.; Gao, Y.; Wang, Y. Physiological and molecular changes in large yellow croaker (*P seudosciaena crocea R)* with high-fat diet-induced fatty liver disease. *Aquac. Res.* **2015**, *46*, 272–282. [CrossRef]
43. Torfi, M.M.; Omid, S.; Rahim, O.; Mehrjooyan, S.; Najafabadi, M.Z.; Hoseini, S.J.; Saghavi, H.; Monem, J. The effect of salinity on growth performance, digestive and antioxidant enzymes, humoral immunity and stress indices in two euryhaline fish species: Yellowfin seabream (*Acanthopagrus latus*) and Asian seabass (*Lates calcarifer*). *Aquaculture* **2021**, *534*, 736329.
44. Elsayed, Y.; Abdel-Wahab, A.; Nasser, A.; Ebaid, H. Histopathological alterations in the liver and intestine of Nile tilapia *Oreochromis niloticus* exposed to long-term sublethal concentrations of cadmium chloride. *Chin. J. Oceanol. Limnol.* **2015**, *33*, 846–852.
45. Li, B.; Sun, S.; Zhu, J.; Yanli, S.; Wuxiao, Z.; Ge, X. Transcriptome profiling and histology changes in juvenile blunt snout bream (*Megalobrama amblycephala)* liver tissue in response to acute thermal stress. *Genomics* **2019**, *111*, 242–250. [CrossRef]
46. Arulraj, J.S.; Pandurengan, P.; Arasan, S.; Gopalrajan, S.; Paulraj, J. Acute Toxicity of Lamda-Cyhalothrin and the Histopathological Changes of Gill and Liver Tissues of Tilapia (*Oreochromis niloticus*). *J. Coast. Res.* **2019**, *86* (Suppl. 1), 235–238. [CrossRef]
47. Rao, V.J. Sublethal effects of an organophosphorus insecticide (RPR-II) on biochemical parameters of tilapia, *Oreochromis mossambicus*. *Comp. Biochem. Physiol. Part C* **2006**, *143*, 492–498.
48. Meijie, G.; Zhenkun, X.; Hongzhi, Z.; Mei, J.; Xie, J. The Effects of Acute Exposure to Ammonia on Oxidative Stress, Hematological Parameters, Flesh Quality, and Gill Morphological Changes of the Large Yellow Croaker (*Larimichthys crocea*). *Animals* **2023**, *13*, 2534.
49. Weiliang, G.; Wenqian, N.; Xiaobo, W.; Chen, R.; Huang, Z.; Ding, Y.; Qin, X.; Cai, L.; Mao, L. Influences of two transport strategies on AMPK mediated metabolism and flesh quality of shrimp (*Litopenaeus vannamei*). *J. Sci. Food Agric.* **2023**, *104*, 727–736.
50. Yanouk, E.; Claudie, Q.; Fabrice, P.; Pichereau, V.; Corporeau, C. Energy and antioxidant responses of pacific oyster exposed to trace levels of pesticides. *Chem. Res. Toxicol.* **2015**, *28*, 1831–1841.
51. Thomas OR, B.; Swearer, S.E. Otolith Biochemistry—A Review. *Rev. Fish. Sci. Aquac.* **2019**, *27*, 458–489. [CrossRef]
52. Fivelstad, S.; Olsen, B.A.; Åsgård, T.; Baeverfjord, G.; Rasmussen, T.; Vindheim, T.; Stefansson, S. Long-term sublethal effects of carbon dioxide on Atlantic salmon smolts (*Salmo salar L.)*: Ion regulation, haematology, element composition, nephrocalcinosis and growth parameters. *Aquaculture* **2003**, *215*, 301–319. [CrossRef]
53. Shrivastava, J.; Ndugwa, M.; Caneos, W.; De Boeck, G. Physiological trade-offs, acid-base balance and ion-osmoregulatory plasticity in European sea bass (*Dicentrarchus labrax*) juveniles under complex scenarios of salinity variation, ocean acidification and high ammonia challenge. *Aquat. Toxicol.* **2019**, *212*, 54–69. [CrossRef] [PubMed]
54. Meng, X.X.; Wei, Y.L.; Liang, M.Q.; Xu, H.G. Progress in cholesterol nutritional requirements of fish. *Anim. Nutr.* **2021**, *33*, 719–728.
55. Wang, X.; Ma, J.; Wang, J.; Li, B.; Huang, B.; Hao, T. Studies on supplementation of stickwater meal to high plant protein diets of juvenile *Epinephelus fuscoguttatus* ♀× *E. lanceolatus* ♂pearl gentian. *Period. Ocean. Univ. China* **2021**, *51*, 31–43.
56. Konstantinos, F.; Pörtner, H.-O.; Antonopoulou, E.; Michaelidis, B. Synergistic effects of acute warming and low pH on cellular stress responses of the gilthead seabream *Sparus aurata*. *J. Comp. Physiol. B* **2015**, *185*, 185–205.
57. Velázquez, S.J.; Herrejón, P.A.G.; Becerra, A.H. Fish Responses to Alternative Feeding Ingredients under Abiotic Chronic Stress. *Animals* **2024**, *14*, 765. [CrossRef]

Journal of
Marine Science and Engineering

MDPI

Article

The Effects of Acute Ammonia Nitrogen Stress on Antioxidant Ability, Phosphatases, and Related Gene Expression in the Kidney of Juvenile Yellowfin Tuna (*Thunnus albacares*)

Yongyue Sun [1,2,3,4,5], Zhengyi Fu [2,3,4,5,6] and Zhenhua Ma [2,3,4,5,6,*]

[1] College of Fisheries and Life Science, Shanghai Ocean University, Shanghai 201306, China
[2] Key Laboratory of Efficient Utilization and Processing of Marine Fishery Resources of Hainan Province, Sanya Tropical Fisheries Research Institute, Sanya 572018, China
[3] South China Sea Fisheries Research Institute, Chinese Academy of Fishery Sciences, Guangzhou 510300, China
[4] Hainan Engineering Research Center for Deep-Sea Aquaculture and Processing, Sanya 572018, China
[5] International Joint Research Center for Conservation and Application of Fishery Resources in the South China Sea, Sanya 572018, China
[6] College of Science and Engineering, Flinders University, Adelaide 5001, Australia
[*] Correspondence: zhenhua.ma@scsfri.ac.cn

Abstract: This study investigated the effects of acute ammonia nitrogen (NH_3-N) exposure on kidney antioxidant ability and phosphatases and related gene expression in juvenile yellowfin tuna (*Thunnus albacares*). The 180 juvenile yellowfin tuna (260.39 ± 55.99 g, 22.33 ± 2.28 cm) were exposed to ammonia for 6, 24, and 36 h using natural seawater (0 mg/L) as a control and NH_3-N at 5 and 10 mg/L. The lipid peroxidation byproduct malondialdehyde (MDA) and the levels of antioxidant enzymes, such as superoxide dismutase (SOD), catalase (CAT), glutathione peroxidase (GSH-PX), alkaline phosphatase (AKP), and acid phosphatase (ACP), were measured using the colorimetric method in the trunk kidney to determine changes in antioxidant ability and phosphatase activity of juvenile yellowfin tuna exposed to NH_3-N. Results indicated that, at 36 h, MDA, SOD, CAT, and GSH-PX levels rose in the 5 mg/L group versus the control. In the 10 mg/L group, MDA and SOD, CAT, and GSH-PX activities significantly increased after 24 and 36 h exposure compared to the control. Phosphatases play a pivotal role in the immune system. AKP activity significantly increased at 6 h, and ACP activity markedly rose at 36 h in the 5 mg/L group versus the control. Real-time fluorescence quantitative PCR was applied to detect alterations in the antioxidant genes SOD2, CAT, and glutathione peroxidase 1b (GPX1b) and immune cytokines-related genes Interleukin 10 (IL-10) and Interleukin 6 receptor (IL-6r) expression in the head kidney in juvenile tuna. Relative to the control, antioxidant gene expression in the 5 mg/L group significantly rose at 6 and 36 h, and in the 10 mg/L group, SOD2 and GPX1b were significantly elevated at 36 h. Compared to the control group, IL-10 expression in the 5 mg/L group significantly increased at 6 h, whereas IL-6r expression decreased. In the 10 mg/L group, both IL-10 and IL-6r levels were observed to be lower. Low ammonia nitrogen concentrations boost antioxidant defenses, phosphatase activities, and gene expression levels, whereas higher levels may induce suppressive effects. In yellowfin tuna juvenile farming, NH_3-N concentration significantly affects the health of the juveniles. When the NH_3-N concentration is between 5–10 mg/L, the stress duration should be limited to 24 h; if the concentration is below 5 mg/L, the stress duration can be extended to 36 h.

Keywords: NH_3-N; head kidney; trunk kidney; antioxidant enzymes; immunological enzyme; antioxidant genes; immune genes

Citation: Sun, Y.; Fu, Z.; Ma, Z. The Effects of Acute Ammonia Nitrogen Stress on Antioxidant Ability, Phosphatases, and Related Gene Expression in the Kidney of Juvenile Yellowfin Tuna (*Thunnus albacares*). *J. Mar. Sci. Eng.* **2024**, *12*, 1009. https://doi.org/10.3390/jmse12061009

Academic Editor: Nguyen Hong Nguyen

Received: 13 April 2024
Revised: 8 June 2024
Accepted: 17 June 2024
Published: 18 June 2024

1. Introduction

Ammonia nitrogen (NH_3-N) represents a prevalent stressor within the aquaculture environment, emerging as the final byproduct in the decomposition of nitrogenous wastes from a variety of sources, such as the excrement of aquatic creatures, the breakdown of leftover feed, and the discharge from industrial and household wastewater processes [1]. In fact, there are two forms of ammonia nitrogen: ionizated (NH_4^+) and non-ionizated (NH_3), with the latter being much more toxic and more likely to enter aquatic animals, causing damage and even death to the organisms [2,3]. Elevated levels of ammonia nitrogen pose a direct threat to the liver and kidney tissues of fish, leading to conditions such as congestion, edema, hepatic coma, and, in severe cases, death [4]. Additionally, ammonia nitrogen exposure can provoke neurological responses in fish, manifesting in abnormal behaviors including reduced appetite, stunted growth, and diminished movement [5–7]. Therefore, mitigating ammonia nitrogen pollution and optimizing aquaculture conditions are crucial steps toward enhancing the efficiency of aquaculture practices.

When subjected to environmental stressors, fish respond by generating an excess of reactive oxygen species (ROS), which leads to heightened levels of malondialdehyde (MDA), a marker of oxidative stress. This escalation compromises their immune defenses and can potentially lead to mortality. In response to such oxidative threats, fish mobilize their antioxidant and immune systems for protection. The core components of the antioxidant defense mechanism include enzymes such as superoxide dismutase (SOD), catalase (CAT), and glutathione peroxidase (GSH-PX), which effectively neutralize ROS, safeguarding the fish against oxidative harm [8]. Additionally, enzymes such as acid phosphatase (ACP) and alkaline phosphatase (AKP), vital elements of the lysosomal system, play crucial roles in bolstering the fish's immune response [9]. Research findings reveal that the cuttlefish (*Sepia pharaonis*) displays a decrease in the activities of SOD and CAT, coupled with an elevation in MDA levels, when subjected to ammonia nitrogen stress [10]. Similarly, Pacific white shrimp (*Litopenaeus vannamei*) exposed to ammonia stress show elevated activities of AKP and ACP, highlighting the impact of environmental stress on marine organisms' biochemical pathways [11].

In response to the stress induced by ammonia toxicity, fish can activate their antioxidant defense mechanisms through the expression of specific antioxidant genes [12]. This phenomenon has been observed in hybrid groupers (*Epinephelus fuscoguttatus*♀ × *E. lanceolatus*♂), where genes responsible for SOD, CAT, and GSH-PX are significantly upregulated under ammonia-nitrogen stress [13]. Ammonia stress not only influences inflammatory cytokines in fish but also triggers inflammatory responses and tissue damage [14]. The increase in pro-inflammatory cytokines, notably Interleukin 6 (IL-6) [15], triggers such inflammatory responses. Conversely, a rise in the anti-inflammatory molecule, Interleukin 10 (IL-10), serves to mitigate these inflammatory processes [16]. Research into these cytokines across different fish species, including Atlantic halibut (*Hippoglossus hippoglossus* L.), orange-spotted grouper (*Epinephelus coioides*), Senegalese sole (*Solea senegalensis*), golden pomfret (*Trachinotus ovatus*), Atlantic cod (*Gadus morhua* L.), and large-bellied seahorse (*Hippocampus abdominalis*) [17–22], has provided insight into the nuanced interplay of enzyme activity, gene expression, and the resulting antioxidant and immune responses. This comprehensive understanding aids in deciphering the complex biological responses of fish to environmental stressors, particularly those related to ammonia exposure.

Yellowfin tuna (*Thunnus albacares*), a member of the Scombridae family within the Scombroid suborder of the order Perciformes, thrives in warm tropical and subtropical waters worldwide. Celebrated for its high protein and low fat content, yellowfin tuna is a favored selection among seafood lovers [23]. The species is abundant in the South China Sea, and advancements in aquaculture techniques, such as indoor recirculating water systems and offshore deep-water net-pan farming, have been notably successful at the Tropical Aquatic Research and Development Centre in Xincun Town, Lingshui County, Hainan Province [24]. While yellowfin tuna aquaculture is also undertaken in regions such as Mexico, Panama, and Indonesia, the industry faces challenges related to the insufficient supply of wild fry [25]. Research on yellowfin tuna has covered various aspects, including stock assessment, fisheries, reproductive biology, and genomics [26–30]. Despite these extensive studies, the physiological response of yellowfin tuna to ammonia and nitrogen levels in artificial aquaculture environments remains largely unexplored.

The influence of ammonia nitrogen stress on the kidney health of marine fish, with a particular focus on juvenile yellowfin tuna, remains an underexplored area of research. This study seeks to examine the impact of short-term ammonia nitrogen exposure on juvenile yellowfin tuna, specifically looking at the activity of antioxidant enzymes and phosphatase, as well as antioxidant and immune-related gene expression. Our research aims to shed light on how yellowfin tuna respond to changes in ammonia nitrogen levels in their environment. The findings from this study are expected to inform improvements in aquaculture methods for yellowfin tuna, whether in recirculating water systems or cage cultures, by focusing on short-term adaptations. Through the development of well-founded and efficient management approaches, we intend to counteract the negative effects of ammonia nitrogen, thus promoting a more efficient and sustainable approach to the aquaculture of yellowfin tuna.

2. Materials and Methods

2.1. Experimental Fish and Design

Juvenile yellowfin tuna were supplied by the Sanya Tropical Fisheries Research Institute located in Lingshui County, Hainan Province, China. These juveniles had an average weight of 260.39 ± 55.99 g and measured an average length of 22.33 ± 2.28 cm. At the experiment's outset, 180 juvenile yellowfin tuna were transferred to an indoor recirculating tank and evenly distributed across nine 5000-L recirculating water tanks, ensuring 20 fish per tank for a week-long acclimation period. During this time, their diet consisted of chilled mixed fish species, and the water quality was meticulously maintained at optimal conditions for their growth: a temperature of 29 ± 1°C, ammonia nitrogen levels under 0.01 mg/L, dissolved oxygen at 7.00 ± 0.50 mg/L, a pH of 8.10, salinity at 32‰, and nitrite levels below 0.01 mg/L.

Based on prior studies [31–33], 0 mg/L (natural seawater) was used as the control group, the ammonia nitrogen concentration was set at 5 and 10 mg/L, respectively, and the experiment was set up with three replicates, where analytically pure ammonium chloride (NH_4Cl) (99.5% purity, Xilong Chemical Co., Ltd., Foshan, China) was added to seawater, dissolved, and configured into a masterbatch of 10 g/L as the ammonia nitrogen source. The relevant calculation formulas are as follows:

Supplementing ammonia nitrogen in terms of mass (mg) = The volume of water (L) × The difference in ammonia nitrogen concentration (mg/L).
The molar quantity of ammonium chloride (mol) = Supplementing ammonia nitrogen in terms of mass (mg) / The molar mass of ammonium chloride (g/mol). (The molar mass of ammonium chloride is approximately 53.49 g/mol).
The mass of ammonium chloride (g) = The molar quantity of ammonium chloride (mol) × The molar mass of ammonium chloride (g/mol).
The actual mass of ammonium chloride (g) = The mass of ammonium chloride (g)/0.995.

The fish were not fed for 24 h before and throughout the duration of the experiment. Utilizing a portable water quality detector (Wuxi Oktan Biotechnology Co. Ltd., Wuxi, China), the concentration of ammonia nitrogen is measured every two hours. Based on the results, ammonia nitrogen is calculated and replenished in a timely manner to maintain the required concentration according to the following formula. Three fish were randomly selected from each tank at 6, 24, and 36 h into the experiment for further analysis.

2.2. Sample Collection

Following the administration of eugenol anesthesia to ensure humane handling, the juvenile yellowfin tuna underwent measurement for body mass and length. Subsequently, a precise dissection was performed to extract the head and trunk kidneys, which were then meticulously placed into clearly labeled cryopreservation tubes. These samples were immediately flash-frozen in liquid nitrogen to preserve their integrity. The following day, the samples were transferred to a −80 °C freezer, setting the stage for further detailed analyses and determinations.

2.3. Measurement of Physiological Indicators

First, 0.1 to 0.2 g of trunk kidney tissue was weighed into a 2-milliliter centrifuge tube, adhering to the reagent instructions. Then, 0.9% saline was added in a ratio of 1:9 (weight in grams to volume in milliliters). Then, 2–3 steel balls were added. The homogenate was then processed using a tissue homogenizer (Hangzhou Aosheng Instrument Co., Ltd., Hangzhou, China). Following homogenization, the mixture was centrifuged at 3500 rpm for 10 min at 4 °C using a centrifuge (Heraeus Corporation, Hanau, Germany). The supernatant was then separated from the tissue pellet so it could be aspirated and dispensed as needed. This supernatant was promptly stored at −80 °C in preparation for the measurement of antioxidant enzyme activity. The evaluation of enzyme activity was conducted utilizing the optical densities obtained with a hybrid microplate reader (BioTek Instruments, Inc., Winooski, VT, USA) and a spectrophotometer (Mepore Instruments, Shanghai, China), ensuring accurate quantification of the enzymes.

The kit for the kidney test was purchased from the Nanjing Jiancheng Bioengineering Institute, Nanjing, China. The MDA assay kit (catalog number A003-1) was employed to quantify MDA, a byproduct of lipid peroxidation that reacts with thiobarbituric acid (TBA) to produce a red compound with a peak absorbance at 532 nm, which was then recorded. SOD activity was evaluated using a water-soluble tetrazolium salt (WST-1)-based method (Catalogue No. A001-3). CAT activity was assessed through the ammonium molybdate method, wherein the reaction between CAT and hydrogen peroxide (H_2O_2) is halted by ammonium molybdate. The remaining H_2O_2 forms a yellow complex with ammonium molybdate, the concentration of which is measured at 405 nm to determine CAT activity (Catalogue No. A007-1). GSH-PX levels were determined by measuring the consumption of reduced glutathione in the enzymatic reaction (Catalogue No. A005-1).

ACP (Catalogue No. A060-2) and AKP (Catalogue No. A059-2) activities were measured by their ability to hydrolyze disodium benzene phosphate, releasing free phenol. This phenol reacts with potassium ferricyanide in an alkaline medium and 4-aminoantipyrine to form a red quinone derivative. The intensity of the red color, indicative of enzyme activity, was then quantitatively assessed, allowing for the precise determination of ACP and AKP levels.

2.4. Expression of Antioxidant-Related and Immune-Related Genes

The head kidney was selected for the assessment of RNA gene relative expression. RNA was extracted from head kidney samples using a Trizol reagent (Lanjieke Technology Co., Ltd., Hefei, China) to isolate the total RNA. The integrity of the RNA was verified through 1% agarose gel electrophoresis (Lanjieke Technology Co., Ltd.), ensuring the RNA was not degraded. The concentration of the isolated RNA was quantified using a Nano-300 spectrophotometer (Ausheng Instrument Co., Ltd., Hangzhou, China), with the extraction deemed successful if the optical density (OD) ratios at 260 nm to 280 nm fell within the range of 1.8 to 2.0. Following RNA extraction, cDNA synthesis was carried out, as per the protocol provided by the One-Step gDNA Removal and cDNA Synthesis SuperMix kit (Quan's Gold Biotechnology Limited, Beijing, China) [34], laying the groundwork for subsequent gene expression analysis.

The gene sequences analyzed in this study of yellowfin tuna were obtained from the unpublished transcriptome data collected by our research team. Primers for the study were meticulously designed with the aid of Primer Premier 5 software (Premier Inc., Charlotte, NC, Canada), and their sequences are detailed in Table 1. These primers were then synthesized by Sangyo Bioengineering Co., Ltd. (Shanghai, China). β-actin [35] was selected as the reference gene to ensure accurate gene expression analysis.

Table 1. RT-PCR primer sequence.

Gene	Acronym		Primer sequences	Mollification Size
Superoxide dismutase 2	SOD2 SOD2	F R	CGGGACTTTGGTTCCTTCCA GCACAAGCAGCGATACGAAG	128
Catalase	CAT CAT	F R	CAGGCAACAACACCCCCA CCAGAAGTCCCACACCAT	122
Glutathione peroxidase 1b	GPX1b GPX1b	F R	GACCACCAGGGATTACAC GGACGGACATACTTCAGA	150
Interleukin 6 receptor	IL-6r IL-6r	F R	TTGTCAGTCATTTTGGCT CTCTGGAGATGTTGGGGT	132
Interleukin 10	IL-10 IL-10	F R	CAGCAAGATACCAACAAG CGACAAGAGAACCAGGAC	190
β-actin	β-actin β-actin	F R	CGCCCTCGTTGTTGAC CCCTTTTGCTCTGTGCC	170

The SYBR Green RealUniversal Colour Fluorescence Quantification Kit (Beijing Tiangen Biochemical Technology Co., Ltd., Beijing, China) was employed for cDNA quantification. The reaction mixture totaled 20 μL, comprising 10 μL of 2 × RealUniversal PreMix, 0.6 μL forward primer (10 μM concentration), 0.6ul reverse primer (10um concentration), 6.8 μL of RNase-free ddH_2O, and 2 μL of the cDNA template (the cDNA contains 1000 nanograms.) [31,32]. Real-time PCR analyses were conducted using a fluorescence quantitative PCR instrument (Langji Scientific Instruments Co., Ltd., Hangzhou, China), with the specific parameters of the reaction outlined in Table 2. Relative expression levels of the target genes were calculated employing the $2^{-\Delta\Delta Ct}$ [36] method, allowing for the precise quantification of gene expression changes.

Table 2. Fluorescence quantitative reaction procedure.

Stage	Cycle	Temperature	Times	Elements
permutability	1×	95 °C	15 min	permutability
PCR reaction	40×	95 °C	10 s	denaturation
		50–60 °C	20 s	annealing (metallurgy)
		72 °C	20–32 s	extend

2.5. Statistical Analysis

The results of the treatment and control groups are presented as mean ± standard deviation (Mean ± SD), with the difference value between the treatment and control groups presented as the treatment-control mean. Subsequently, a one-way analysis of variance (ANOVA) was conducted to evaluate the results of the treatment and control groups. Prior to conducting the ANOVA, we performed a preliminary Chi-square test to check for normal distribution and homogeneity of variances. In instances where significant differences were identified ($p < 0.05$), Duncan's multiple comparison test was employed to discern the specific differences between groups. For visual representation, the experimental data were graphically plotted using Origin 2022 software (2022 edition).

3. Results

3.1. Effect of Acute Ammonia Nitrogen Stress on the Antioxidant Ability of the Trunk Kidney of Juvenile Yellowfin Tuna

Over time, both the 5 mg/L and 10 mg/L treatment groups exhibited an upward trend in MDA concentration, with the most pronounced changes observed at the 36-h mark. Specifically, the 5 mg/L treatment group experienced an increase of 609.38 ± 70.18 nmol/mgprot, while the 10 mg/L treatment group saw an increase of 1003.97 ± 6.45 nmol/mgprot

(Figure 1a). In the 5 mg/L ammonia nitrogen treatment group, the MDA concentration in the kidney of juvenile yellowfin tuna was significantly lower than that of the control group at both the 6-h and 24-h time points ($p = 0.001$), yet it notably surpassed the control group's level at the 36-h mark ($p = 0.001$, Figure 1b). Within the 10 mg/L treatment group, the MDA concentration was significantly reduced compared to the control group at the 6-h mark ($p = 0.001$), but it was significantly elevated at both the 24-h and 36-h time points ($p = 0.001$).

The SOD activity of both the 5 mg/L and 10 mg/L treatment groups exhibited an upward trend over time, with the most significant increase observed at 36 h. Specifically, the 5 mg/L treatment group saw an increase of 13.35 ± 0.86 U/mgprot, while the 10 mg/L treatment group experienced an increase of 12.08 ± 0.72 U/mgprot (Figure 1c). In the 5 mg/L ammonia nitrogen treatment group, the SOD activity in juvenile yellowfin tuna was significantly lower than that of the control group at both the 6-h and 24-h time points ($p = 0.001$), but it notably surpassed the control group's level at the 36-h mark ($p = 0.001$, Figure 1d). Within the 10 mg/L treatment group, there was no significant difference in SOD activity compared to the control group at the 6-h mark ($p = 0.425$), but at both the 24-h and 36-h time points, the SOD activity in this group was significantly higher than that of the control group ($p = 0.001$).

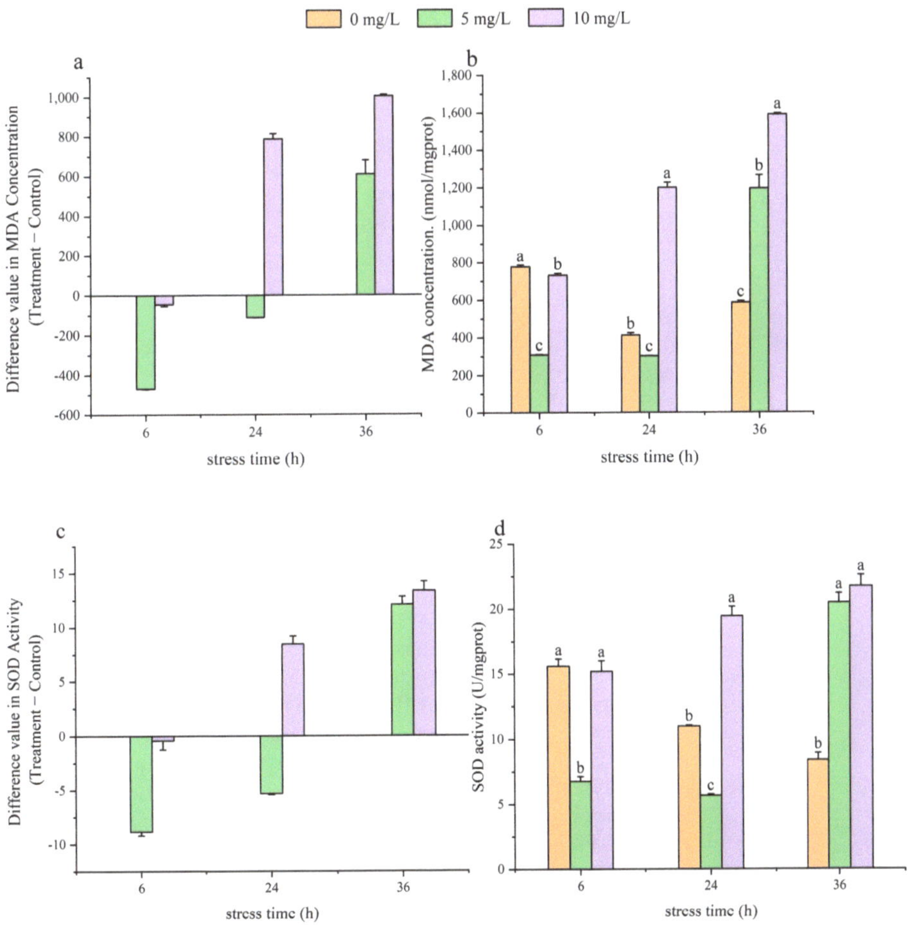

Figure 1. *Cont.*

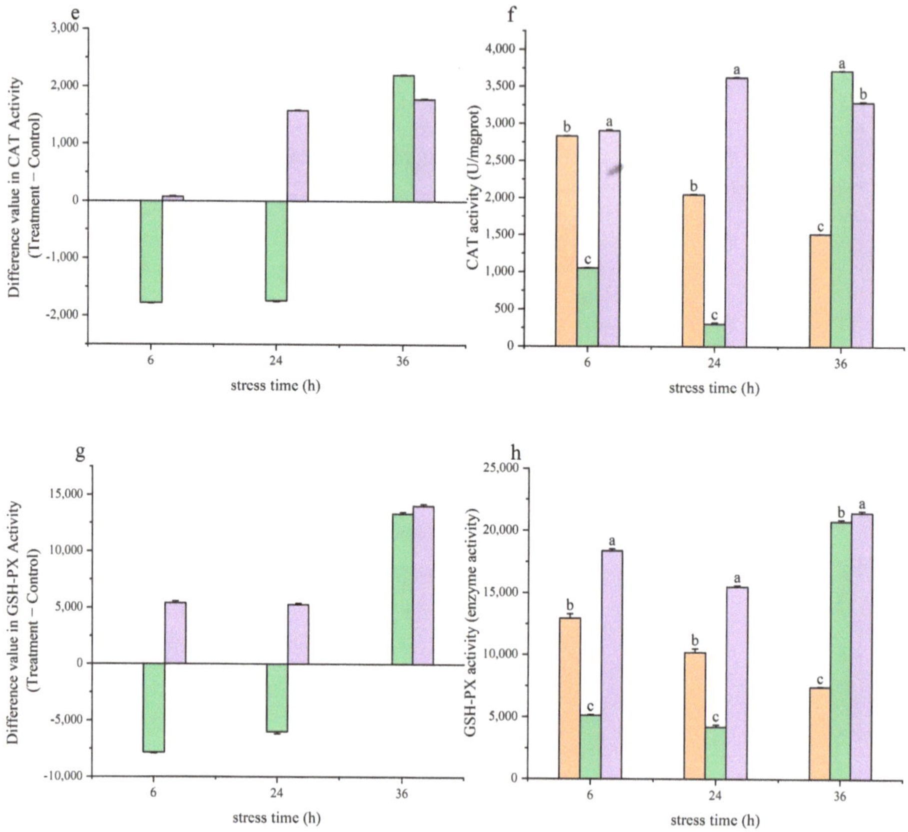

Figure 1. Effect of acute ammonia nitrogen stress on the antioxidant ability in the trunk kidney of juvenile yellowfin tuna (n = 9). (**a**) Difference value in malondialdehyde (MDA) concentration (treatment-control), (**b**) MDA concentration in treatment and control groups, (**c**) difference value in superoxide dismutase (SOD) activity (treatment-control), (**d**) SOD activity in treatment and control groups, (**e**) difference value in catalase (CAT) activity (treatment-control), (**f**) CAT activity in treatment and control groups, (**g**) difference value in glutathione peroxidase (GSH-PX) activity (treatment-control), (**h**) GSH-PX activity in treatment and control groups. The difference value = treatment value − control mean value. Different letters signify the significance of varying ammonia concentrations at the same time point ($p < 0.05$).

The CAT activity within the experimental groups displayed a consistent upward trajectory over time, with the most pronounced increase observed at the 36-h mark. Specifically, the activity rose by 2203.90 ± 4.21 U/mgprot in one group and by 11776.05 ± 11.79 U/mgprot in another (Figure 1e). In the 5 mg/L ammonia nitrogen treatment group, the CAT activity in juvenile yellowfin tuna was significantly lower than that of the control group at both the 6-h and 24-h time points (p = 0.001), but it was notably higher at the 36-h mark (p = 0.001, Figure 1f). Within the 10 mg/L treatment group, CAT activity was significantly elevated above the control group levels throughout the 6 to 36-h period (p = 0.001).

The GSH-PX activity of the 5 mg/L treatment group exhibited an upward trend over time, with the most significant increase observed at 36 h. The enzyme activity increased by 13,325.09 ± 141.68 units (Figure 1g). In the 5 mg/L ammonia nitrogen treatment group, the GSH-PX activity in juvenile yellowfin tuna was significantly lower than that of the control group at both the 6-h and 24-h time points (p = 0.001), yet it was notably higher at the 36-h mark (p = 0.001, Figure 1h). The GSH-PX activity of the 10 mg/L treatment group showed an upward trend following a 24-h stress period, with the largest increase in activity observed at 36 h, increasing by 14,030.91 ± 150.08 enzyme activity units. Within the 10 mg/L treatment group, the GSH-PX activity was consistently elevated above the control group levels throughout the 6- to 36-h period ($p = 0.001$).

3.2. Effect of Acute Ammonia Nitrogen Stress on the Activity of Phosphatase in the Trunk Kidney of Juvenile Yellowfin Tuna

AKP activity in the 5 mg/L treatment group showed a decreasing and then increasing trend, with the greatest increase in AKP activity at 6 h, 0.063309434 ± 0.02163058 King units/gprot (Figure 2a). In the 5 mg/L ammonia nitrogen treatment group, AKP activity was significantly increased at the 6-h mark compared to the control group ($p = 0.007$), while, at the 24-h mark, it was found to be lower than the control level ($p = 0.001$). At the 36-h mark, no significant difference was observed compared to the control group ($p = 0.095$, Figure 2b). The AKP activity in the 10 mg/L treatment group exhibited a decreasing trend over time, with the largest reduction observed at 36 h, amounting to a decrease of 0.01 ± 0.01 King unit/gprot. Conversely, in the 10 mg/L treatment group, no significant differences were detected in AKP activity compared to the control group during the 6- to 36-h period ($p = 0.397$ at 6 h, $p = 0.934$ at 24 h, and $p = 0.225$ at 36 h).

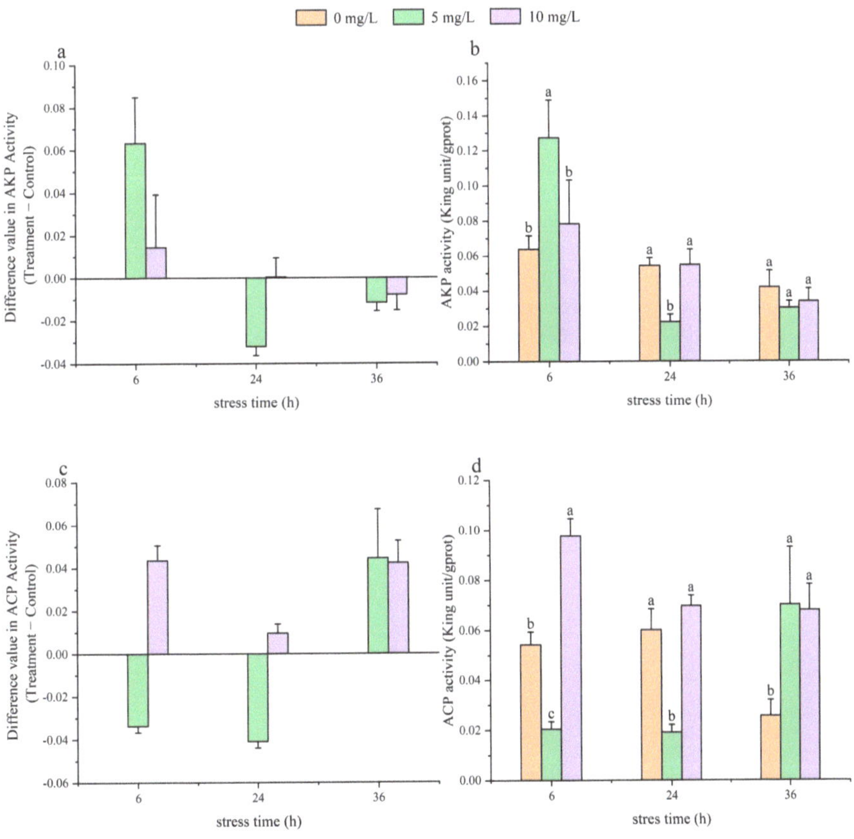

Figure 2. Effect of acute ammonia nitrogen stress on the activity of phosphatase in the trunk kidney of juvenile yellowfin tuna ($n = 9$). (**a**) Difference value in alkaline phosphatase (AKP) activity (treatment-control), (**b**) AKP activity in treatment and control groups, (**c**) difference value in acid phosphatase (ACP) activity (treatment-control), (**d**) ACP activity in treatment and control groups. The difference value = treatment value − control mean value. Different letters signify the significance of varying ammonia concentrations at the same time point ($p < 0.05$).

The ACP activity in both the 5 mg/L and 10 mg/L ammonia nitrogen treatment groups exhibited a pattern of initial decline followed by an increase over time, with the greatest decrease observed at 24 h. Specifically, the 5 mg/L group experienced a reduction of 0.041 ± 0.01 King unit/gprot, while the 10 mg/L group saw a decrease of 0.01 ± 0.01 King unit/gprot (Figure 2c). In the 5 mg/L ammonia nitrogen treatment group, ACP activity was significantly lower than that of the control group at both 6 and 24 h (p = 0.001, Figure 2d), yet it was notably higher at 36 h (p = 0.011). For the 10 mg/L treatment group, ACP activity was markedly elevated above the control levels at both 6 and 36 h (p = 0.001 at 6 h, p = 0.013 at 36 h), while, at 24 h, no significant difference was observed compared to the control group (p = 0.083).

3.3. *Effects of Acute Ammonia Nitrogen Stress on Antioxidant Genes in the Head Kidney of Juvenile Yellowfin Tuna*

In the 5 mg/L treatment group, the expression levels of SOD2 initially declined and subsequently demonstrated an upward trajectory over the course of the study (Figure 3a). In the 5 mg/L ammonia nitrogen treatment group, the relative expression of the SOD2 gene in juvenile yellowfin tuna was comparable to that of the control group at the 6-h mark (p = 0.201, Figure 3b), but it significantly decreased at the 24-h mark (p = 0.001) and then surpassed the control group's level at the 36-h mark (p = 0.009). In the 10 mg/L treatment group, the expression of SOD2 exhibited a consistent increase over the temporal progression. In the 10 mg/L treatment group, the SOD2 gene expression showed no significant difference from the control group at the 6-h mark (p = 0.206), was lower than the control at the 24-h mark (p = 0.004), and finally significantly exceeded the control group at the 36-h mark (p = 0.05).

The expression of the CAT gene in the 5 mg/L treatment group displayed a trend of initial decrease followed by a subsequent increase over time (Figure 3c). In the 5 mg/L ammonia nitrogen treatment group, the relative expression of the CAT gene in juvenile yellowfin tuna showed no significant difference from the control group during the 6- to 36-h period (at 6 h, p = 0.159; at 24 h, p = 0.189; at 36 h, p = 0.757; Figure 3d). In the 10 mg/L treatment group, the expression of CAT exhibited a sustained upward trend over time. Similarly, in the 10 mg/L ammonia nitrogen treatment group, the expression level of the CAT gene did not significantly differ from the control group over the same 6 to 36-h period (at 6 h, p = 0.204; at 24 h, p = 0.56; at 36 h, p = 0.936).

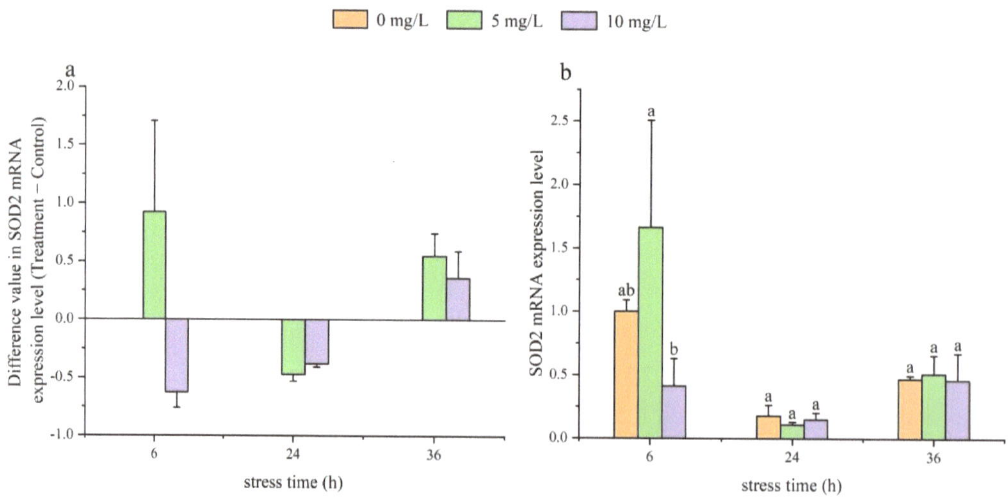

Figure 3. *Cont.*

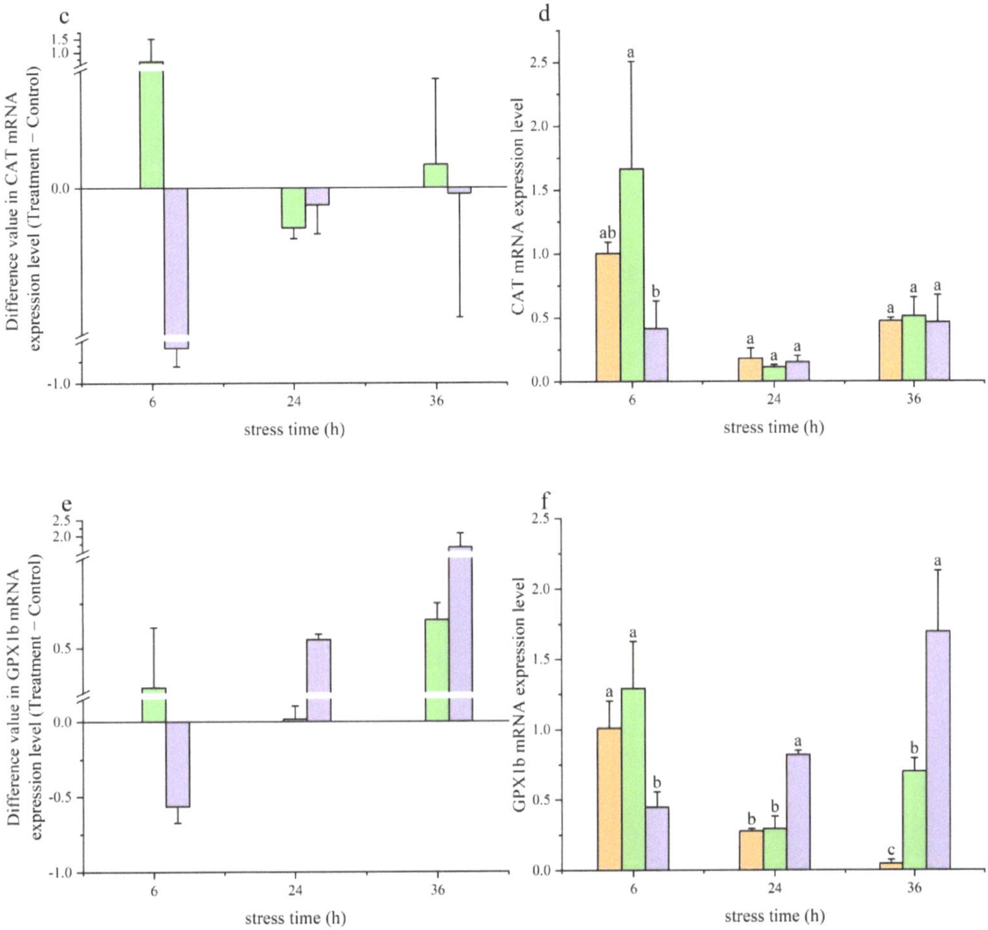

Figure 3. Effects of acute ammonia nitrogen stress on antioxidant genes in the head kidney of juvenile yellowfin tuna (n = 9). (**a**) Difference value in superoxide dismutase 2 (SOD2) mRNA expression level (treatment-control), (**b**) SOD2 mRNA expression level in treatment and control groups, (**c**) difference value in catalase (CAT) mRNA expression level (treatment-control), (**d**) CAT mRNA expression level in treatment and control groups, (**e**) difference value in glutathione peroxidase 1b (GPX1b) mRNA expression level (treatment-control), (**f**) GPX1b mRNA expression level in treatment and control groups. The difference value = treatment value − control mean value. Different letters signify the significance of varying ammonia concentrations at the same time point ($p < 0.05$).

In the 5 mg/L group, GPX1b gene expression decreased and then increased over time (Figure 3e). In the 5 mg/L treatment group, the relative expression level of the GPX1b gene in juvenile yellowfin tuna showed no significant difference from the control group at both the 6-h and 24-h time points (at 6 h, $p = 0.189$, at 24 h, $p = 0.712$, Figure 3f), but it was significantly elevated at the 36-h mark ($p = 0.02$). Expression of the GPX1b gene in the 10 mg/L treatment group showed an increasing trend with time. In the 10 mg/L treatment group, the expression of the GPX1b gene was significantly lower than that of the control group at the 6-h mark ($p = 0.024$), yet it was notably higher than the control group at both the 24-h and 36-h time points ($p = 0$).

3.4. Effects of Acute Ammonia Nitrogen Stress on Immune-Related Genes in the Head Kidney of Juvenile Yellowfin Tuna

IL-10 showed a trend of first decreasing and then increasing over time in the treatment group with a concentration of 5 mg/L (Figure 4a). In the 5 mg/L treatment group, the relative expression level of the IL-10 gene in juvenile yellowfin tuna was significantly higher than that of the control group at the 6-h mark ($p = 0.001$), but it was significantly lower at the 24-h mark ($p = 0.001$). By 36 h, there was no significant difference in expression levels compared to the control group ($p = 0.944$, Figure 4b). IL-10 demonstrated an increasing trend over time in the treatment group with a concentration of 10 mg/L. In the 10 mg/L treatment group, the IL-10 gene expression was significantly lower than the control group at both the 6-h ($p = 0.002$) and 24-h ($p = 0.001$) time points, but by 36 h, there was no significant difference from the control group ($p = 0.955$).

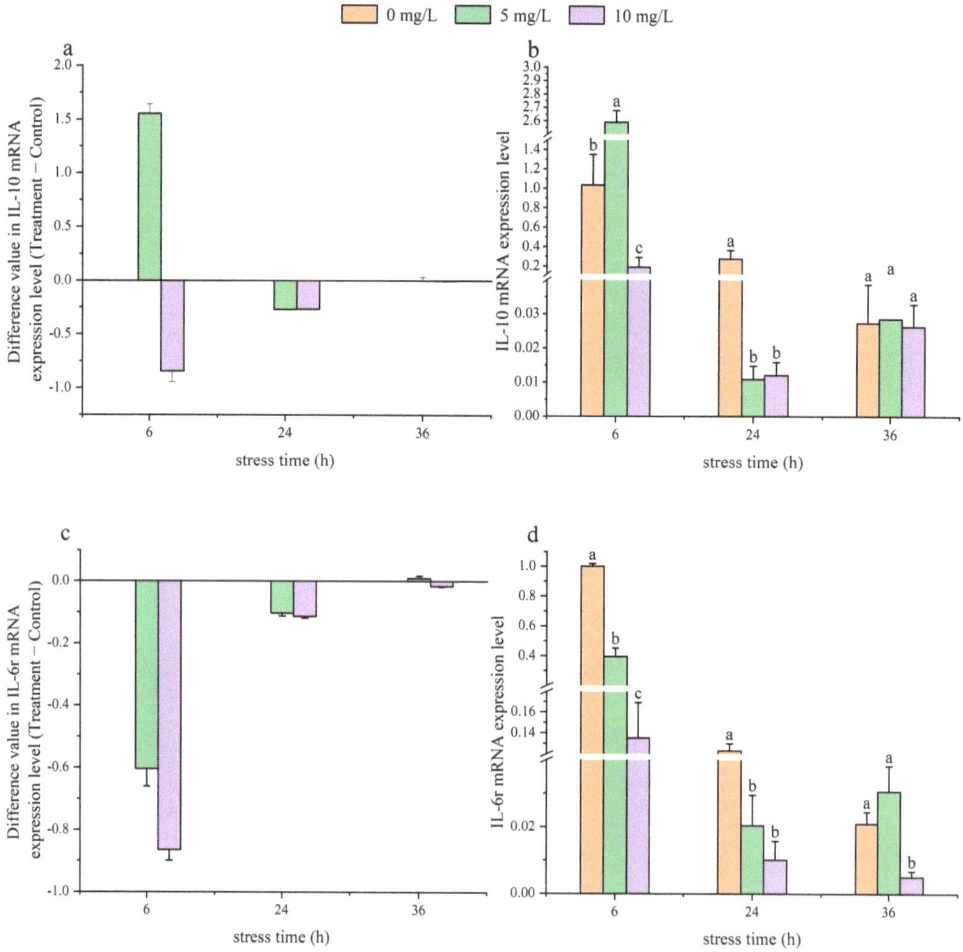

Figure 4. Effects of acute ammonia nitrogen stress on immune-related genes in the head kidney of juvenile yellowfin tuna ($n = 9$). (**a**) Difference value in Interleukin 10 (IL-10) mRNA expression level (treatment-control), (**b**) IL-10 mRNA expression level in treatment and control groups, (**c**) difference value in Interleukin 6 receptor (IL-6r) mRNA expression level (treatment-control), (**d**) IL-6r gene expression levels in treatment and control groups. The difference value = treatment value − control mean value. Different letters signify the significance of varying ammonia concentrations at the same time point ($p < 0.05$).

In the experimental groups treated with 5 and 10 mg/L of ammonia nitrogen, the expression of IL-6r exhibited an upward trend over time (Figure 4c). In the 5 mg/L treatment group, the expression of the IL-6r gene in juvenile yellowfin tuna was significantly lower than that of the control group at both the 6-h and 24-h time points ($p = 0.001$). By 36 h, there was no significant difference in expression levels compared to the control group ($p = 0.053$, Figure 4d). In the 10 mg/L treatment group, the IL-6r gene expression remained consistently lower than the control group from 6 h to 36 h ($p = 0.001$ at 6 and 24 h, $p = 0.007$ at 36 h).

4. Discussion

The kidneys serve as the primary detoxification and target organ for toxic substances [37]. In fish, the kidneys are situated on the ventral side, beneath the spine, and are divided into the head kidney and the trunk kidney [38]. The trunk kidney, which acts as the urinary organ, handles excretory functions [39]. Numerous studies have documented the impact of harmful environmental conditions on fish kidneys. For instance, exposure to ammonia nitrogen has been found to variably influence the activity of antioxidant enzymes in the kidneys of both amberjack (*Seriola dumerili*) and medaka (*Oryzias dancena*) [39]. Additionally, exposure to cadmium has been shown to impair both the structure and function of the kidneys in marine catfish (*Arius arius*) [39]. Ammonia nitrogen is a critical environmental factor in aquaculture, with high concentrations leading to oxidative stress in fish. This stress results in tissue damage and can severely impair fish growth, compromise their health, or even cause mortality [40]. In an experiment involving yellowfin tuna, changes were observed in the trunk kidney's MDA concentration, and the activities of SOD, GSH-PX, and CAT were monitored over time. These parameters all peaked at 36 h during a 5 mg/L ammonia nitrogen treatment. Similar trends were observed in Japanese seabass (*Lateolabrax japonicus*) [41]. The initial increase in immune response at 6 and 24 h in the 5 mg/L treatment likely facilitated the clearance of a substantial amount of ROS, reducing oxidative damage and consequently MDA production. This reduction might have led to lower activities of SOD, CAT, and GSH-PX compared to the control. As stress duration increased, the accumulation of ROS in the body heightened, causing more oxidative stress and MDA production, which in turn induced an increase in antioxidant enzyme activities due to the accumulation of H_2O_2. In bighead carp (*Hypophthalmythys*), low concentrations of ammonia stress activate its antioxidant system, leading to a significant increase in the activity of SOD and CAT. Conversely, high concentrations of ammonia stress cause damage to the antioxidant system, thereby suppressing the activity of these enzymes [42]. Conversely, in this experiment with juvenile yellowfin tuna, the 10 mg/L ammonia nitrogen treatment group exhibited significantly higher MDA concentrations and SOD, CAT, and GSH-PX activities at 6 h and 24 h compared to the 5 mg/L group. The higher concentration of ammonia nitrogen might have caused a gradual increase in ammonia accumulation in the organisms over time, leading to lipid peroxidation, a rise in MDA concentrations, and subsequent increases in SOD activity due to the production of large amounts of H_2O_2. Concurrently, CAT activity decreased while GSH-PX activity rose at 36 h, which resulted in inefficient clearance of H_2O_2, further exacerbating lipid peroxidation and increasing MDA concentrations. These processes indicate more severe damage to the organisms as the MDA concentration increases [43]. Moderate exposure to ammonia nitrogen appears to act as an adaptive challenge, stimulating the antioxidant defense mechanisms of juvenile yellowfin tuna and thereby enhancing their resistance to oxidative stress. During this process, the antioxidant capacity of juvenile yellowfin tuna was significantly improved when exposed to 5 mg/L of ammonia nitrogen for 36 h, or to 10 mg/L for 24 h. To minimize the impacts of ammonia nitrogen concentration on their antioxidant capabilities, it is crucial to strictly manage both the ammonia nitrogen levels and the duration of exposure.

ACP and AKP are critical enzymes for phosphate group transfer and transformation, playing a key role in the metabolism associated with an animal's defense mechanisms. These enzymes facilitate the absorption and transfer of substances, forming a hydrolytic enzyme system to eliminate foreign materials and bolster the body's defense [9,44]. In juvenile barramundi (*Lates calcarifer*), ACP levels showed an increase from 24 to 72 h under ammonia nitrogen stress [45], while plasma AKP levels in Hsu's ping-pong (*Sebastes schlegelii*) increased initially and then decreased under similar stress conditions [46]. In this study, AKP activity in the 5 mg/L treatment group was elevated at 6 h, and ACP activity increased at 36 h, both compared to the control group. This may be due to the 5 mg/L ammonia nitrogen concentration enhancing oxidative stress and non-specific immunity in yellowfin tuna, leading to decreased MDA levels. The changes in AKP, related to metabolic transport, suggest increased lipid metabolism and reduced lipid oxidation. Over time, the prolonged stress leads to an excessive accumulation of ammonia nitrogen, diminishing the metabolic activities and AKP activity in juvenile yellowfin tuna, resulting in lipid oxidation. In response, antioxidant enzymes and various lysosomes, including ACP—an enzyme predominant in lysosomes involved in phagocytic processes—activate to eliminate ROS, thereby driving the increase in ACP concentration. In the context of alkalinity stress, both ACP and AKP activities first increased in low alkalinity conditions and then decreased, whereas they increased significantly under high alkalinity stress [47]. In this study, the 10 mg/L ammonia nitrogen treatment group showed significantly lower AKP activity compared to the 5 mg/L group at 6 h, while ACP activity was significantly higher in the 10 mg/L group at both 6 and 24 h. Ammonia nitrogen has a pronounced inhibitory effect on the AKP activity in juvenile yellowfin tuna. However, at lower concentrations, ammonia nitrogen exhibits a positive stimulatory effect on the ACP activity in these juvenile fish. Conversely, when ammonia nitrogen levels rise to higher concentrations, it significantly inhibits the ACP activity in the juvenile yellowfin tuna. This discovery underscores the intricate influence of ammonia nitrogen levels on the phosphatase activity in juvenile yellowfin tuna, offering a scientific basis for optimizing the aquaculture environment for these species.

The head kidney in fish is a crucial hematopoietic tissue and a site for immune cell differentiation and proliferation. In gilded snapper (*Sparus aurata* L.), researchers observed a down-regulation of some immune genes in the kidney when exposed to highly polluted marine sediments [36]. Sinha et al. reported stable SOD gene expressions but reduced CAT gene expressions in the liver of European black bass (*Dicentrarchus labrax*) under acute hypotonic stress [48]. Conversely, antioxidant genes such as SOD and CAT saw increased expression in the liver of puffer fish (*Takifugu obscurus*) subjected to high- or low-temperature stress for 1 h [49]. Ammonia stress is known to induce oxidative stress in fish, altering the expression of genes related to antioxidants [50]. In this experiment, the expression levels of antioxidant genes in the 5 mg/L ammonia-treated juvenile yellowfin tuna were not significantly higher than the control at 6 h. This pattern suggests that initial ammonia-induced oxidative stress prompted an upregulation of antioxidant genes to bolster the organism's antioxidant defenses and to efficiently remove excessive ROS, thereby maintaining metabolic balance. As the stress duration increased, the yellowfin tuna gradually adapted to the 5 mg/L ammonia concentration, but a notable increase in ammonia accumulation at 36 h led to a renewed increase in antioxidant gene expression to mitigate oxidative stress. In the 10 mg/L treatment group, the expression of GPX1b increased at 36 h compared to the control group, possibly due to the higher ammonia concentration suppressing SOD2 and CAT expressions while enhancing the immune response, leading to the overexpression of GPX1b. It is documented that the GPX1 gene is often overexpressed in many cancers, influencing tumor cell proliferation, invasion, migration, apoptosis, and immune responses, and is considered a tumor promoter [51]. The disparity between gene expression and antioxidant enzyme activity might be attributed to additional regulatory mechanisms, such as post-transcriptional and post-translational modifications, which can lead to differences between gene expression levels and their corresponding enzyme activities at the protein level [52]. Initially, low concentrations of ammonia nitrogen elevated the relative expression levels of antioxidant genes in juvenile yellowfin tuna. However, as the stress period extended, these levels began to decline. Conversely, high concentrations of ammonia nitrogen exerted varied impacts on the relative expressions of these antioxidant genes.

IL-10, an anti-inflammatory cytokine, plays a crucial role in mitigating inflammation and autoimmune disorders by counteracting pro-inflammatory factors [53]. In contrast, IL-6r is a pro-inflammatory cytokine essential for promoting the expression of proteins linked with acute inflammation [54]. Studies have indicated that catfish (*Lophiosilurus alexandri*) exhibit significantly higher plasma IL-6r levels than the control group during 24 and 72 h of hypoxic stress [55]. Ammonia stress in fish can trigger an inflammatory response, altering the expression of associated genes [14]. In this study, within the 5 mg/L ammonia-treated group, IL-10 gene expression was up-regulated, while IL-6r gene expression was down-regulated at 6 h compared to the control group. This suggests that, early in the stress response, oxidative stress induced inflammation in juvenile yellowfin tuna, which led to an increase in IL-10 expression. This cytokine, by enhancing autoimmunity, helps prevent excessive inflammatory reactions that could damage cells and tissues. Consequently, IL-10 suppressed the expression of the pro-inflammatory cytokine IL-6r, thus exerting a regulatory effect. At 36 h, there were no significant differences in IL-10 and IL-6r levels between the ammonia nitrogen experimental group and the control group, possibly reflecting the juvenile yellowfin tuna's gradual adaptation to the ammonia nitrogen environment and the kidneys' timely regulation of genes related to the inflammatory response [56]. The interaction between IL-10 and IL-6r helped maintain a stable level that mitigated the inflammatory effects of ammonia nitrogen on the kidneys. Initially, low concentrations of ammonia nitrogen stimulated the expression of genes related to IL-10, but as time progressed, this expression was suppressed, and ammonia nitrogen exerted an inhibitory effect on IL-6r expression.

5. Conclusions

This study investigated the effects of acute ammonia nitrogen stress on the antioxidant ability and phosphatase activity in the trunk kidneys of juvenile yellowfin tuna, as well as the expression of antioxidant and immune genes in the head kidney. The results showed that ammonia exposure enhanced the antioxidant ability of trunk kidneys to varying degrees and inhibited phosphatase activity. Furthermore, ammonia nitrogen exposure reduces the expression of antioxidant and immune genes in the head kidney. These results suggest that the antioxidant system can effectively neutralize excess free radicals and oxidative intermediates, thereby maintaining the dynamic balance of the intracellular antioxidant system within the range of ammonia-nitrogen concentration changes. However, the occurrence of oxidative stress and immune-inflammatory responses in the head kidney influences the transcription levels of these genes. Based on these observations, it is recommended that ammonia nitrogen levels be strictly controlled below 10 mg/L and that the stress duration should not exceed 24 h. For ammonia nitrogen mass concentrations below 5 mg/L, the stress duration should preferably be controlled within 36 h. This study contributes to a better understanding of the physiological changes in yellowfin tuna under ammonia nitrogen stress, establishing a foundation for subsequent artificial cultivation and offering insights that can enhance the culture efficiency of yellowfin tuna.

Author Contributions: Conceptualization, Y.S.; methodology, Y.S.; software, Z.F.; validation, Z.F.; formal analysis, Y.S. and Z.F.; investigation, Y.S. and Z.F.; resources, Z.F. and Z.M.; data curation, Y.S.; writing—original draft preparation, Y.S.; writing—review and editing, Z.F. and Z.M.; visualization, Y.S. and Z.F.; supervision, Z.M.; project administration, Z.M.; funding acquisition, Z.M. All authors have read and agreed to the published version of the manuscript.

Funding: This work was supported by the Hainan Major Science and Technology Project, grant number ZDKJ2021011; the Project of Sanya Yazhou Bay Science and Technology City, grant number SKJC-2022-PTDX-015; the Central Public-Interest Scientific Institution Basal Research Fund, CAFS, grant number 2023TD58, 2024XT04; the Central Public-Interest Scientific Institution Basal Research Fund South China Sea Fisheries Research Institute, CAFS, grant number 2021SD09; the Hainan Provincial Natural Science Foundation of China, grant number 323QN331; the Hainan Province Science and Technology Special Fund, grant number ZDYF2024XDNY247.

Institutional Review Board Statement: The experiment complied with the regulations and guidelines established by the Animal Care and Use Committee of the South China Sea Fisheries Research Institute, Chinese Academy of Fishery Sciences. The approval number is 2020TD55, which was approved on 5 January 2020.

Informed Consent Statement: Not applicable.

Data Availability Statement: The original contributions presented in the study are included in the article. Further inquiries can be directed to the corresponding authors.

Conflicts of Interest: The authors declare no conflicts of interest.

References

1. Lu, J.; Yao, T.; Shi, S.; Ye, L. Effects of acute ammonia nitrogen exposure on metabolic and immunological responses in the Hong Kong oyster *Crassostrea hongkongensis*. *Ecotoxicol. Environ. Saf.* **2022**, *237*, 113518. [CrossRef] [PubMed]
2. Da Silva, F.J.R.; Lima, F.R.S.; Do Vale, D.A.; Do Carmo, M.V. High levels of total ammonia nitrogen as NH_4^+ are stressful and harmful to the growth of Nile tilapia juveniles. *Acta Scientiarum. Biol. Sci.* **2013**, *35*, 475–481.
3. Zhang, C.; Ma, J.; Qi, Q.; Xu, M.; Xu, R. Effects of ammonia exposure on anxiety behavior, oxidative stress and inflammation in guppy (*Poecilia reticulate*). *Comp. Biochem. Physiol. Part. C Toxicol. Pharmacol.* **2023**, *265*, 109539. [CrossRef] [PubMed]
4. Cong, M.; Wu, H.; Cao, T.; Ji, C.; Lv, J. Effects of ammonia nitrogen on gill mitochondria in clam *Ruditapes philippinarum*. *Environ. Toxicol. Pharmacol.* **2019**, *65*, 46–52. [CrossRef] [PubMed]
5. Rodrigues, R.V.; Schwarz, M.H.; Delbos, B.C. Acute toxicity and sublethal effects of ammonia and nitrite for juvenile cobia *Rachycentron canadum*. *Aquaculture* **2007**, *271*, 553–557. [CrossRef]
6. Xu, W.; Zhu, Z.; Ge, F.; Han, Z.; Li, J. Analysis of behavior trajectory based on deep learning in ammonia environment for fish. *Sensors* **2020**, *20*, 4425. [CrossRef] [PubMed]
7. Zhang, W.; Xia, S.; Zhu, J.; Miao, L.; Ren, M.; Lin, Y.; Ge, X.; Sun, S. Growth performance, physiological response and histology changes of juvenile blunt snout bream, *Megalobrama amblycephala* exposed to chronic ammonia. *Aquaculture* **2019**, *506*, 424–436. [CrossRef]
8. Long, S.; Dong, X.; Yan, X.; Liu, H.; Tan, B.; Zhang, S.; Chi, S.; Yang, Q.; Liu, H.; Yang, Y. The effect of oxidized fish oil on antioxidant ability, histology and transcriptome in intestine of the juvenile hybrid grouper (♀*Epinephelus fuscoguttatus* × ♂*Epinephelus lanceolatus*). *Aquac. Rep.* **2021**, *22*, 100921. [CrossRef]
9. Bao, J.; Li, X.; Xing, Y.; Feng, C.; Jiang, H. Effects of hypoxia on immune responses and carbohydrate metabolism in the Chinese mitten crab, *Eriocheir sinensis*. *Aquac. Res.* **2020**, *51*, 2735–2744. [CrossRef]
10. Chai, Y.; Peng, R.; Jiang, M.; Jiang, X.; Han, Q.; Han, Z. Effects of ammonia nitrogen stress on the blood cell immunity and liver antioxidant function of *Sepia pharaonis*. *Aquaculture* **2022**, *546*, 737417. [CrossRef]
11. Chen, S.; Yu, Y.; Gao, Y.; Yin, P.; Tian, L.; Niu, J.; Liu, Y. Exposure to acute ammonia stress influences survival, immune response and antioxidant status of pacific white shrimp (*Litopenaeus vannamei*) pretreated with diverse levels of inositol. *Fish Shellfish Immunol.* **2019**, *89*, 248–256. [CrossRef] [PubMed]
12. Abdel-Latif, H.M.; Shukry, M.; Abd-Elaziz, R.A. Clinico-pathological findings and expression of inflammatory cytokines, apoptosis, and oxidative stress-related genes draw mechanistic insights in Nile tilapia reared under ammonia-N exposure and *Aeromonas hydrophila* challenge. *Fish Shellfish Immunol.* **2022**, *127*, 1–12. [CrossRef] [PubMed]
13. Cao, J.; Mei, J.; Xie, J. Combined effects of hypoxia and ammonia-N exposure on the immune response, oxidative stress, tissue injury and apoptosis of hybrid grouper (*Epinephelus fuscoguttatus*♀× *E. lanceolatus*♂). *Environ. Sci. Pollut. Res.* **2024**, *31*, 845–856. [CrossRef] [PubMed]
14. Collet, B. Innate immune responses of salmonid fish to viral infections. *Dev. Comp. Immunol.* **2014**, *43*, 160–173. [CrossRef] [PubMed]
15. Tabaei, S.; Motallebnezhad, M.; Tabaee, S.S. Systematic review and meta-analysis of association of polymorphisms in inflammatory cytokine genes with coronary artery disease. *Inflamm. Res.* **2020**, *69*, 1001–1013. [CrossRef]
16. Yang, P.; Liu, J.; Xiao, J.; Jian, H.; Chen, H. Associations between seven common cytokine gene polymorphisms and coronary artery disease: Evidence from a meta-analysis. *Int. Arch. Allergy Immunol.* **2020**, *181*, 301–310. [CrossRef]
17. Overgard, A.; Nepstad, I.; Nerland, A.H.; Patel, S. Characterisation and expression analysis of the Atlantic halibut (*Hippoglossus hippoglossus* L.) cytokines: IL-1 beta, IL-6, IL-11, IL-12 beta and IFN gamma. *Mol. Biol. Rep.* **2012**, *39*, 2201–2213. [CrossRef] [PubMed]

18. Furne, M.; Holen, E.; Araujo, P.; Lie, K.K.; Moren, M. Cytokine gene expression and prostaglandin production in head kidney leukocytes isolated from Atlantic cod (*Gadus morhua*) added different levels of arachidonic acid and eicosapentaenoic acid. *Fish Shellfish Immunol.* **2013**, *34*, 770–777. [CrossRef]
19. Ponce, M.; Zuasti, E.; Anguís, V.; Fernández-Díaz, C. Effects of the sulfated polysaccharide ulvan from Ulva ohnoi on the modulation of the immune response in Senegalese sole (*Solea senegalensis*). *Fish Shellfish Immunol.* **2020**, *100*, 27–40. [CrossRef] [PubMed]
20. Tharuka, M.N.; Bathige, S.; Oh, M.; Lee, S.; Kim, M.; Priyathilaka, T.T.; Lee, J. Molecular characterization and expression analysis of big-belly seahorse (*Hippocampus abdominalis*) interleukin-10 and analysis of its potent anti-inflammatory properties in LPS-induced murine macrophage RAW 264.7 cells. *Gene* **2019**, *685*, 1–11. [CrossRef]
21. Peng, Y.; Cai, X.; Zhang, G.; Wang, J.; Li, Y.; Wang, Z.; Wang, B.; Xiong, X.; Wu, Z.; Jian, J. Molecular characterization and expression of interleukin-10 and interleukin-22 in golden pompano (*Trachinotus ovatus*) in response to Streptococcus agalactiae stimulus. *Fish Shellfish Immunol.* **2017**, *65*, 244–255. [CrossRef] [PubMed]
22. Chen, H.; Lin, H.; Foung, Y.; Lin, J.H. The bioactivity of teleost IL-6: IL-6 protein in orange-spotted grouper (*Epinephelus coioides*) induces Th2 cell differentiation pathway and antibody production. *Dev. Comp. Immunol.* **2012**, *38*, 285–294. [CrossRef] [PubMed]
23. Block, B.A.; Keen, J.E.; Castillo, B.; Dewar, H.; Freund, E.V.; Marcinek, D.J.; Brill, R.W.; Farwell, C. Environmental preferences of yellowfin tuna (*Thunnus albacares*) at the northern extent of its range. *Mar. Biol.* **1997**, *130*, 119–132. [CrossRef]
24. Zhang, N.; Yang, R.; Fu, Z.; Yu, G.; Ma, Z. Mechanisms of Digestive Enzyme Response to Acute Salinity Stress in Juvenile Yellowfin Tuna (*Thunnus albacares*). *Animals* **2023**, *13*, 3454. [CrossRef] [PubMed]
25. Zhou, S.; Zhang, N.; Fu, Z.; Yu, G.; Ma, Z.; Zhao, L. Impact of Salinity Changes on the Antioxidation of Juvenile Yellowfin Tuna (*Thunnus albacares*). *J. Mar. Sci. Eng.* **2023**, *11*, 132. [CrossRef]
26. Fonteneau, A. An overview of yellowfin tuna stocks, fisheries and stock status worldwide. In Proceedings of the IOTC 7th Working Party on Tropical Tunas, Phuket-Thailand, Phuket, Thailand, 18–22 July 2005.
27. Grewe, P.M.; Feutry, P.; Hill, P.L.; Gunasekera, R.M.; Schaefer, K.M.; Itano, D.G.; Fuller, D.W.; Foster, S.D.; Davies, C.R. Evidence of discrete yellowfin tuna (*Thunnus albacares*) populations demands rethink of management for this globally important resource. *Sci. Rep.* **2015**, *5*, 1–9. [CrossRef] [PubMed]
28. Andamari, R.; Hutapea, J.H.; Prisantoso, B.I. Reproduction aspects of the yellowfin tuna (*Thunnus albacares*). *J. Ilmu Teknol. Kelaut.* **2012**, *4*, 1.
29. Estess, E.E.; Klinger, D.H.; Coffey, D.M.; Gleiss, A.C.; Rowbotham, I.; Seitz, A.C.; Rodriguez, L.; Norton, A.; Block, B.; Farwell, C. Bioenergetics of captive yellowfin tuna (*Thunnus albacares*). *Aquaculture* **2017**, *468*, 71–79. [CrossRef]
30. Pecoraro, C.; Babbucci, M.; Franch, R.; Rico, C.; Papetti, C.; Chassot, E.; Bodin, N.; Cariani, A.; Bargelloni, L.; Tinti, F. The population genomics of yellowfin tuna (*Thunnus albacares*) at global geographic scale challenges current stock delineation. *Sci. Rep.* **2018**, *8*, 13890. [CrossRef]
31. He, Y.; Fu, Z.; Dai, S.; Yu, G.; Ma, Z. Dietary curcumin supplementation can enhance health and resistance to ammonia stress in the greater amberjack (*Seriola dumerili*). *Front. Mar. Sci.* **2022**, *9*, 961783. [CrossRef]
32. Wang, W.; Dai, S.; Liu, L.; Fu, Z.; Yang, R.; Yu, G.; Ma, Z.; Zong, H. Daily Rhythmicity of Muscle-Related and Rhythm Genes Expression in Mackerel Tuna (*Euthynnus affinis*). *Biology* **2023**, *12*, 1211. [CrossRef]
33. Fu, W. Comparison of Histological and Transcriptome Studies of the Gills between Juvenile and Adult Yellowfin Tuna (*Thunnus albacares*) from the South China Sea. Master's Thesis, Hainan University, Haikou, China, 2023.
34. Livak, K.J.; Schmittgen, T.D. Analysis of relative gene expression data using real-time quantitative PCR and the $2^{-\Delta\Delta Ct}$ method. *Methods* **2001**, *25*, 402–408. [CrossRef]
35. Hamad, A.R. Physiological and Cellular Responses of Marine Medaka Fish (*Oryzias dancena*) to Ammonia Stress Treatment. Ph.D. Dissertation, Pukyong National University, Busan, Republic of Korea, 2018.
36. Hamed, S.B.; Guardiola, F.; Cuesta, A.; Martínez, S.; Martínez-Sánchez, M.J.; Pérez-Sirvent, C.; Esteban, M.Á. Head kidney, liver and skin histopathology and gene expression in gilthead seabream (*Sparus aurata* L.) exposed to highly polluted marine sediments from Portman Bay (Spain). *Chemosphere* **2017**, *174*, 563–571. [CrossRef]
37. Zwollo, P.; Cole, S.; Bromage, E.; Kaattari, S. B cell heterogeneity in the teleost kidney: Evidence for a maturation gradient from anterior to posterior kidney. *J. Immunol.* **2005**, *174*, 6608–6616. [CrossRef] [PubMed]
38. Hong, J.; Fu, Z.; Hu, J.; Zhou, S.; Yu, G.; Ma, Z. Dietary curcumin supplementation enhanced ammonia nitrogen stress tolerance in greater amberjack (*Seriola dumerili*): Growth, serum biochemistry and expression of stress-related genes. *J. Mar. Sci. Eng.* **2022**, *10*, 1796. [CrossRef]
39. Mani, R.; Rose, S.; Suresh, A.; Sambantham, S.; Anandan, B.; Ibrahim, M.; Meena, B. Cellular alterations and damage to the renal tissue of marine catfish *Arius arius* following Cd exposure and the possible sequestrant role of Metallothionein. *Mar. Pollut. Bull.* **2021**, *163*, 111930. [CrossRef]
40. Ching, B.; Chew, S.F.; Wong, W.P.; Ip, Y.K. Environmental ammonia exposure induces oxidative stress in gills and brain of *Boleophthalmus boddarti* (mudskipper). *Aquat. Toxicol.* **2009**, *95*, 203–212. [CrossRef] [PubMed]
41. Guo, M.; Yan, Q.; Dong, Y.; Ding, Z.; Mei, J.; Xie, J. Apoptotic Changes, Oxidative Stress and Immunomodulatory Effects in the Liver of Japanese Seabass (*Lateolabrax japonicus*) Induced by Ammonia-Nitrogen Stress during Keep-Live Transport. *Biology* **2023**, *12*, 769. [CrossRef] [PubMed]

42. Sun, H.; Lü, K.; Minter, E.J.; Chen, Y.; Yang, Z.; Montagnes, D.J. Combined effects of ammonia and microcystin on survival, growth, antioxidant responses, and lipid peroxidation of bighead carp *Hypophthalmythys nobilis* larvae. *J. Hazard. Mater.* **2012**, *221*, 213–219. [CrossRef]
43. Zhang, Z.; Liu, Q.; Cai, J.; Yang, J.; Shen, Q.; Xu, S. Chlorpyrifos exposure in common carp (*Cyprinus carpio* L.) leads to oxidative stress and immune responses. *Fish Shellfish. Immunol.* **2017**, *67*, 604–611. [CrossRef]
44. Rahman, M.F.; Siddiqui, M. Biochemical effects of vepacide (from *Azadirachta indica*) on Wistar rats during subchronic exposure. *Ecotoxicol. Environ. Saf.* **2004**, *59*, 332–339. [CrossRef] [PubMed]
45. Liu, Y.; Hu, J.; Zhou, S.; Yang, R.; Qin, J.G.; Ma, Z.; Yang, Q. Effect of acute ammonia stress on antioxidant enzymes and digestive enzymes in barramundi *Lates calcarifer* larvae. *Isr. J. Aquac.-Bamidgeh* **2018**, *70*, 70–2018. [CrossRef]
46. Gao, Y.; Gao, Y.; Wang, J.; Li, M.; Jia, Y.; Mend, Z. Effect of acute ammonia stress on the plasma biochemicalindexes of *Sebastes schlegelii*. *Mar. Sci.* **2023**, *47*, 49–59.
47. Li, M.; Wang, S.; Zhao, Z.; Luo, L.; Zhang, R.; Guo, K.; Zhang, L.; Yang, Y. Effects of alkalinity on the antioxidant capacity, nonspecific immune response and tissue structure of Chinese Mitten Crab *Eriocheir sinensis*. *Fishes* **2022**, *7*, 206. [CrossRef]
48. Sinha, A.K.; AbdElgawad, H.; Zinta, G.; Dasan, A.F.; Rasoloniriana, R.; Asard, H.; Blust, R.; De Boeck, G. Nutritional status as the key modulator of antioxidant responses induced by high environmental ammonia and salinity stress in European sea bass (*Dicentrarchus labrax*). *PLoS ONE* **2015**, *10*, e135091. [CrossRef] [PubMed]
49. Cheng, C.; Guo, Z.; Luo, S.; Wang, A. Effects of high temperature on biochemical parameters, oxidative stress, DNA damage and apoptosis of pufferfish (*Takifugu obscurus*). *Ecotoxicol. Environ. Saf.* **2018**, *150*, 190–198. [CrossRef] [PubMed]
50. Hegazi, M.M.; Attia, Z.I.; Ashour, O.A. Oxidative stress and antioxidant enzymes in liver and white muscle of Nile tilapia juveniles in chronic ammonia exposure. *Aquat. Toxicol.* **2010**, *99*, 118–125. [CrossRef] [PubMed]
51. Zhao, Y.; Wang, H.; Zhou, J.; Shao, Q. Glutathione peroxidase GPX1 and its dichotomous roles in cancer. *Cancers* **2022**, *14*, 2560. [CrossRef] [PubMed]
52. Liang, X.; Wang, F.; Li, K.; Nie, X.; Fang, H. Effects of norfloxacin nicotinate on the early life stage of zebrafish (*Danio rerio*): Developmental toxicity, oxidative stress and immunotoxicity. *Fish Shellfish Immunol.* **2020**, *96*, 262–269. [CrossRef]
53. Saraiva, M.; O'Garra, A. The regulation of IL-10 production by immune cells. *Nat. Rev. Immunol.* **2010**, *10*, 170–181. [CrossRef]
54. Uciechowski, P.; Dempke, W. Interleukin-6: A masterplayer in the cytokine network. *Oncology* **2020**, *98*, 131–137. [CrossRef] [PubMed]
55. Baldissera, M.D.; Souza, C.F.; Boaventura, T.P.; Nakayama, C.L.; Baldisserotto, B.; Luz, R.K. Purinergic signaling as a potential target of hypoxia stress-induced impairment of the immune system in freshwater catfish *Lophiosilurus alexandri*. *Aquaculture* **2018**, *496*, 197–202. [CrossRef]
56. Li, M.; Lai, H.; Li, Q.; Gong, S.; Wang, R. Effects of dietary taurine on growth, immunity and hyperammonemia in juvenile yellow catfish *Pelteobagrus fulvidraco* fed all-plant protein diets. *Aquaculture* **2016**, *450*, 349–355. [CrossRef]

Journal of
Marine Science and Engineering

MDPI

Article

Effect of Licorice on Gene Expression Related to the Growth of Asian Seabass *Lates calcarifer*

Rui Yang [1,2,3,4,†], Wang Zhao [1,2,3,4,†], Yifu Wang [1,2,3,4], Zhengyi Fu [1,2,3,4,5], Jing Hu [1,2,3,4], Shengjie Zhou [1,2,3,4], Minghao Li [1,2,3,4] and Zhenhua Ma [1,2,3,4,5,*]

1 Key Laboratory of Efficient Utilization and Processing of Marine Fishery Resources of Hainan Province, Sanya Tropical Fisheries Research Institute, Sanya 572018, China
2 South China Sea Fisheries Research Institute, Chinese Academy of Fishery Sciences, Guangzhou 510300, China
3 Hainan Engineering Research Center for Deep-Sea Aquaculture and Processing, Sanya 572018, China
4 International Joint Research Center for Conservation and Application of Fishery Resources in the South China Sea, Sanya 572018, China
5 College of Science and Engineering, Flinders University, Adelaide 5001, Australia
* Correspondence: zhenhua.ma@scsfri.ac.cn
† These authors contributed equally to this work.

Abstract: The Asian seabass (*Lates calcarifer*) has high economic value and is the primary aquaculture species in China. Licorice (*Glycyrrhiza uralensis)* as a feed additive has demonstrated significant immunological benefits in aquaculture. However, its effects on the growth of aquatic animals are largely unexplored. This study explored the influence of licorice on the level of growth-related genes in Asian seabass by conducting an experiment using artificial feed with 0%, 1%, 3%, and 5% licorice. The impact on growth performance and the expression of several genes, including growth hormone–releasing hormone (*GHRH*), growth hormone (*GH*), growth hormone receptor (*GHR*), insulin-like growth factor 1 (*IGF1*), *IGF2*, IGF2 receptor (*IGF2R*), myostatin 1 (*MSTN1*), and myostatin 2 (*MSTN2*), were studied over 56 d. According to the results, the 3% and 5% licorice-supplemented diets significantly improved survival rates and weight gain compared to the control group. Licorice affected the level of growth-associated genes in Asian seabass and significantly increased the levels of *GHR* and *IGF1* in the liver. However, a 5% licorice diet downregulated the expression of *IGF2*. As the licorice content in the diet increased, the levels of *IGF2R* and *MSTN1* in the muscle tissue first decreased and then increased, and licorice addition inhibited the *MSTN2* expression. The inclusion of licorice in the feed led to a significant downregulation of the *GH* and *GHRH* expression ($p < 0.05$). In summary, adding a certain proportion of licorice to the diet can improve the survival rate of the Asian seabass. Moreover, a proper proportion of licorice can increase the expression of related growth genes of fish, effectively increasing their weight gain rate and specific growth rate.

Keywords: traditional Chinese herbs; plant source additive; growth performance; growth-related genes; muscle growth-related genes

Citation: Yang, R.; Zhao, W.; Wang, Y.; Fu, Z.; Hu, J.; Zhou, S.; Li, M.; Ma, Z. Effect of Licorice on Gene Expression Related to the Growth of Asian Seabass *Lates calcarifer*. *J. Mar. Sci. Eng.* **2024**, *12*, 1036. https://doi.org/10.3390/jmse12071036

Academic Editors: Roberto Carlucci and Ka Hou Chu

Received: 5 May 2024
Revised: 3 June 2024
Accepted: 18 June 2024
Published: 21 June 2024

1. Introduction

The Asian seabass is a euryhaline fish found mainly in the Western Pacific, being widely farmed in Australia [1]. Renowned for its taste, rapid growth, and large size, the Asian seabass has a high economic value and is the primary aquaculture species in China [2]. Extensive studies were conducted on the dietary requirements of the Asian seabass, including proteins, fats, carbohydrates, and minerals, as well as the need for additives such as probiotics and antibiotics [3,4]. Studies have shown that some traditional Chinese herbs or extracts, such as *Allium sativum* [5], *Cissus quadrangularis* [6], *Mentha piperita* [7], *Polygonum chinense* [8], can improve the growth performance of Asian seabass. Traditional Chinese herbs, which are rich in nutrients and medicinal components, are characterized by their abundance, multi-targeting properties, diverse pharmacological effects, low toxicity, and minimal residues, making them ideal green additives for aquaculture feed [9–11]. These additives promote growth, enhance immunity, reduce costs, and improve profitability [12].

Licorice, a traditional medicinal herb, contains various components, including triterpene saponins, flavonoids, alkaloids, polysaccharides, and amino acids [13–17]. Its pharmacological actions mainly involve antitumor, anti-arrhythmic, antispasmodic, and antitussive effects, mediated by flavonoids; anti-inflammatory, antiviral, antitumor, and detoxifying effects, mediated by glycyrrhizic acid; and immunomodulatory, antiviral, and antitumor effects, mediated by polysaccharides [18–25]. Recently, licorice (*Glycyrrhiza uralensis)* has been studied as a feed additive in aquaculture and has been shown to significantly improve stress resistance, pathogen resistance, survival rate, and antioxidant capacity in fish [12,13]. For instance, adding fermented licorice to the feed of orange-spotted groupers (*Epinephelus coioides*) can reduce liver tissue damage and enhance antioxidative capacity, thereby increasing the survival rates under nitrite stress conditions [26]. In addition, licorice significantly improved the stress resistance of goldfish (*Carassius auratus*) and enhanced the resistance to *Aeromonas hydrophila* in goldfish and Chinese soft-shelled turtles (*Pelodiscus sinensis*) [20,27]. In terms of growth, licorice has a certain promotional effect on freshwater species such as tilapia, sturgeon, grass carp, and koi carp [28–31]. Previous studies have found that licorice can promote the growth of Asian seabass, but no further studies have been conducted [32]. In particular, whether licorice can really promote the expression of growth-related genes is still unexplored.

In this study, juvenile Asian seabass were fed with feed additives containing licorice, and then their growth performance was measured. The qrt-PCR method was used to study the effects on the level of growth-related genes and muscle growth-related genes. The present research aimed to reveal the mechanisms by which licorice influences the growth of Asian seabass at the molecular level, providing a scientific basis for its application in the aquaculture production of Asian seabass and acting as an important reference for the use of traditional Chinese herbal resources.

2. Materials and Methods

2.1. Feed Formulation and Experimental Design

Fish and soybean meals were the main protein sources used. Fish oil and wheat flour were the main fat and carbohydrate sources, respectively. A single-factor concentration gradient method was employed to design four levels of licorice feed with mass fractions of 0%, 1%, 3%, and 5%. The basic feed without licorice served as a control group. According to the nutritional requirements of Asian seabass [33], the experimental feeds were formulated to have a protein level of 41% and a lipid level of 17%, with consistency maintained across all experimental groups. The raw materials were ground and sieved through a 40-mesh sieve, mixed thoroughly using a mixer, pelletized using a small pellet machine (pellet diameter 2.0 mm), and stored at −20 °C for later use. The feed ingredients and nutritional components are listed in Table 1.

2.2. Experimental Method

Juvenile Asian seabass were bred by the Sanya Tropical Fisheries Research Institute Lingshui Experimental Center (Lingshui, Hainan, China). The fish weighed 13.93 ± 0.87 g and had a length of 8.78 ± 0.39 cm. Three hundred and sixty healthy, active, and responsive Asian seabass with smooth skin, no injuries, and good feeding behavior were randomly classified into four experimental groups, with three replicates per gradient and 30 fish per replicate. The feeding trial was conducted in flow-through seawater culture tanks (800 L) at the Sanya Tropical Fisheries Research Institute Lingshui Experimental Center (Lingshui, Hainan, China). During the experiment, the feeding behavior and mortality of the fish, as well as the water quality were monitored and recorded. The key water quality parameters were 26–29 °C, pH 7.3–7.8, and nitrite < 0.02 mg/L. Feeding was performed at 9:00 AM and 3:00 PM every day, using a satiation feeding method until the fish stopped feeding.

Approximately 1 h after feeding, the tanks were siphoned to remove feces and prevent water pollution. The water was changed at 4:30 PM, with two-thirds of the water being replaced. The experiment lasted for 56 d, after which the fish were weighed for calculated growth performance.

Table 1. Feed formula and the list of ingredients.

Ingredients	Diets			
	Control Group (0% *G.*)	**Test Group (1% *G.*)**	**Test Group (3% *G.*)**	**Test Group (5% *G.*)**
Fish meal (Fm)	50	50	50	50
Wheat flour (Wf)	23	22	20	18
Soybean meal (Sm)	12.9	12.9	12.9	12.9
Vitamin premix (Vp) (1)	0.5	0.5	0.5	0.5
Mineral premix (Mp) (2)	0.5	0.5	0.5	0.5
Fish oil (Fo)	13	13	13	13
Glycyrrhiza meal (*G.* m)	0	1	3	5
Choline chloride (Cc)	0.1	0.1	0.1	0.1
Dry ingredients (%)				
Crude protein (Cp)	41.44	41.31	41.06	40.81
Crude lipid (Cl)	17.53	17.51	17.46	17.41
Crude ash (Ca)	9.26	9.22	9.13	9.05
Total energy (Te)	20.28	20.12	19.79	19.46

Notes: (1) Vp: VA 900,000 IU, VB_1 320 mg, VB_2 1090 mg, VB_5 2000 mg, VB_6 500 mg, VB_{12} 116 mg, VC 5000 mg, VD 250,000 IU, VE 50 IU, VK_3 60 IU, niacin 40 mg, folic acid 5 mg, phaseomannite 150 mg, calcium pantothenate 20 mg, biotin 0.2 mg. (2) Mp: $MgSO_4 \cdot 7H_2O$ 3.0 $g \cdot 100\ g^{-1}$, KCl 0.7 $g \cdot 100\ g^{-1}$, KI 0.015 $g \cdot 100\ g^{-1}$, $ZnSO_4 \cdot 7H_2O$ 0.14 $g \cdot 100\ g^{-1}$, $MnSO_4 \cdot 4H_2O$ 0.03 $g \cdot 100\ g^{-1}$, $CuCl_2$ 0.05 $g \cdot 100\ g^{-1}$, $CoCl \cdot 6H_2O$ 0.005 $g \cdot 100\ g^{-1}$, $FeSO_4 \cdot 7H_2O$ 0.15 $g \cdot 100\ g^{-1}$, $KH_2PO_4 \cdot H_2O$ 45.0 $g \cdot 100\ g^{-1}$, $CaCl_2$ 28.0 $g \cdot 100\ g^{-1}$. The dietary energy was calculated as protein (23.64 $MJ \cdot kg^{-1}$), carbohydrate (17.15 $MJ \cdot kg^{-1}$), and lipid (39.54 $MJ \cdot kg^{-1}$). Source of materials: fish meal (Changsheng fishmeal factory, Cangzhou, China), wheat flour (Shandong developed face industry Co., Ltd., Dezhou, China), soybean meal (China Textile grain and oil Co., Ltd., Rizhao, China), vitamin premix (Henan Fangmu Shanze biological technology Co., Ltd., Nanyang, China), mineral premix (Henan Fangmu Shanze biological technology Co., Ltd., Nanyang, China), fish oil (Rongcheng City sea source fish oil aquatic products Co., Ltd., Rongcheng, China), glycyrrhiza meal (Longzhilin Medicine Store, Lingshui, China), Choline chloride (Taian Havay Group Co., Ltd., Tai'an, China).

2.3. *Experiment Sampling*

At the end of the feeding trial, three fish were sampled from each tank (i.e., nine fish per treatment) and placed in seawater containing 7 mg/L eugenol (provided by Changshu Shangchi Dental Materials Co., Ltd., Changshu, China) for anesthesia. Once anesthetized, the fish were quickly dissected to collect the brain tissue for *GHRH* and *GH*; the liver tissue for *GHR*, *IGF1*, and *IGF2*; and the muscle tissue for *IGF2R*, myostatin 1 (*MSTN1*), and Myosta2tin 2 (*MSTN2*). The tissues were then placed in cryogenic vials, flash-frozen, and stored at −80 °C.

2.4. *RNA Extraction*

The stored tissues were ground, and the total RNA was extracted using the previous method [34]. The concentration of the extracted RNA was detected using an ND 5000 micro-volume spectrophotometer (Beijing Baitake Biotechnology Co., Ltd., Beijing, China) at 260 and 280 nm to analyze the RNA integrity and purity.

2.5. *qrtPCR Experiment*

The extracted RNA was used for cDNA synthesis based on the PrimeScript™ Mix (Takara Bio Engineering Company) protocol and was then stored at −20 °C. The reaction mix for reverse transcription included 2 μL RT Master Mix. The conditions for the reverse transcription reaction were 37 °C for half an hour, followed by 85 °C for 5 s. The growth-related gene sequences of Asian seabass, including *GHRH*, *GH*, *GHR*, *IGF1*, *IGF2*, *IGF2R*, *MSTN1*, and *MSTN2* were obtained from the NCBI database, and primers were designed with a reference gene as the control (Table 2). The qPCR was performed on an rt-PCR device (Langji Scientific Company). The 20 μL reaction mix included 10 μL 2 × Real PreMix, 0.6 μL 10 μM primers, and 2 μL diluted cDNA. The PCR program was as follows: 95 °C for 15 min; 95 °C for 10 s; 58 °C for 20 s; 72 °C for 30 s; 40 cycles. A melt curve was applied to ensure the specificity of the products and the absence of primer–dimer formation. A no-DNA template control was used to confirm the absence of contamination during PCR.

Table 2. Primer of growth-related genes in barramundi were used in qPCR.

Gene Classification	Gene	Sample	Sequence (5′-3′)	Amplicon Size (bp)	Accession No.
Growth-related gene	*GHRH*	brain	F: GCGTGTTGTTGCACAGGCC R: CTACAGGCCGGTGTTGTTTA	121	XM018681526
	GH		F: AGGTGTTGTTGACAGGCAC R: AACTCCCAGGTGTTGTTCAA	86	X59378
	GHR	liver	F: AAGGTGTTGTTAACAGGCAGC R: GCACGTGTTGTTGACAGGCGC	206	XM_018702498
	IGF1		F: TGACAGGCGGTGTGTTGTTCT R: TGGTGTTGTTTACTAACCT	144	EU136176
	IGF2		F: AGACAGGCAAGTGTTGTTGTG R: GAAGATAACCTGCTCCTGTG	131	XM_018664155
	IGF2R	muscle	F: AGCTGGAAACCCCGAATT R: GAGCGAGACAGGCTGGATA	150	XM_018687313.1
	MSTN1		F: AACTGCGAATGAAAGAAGCTC R: CTTGGACGATGGACTCAGGT	204	XM_018696695
	MSTN2		F: GTCTGTTCAGCCTCAGTCCA R: CGGGTGTTGTTTCCCTCTTT R: GACGTCCAATGGGCTTTCT R: CAAACAGGGTGATGGGGTA	145	XM_018661271
	β-actin		F: AACCAAACGCCCAACAACT R: ATAACTGAAGCCATGCCAATG	112	XM_018667666

Notes: Hormone–releasing hormone (*GHRH*), growth hormone (*GH*), growth hormone receptor (*GHR*), insulin-like growth factor 1 (*IGF1*), insulin-like growth factor 2 (*IGF2*), *IGF2* receptor (*IGF2R*), myostatin 1 (*MSTN1*), and myostatin 2 (*MSTN2*). The PCR efficiency of the primers listed in the table has been verified to be 90–110%.

2.6. Calculation and Statistical Analysis

The calculation formulas of survival rate (SR), weight gain (WG), feed intake (FI), and specific growth rate (SGR) are as follows:

SR (%) = 100 × number of fish at the end of the test/number of fish at the beginning of the test:

$$\text{WG (g fish}^{-1}) = W_t - W_0$$

$$\text{FI (g fish}^{-1}\text{ d}^{-1}) = \text{(feed consumed per tank/fish)}/t$$

$$\text{SGR (\%/d)} = 100 \times [(\text{Ln}W_t - \text{Ln}W_0)]/t$$

W_t—final average weight (g); W_0—initial average weight (g); t—the number of days the feeding text lasted.

The relative levels of gene mRNA were detected based on the $2^{-\Delta\Delta Ct}$ approach, with the *β-actin* as the reference gene for normalization. Experimental data were presented as mean ± SD. The SPSS 19.0 tool was applied for statistical treatment data. The Shapiro–Wilk test was employed to check the data for normal distribution, and the Levene test was used to assess the homogeneity of variance. Comparisons among the groups were conducted using the ANOVA (LSD test), with a statistical level of 0.05.

3. Results

3.1. Effects of Licorice in Feed on the Growth Performance of Asian Seabass

After adding different levels of licorice to the feed and feeding for 56 days, the survival rate, weight gain rate, body-length growth rate, and specific growth rate of the seabass are shown in Figure 1. The survival rate of the 3% and 5% licorice-supplemented diet group was significantly higher than that of the control group ($p < 0.05$) (Figure 1A). The survival rate reached the maximum in the 3% licorice-supplemented diet group, and the value was (98.89 ± 1.93) %. The WG was significantly increased in all the licorice treatment groups compared to the control group ($p < 0.05$) (Figure 1B). The FI was not significant among the groups ($p > 0.05$) (Figure 1C). The SGR of the 5% licorice-supplemented diet group was significantly higher than that of the control group ($p < 0.05$) (Figure 1D).

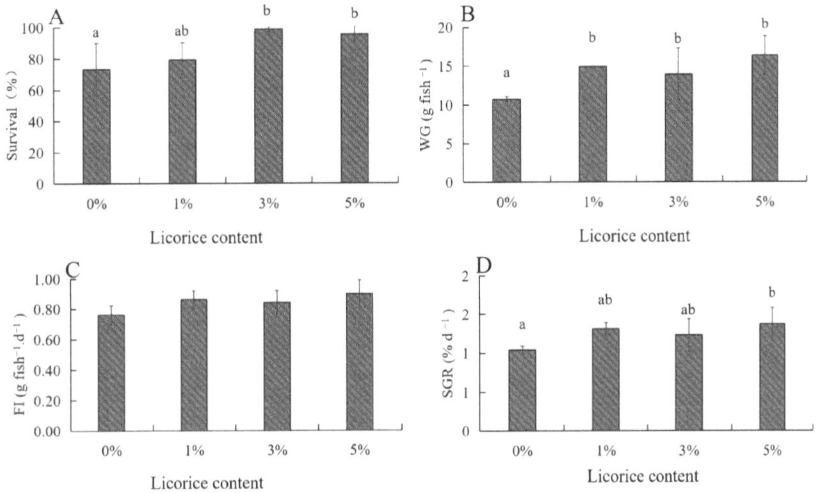

Figure 1. Effects of licorice on survival rate and growth of Asian seabass. (**A**) survival, (**B**) weight gain (WG), (**C**) feed intake (FI), (**D**) specific growth rate (SGR); means within rows with the same superscript are not significantly different ($p > 0.05$), while the different letters mean significant differences ($p < 0.05$).

3.2. Effects of Licorice on the Expression of Growth-Related Genes in the Asian Seabass Liver

As shown in Figure 2, adding licorice to the feed significantly affected the level of the *GHR* and IGF genes in the livers of Asian seabass. According to the comparison result, the relative expression level of the *GHR* gene in the liver increased significantly after the addition of licorice ($p < 0.05$), and the influence of licorice on the *GHR* gene's relative expression level varied with its concentration, with the highest increase of 362% observed in the 5% group, followed by the 1% group, and the lowest in the 3% group. The relative expression level of the *IGF1* gene in the livers of the experimental groups significantly increased with the increase in licorice content in the feed ($p < 0.05$), with the 5% group showing the max level of the *IGF1* gene, which was 751% of the control group's level. The *IGF2* gene level showed an initial increase, followed by a decrease with increasing licorice content, with significant differences between the two groups ($p < 0.05$). The 1% group had the highest level, followed by the 3% group, with both levels significantly increased ($p < 0.05$). The *IGF2* level in the 5% group was significantly lower ($p < 0.05$).

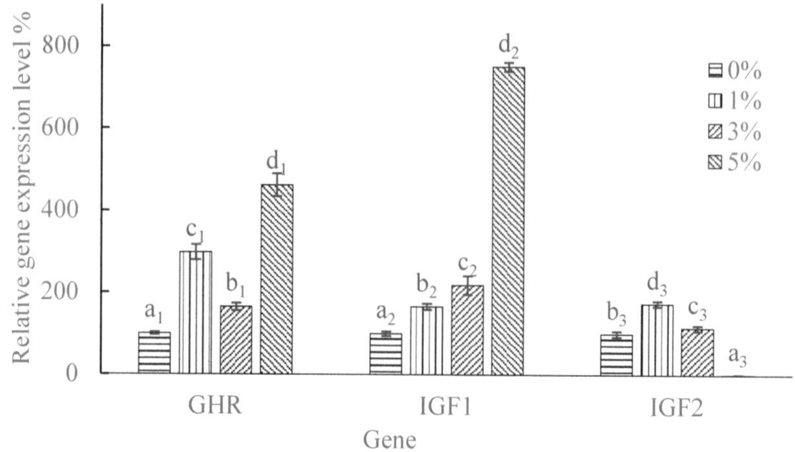

Figure 2. The relative effect of licorice on the level of growth-related genes in the liver tissue. Note: Different letters indicate significant differences. Subscripts 1, 2, and 3 represent *GHR*, *IGF1*, and *IGF2*, respectively.

3.3. Effects of Licorice on Level of Muscle Growth-Related Genes in Asian Seabass

As indicated in Figure 3, licorice significantly affected the level of the muscle growth-associated genes *IGF2R*, *MSTN1*, and *MSTN2* in Asian seabass. With an increase in the licorice content in the feed, the expression level of the *IGF2R* gene initially decreased and then increased; however, the expression levels of *IGF2R* in all experimental groups were significantly lower than those in the other group ($p < 0.05$). Similar to the results of the *IGF2R* gene, the *MSTN1* gene level also initially decreased and then increased, with the lowest level observed in the 1% group. The level increased in the 3% group and was significantly higher compared to the other groups ($p < 0.05$) and continued to increase in the 5% group, reaching 333% of that in the control group. The addition of licorice to the feed caused a significant reduction in the *MSTN2* gene level at all levels ($p < 0.05$).

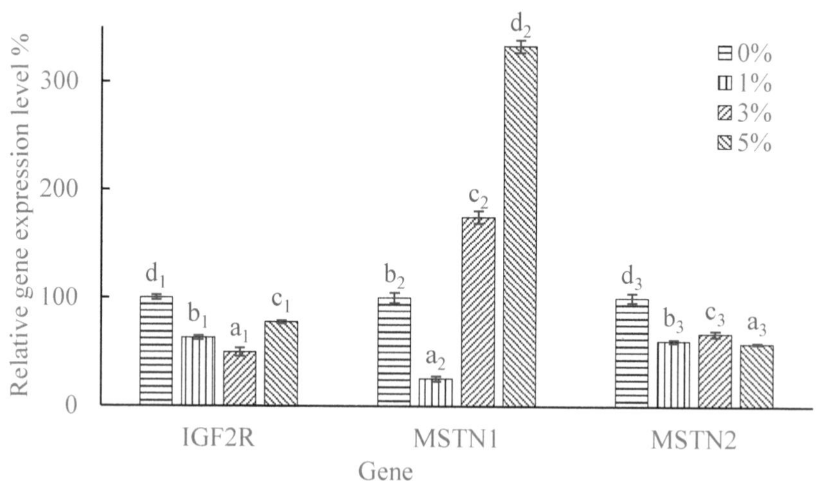

Figure 3. The relative effect of *Glycyrrhiza uralensis* on the level of growth-associated genes in the muscle tissue. Note: Different letters indicate significant differences. Subscripts 1, 2, and 3 represent *IGF2R*, *MSTN1*, and *MSTN2*, respectively.

3.4. Effects of Licorice on Level of Growth-Related Genes in Brain Tissue of Asian Seabass

The influence of licorice on the level of the growth-related genes *GH* and *GHRH* in the brain tissue of Asian seabass is illustrated in Figure 4. After adding licorice, the level of *GH* decreased significantly ($p < 0.05$). The level of the *GHRH* gene exhibited a trend of initially decreasing and then increasing with increasing licorice content. The lowest level appeared in the 3% licorice group, followed by the 5% licorice group, with the highest level observed in the 1% licorice group. However, the expression levels of *GHRH* in all experimental groups were significantly lower than those in the control group ($p < 0.05$).

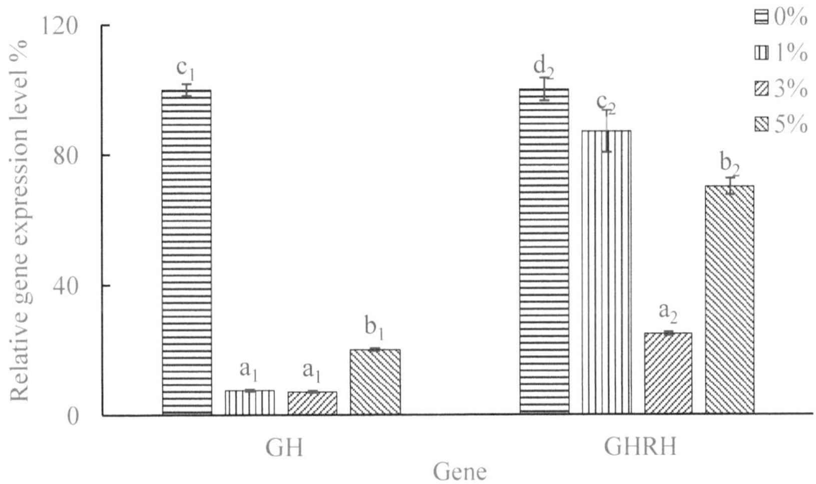

Figure 4. The relative effect of *Glycyrrhiza uralensis* on the level of growth-associated genes in the brain tissue. Note: The letters indicate significant differences. Subscripts 1 and 2 represent *GH*, and *GHRH*, respectively.

4. Discussion

We found that adding a proper concentration of licorice in the feed significantly promoted the expression of genes related to *GHR*, *IGF1*, and *IGF2* in the liver of Asian seabass. The growth of teleost fish is primarily adjusted by the *GH*/IGF system and other endocrine factors. *GH* binds to *GHR* on the surface of target organs, stimulating the liver to secrete IGF [35,36]. IGF binds to IGF receptors in target tissues, initiating a series of cellular processes related to growth, such as cell proliferation and differentiation, leading to overall growth [37]. In the present research, it was found that adding licorice to Asian seabass feed significantly upregulated the level of *GHR*. The downstream genes of the *GH*/IGF axis in the liver, *IGF1*, and *IGF2* showed significant changes, confirming that *GH* can exert biological effects through *IGF1* and *IGF2* after binding to *GHR* [38]. When the addition of licorice did not exceed 3%, the levels of *IGF1* and *IGF2* significantly increased, suggesting that a 3% licorice addition could significantly promote processes related to cell growth. The max level of *IGF1* appeared in the 5% licorice group but the significant downregulation of the *IGF2* expression indicates that high levels of licorice in Asian seabass feed might affect the functions related to *IGF2*.

In Nile tilapia (*Oreochromis niloticus*), *GH* is expressed only in the pituitary gland, whereas *GHR* is expressed in various tissues and organs, including muscle, hypothalamus, and thymus, with the highest expression in the liver, indicating that *GH* has multiple physiological functions [39]. In teleosts, *MSTN1* primarily inhibits muscle hyperplasia but not hypertrophy, which is mainly achieved through the downregulation of *MSTN2* expression [40]. This research showed the variable influence of licorice on *MSTN1* expression, indicating the stages of rapid muscle hyperplasia in Asian seabass [41]. The lower expression of *MSTN1* and *MSTN2* with 1% licorice suggests reduced inhibition of muscle hyperplasia and promotion of muscle hypertrophy, while the increased *MSTN1* expression with 3% and 5% licorice indicates an enhanced inhibition of muscle hyperplasia and the promotion of muscle hypertrophy.

This study found that after adding licorice, the level of *GH* in the brain samples of Asian seabass was downregulated, and the levels of *GHR, IGF1,* and *IGF2* in the liver tissue were upregulated, possibly due to a delayed regulatory effect of *GH* on the growth rate and its negative feedback regulation with *IGF1* [42]. The hypothalamus-secreted *GHRH* physiologically regulates the generation and release of *GH* in the pituitary gland, which is an important hormone for regulating fish growth, development, reproduction, and immunity [43]. In the present research, the *GHRH* and *GH* expression was significantly downregulated and did not show a linear relationship, which may be related to the multi-factorial influence on *GH* secretion that maintains a dynamic balance between promoting and inhibiting factors [44].

The beneficial effects of licorice may be attributed to glycyrice polysaccharides (GPS). Dietary supplementation of GPS has been found to enhance growth performance, body size, and the relative expression of the growth-related gene IGF-1 in broilers [45]. Furthermore, it can improve serum and intestinal immune status, promote the expression of immune-related genes in the spleen, and enhance broiler immunity. The optimal supplemental concentration is 600 mg/kg [46,47]. In weaned piglets, GPS supplementation significantly promotes the mRNA expression levels of the IGF-1 gene in the liver, as well as the IGF-1 and IGF-2 genes in the dorsal longus muscle ($p < 0.05$) [48]. Additionally, the dietary addition of GPS improves growth performance, reduces diarrhea rate, enhances humoral immunity, promotes the related growth gene expression, and even exhibits a certain level of resistance against PRRSV infection in piglets. Supplementation with 1000 mg/kg GPS alleviates stress response, reduces diarrhea rate, and improves growth performance by enhancing the intestinal mucosal barrier effect, immune function, and intestinal microflora structure in weaned piglets [49]. Moreover, GPS exhibits inhibitory effects on the TLR4/MyD88/NF-κB signaling pathway, thereby reducing the excessive expression of immune and inflammatory, apoptosis, and tight junction protein genes induced by LPS in IPEC-J2 cells [50]. Additionally, they enhance the cell's antioxidant capacity and decrease ROS accumulation and the apoptosis rate, ultimately alleviating the inflammatory damage caused by LPS in the IPEC-J2 cells. It also enhances antioxidant capacity, reduces ROS accumulation and cell apoptosis rates, and mitigates inflammatory damage induced by LPS in IPEC-J2 cells. GPS serves as an effective immune enhancer to enhance the integrity of the intestinal barrier [51]. The addition of licorice to animal feed can effectively stimulate the growth of terrestrial animals such as chickens and pigs, as well as that of aquatic species such as Asian seabass. Moreover, it exhibits immune-enhancing properties, thereby positioning licorice as a promising feed additive with excellent application prospects.

5. Conclusions

In conclusion, the inclusion of licorice in the diet of Asian seabass significantly enhances growth performance and survival rates. Specifically, diets supplemented with 3% and 5% licorice showed notable improvements in weight gain and survival compared to the control group. Licorice supplementation positively influenced the expression of growth-related genes, particularly increasing the *GHR* and *IGF1* levels in the liver. However, higher levels of licorice (5%) resulted in the downregulation of *IGF2* expression and complex effects on the *IGF2R* and *MSTN1* levels in the muscle tissue. Additionally, licorice inhibited the expression of *GH, GHRH,* and *MSTN2*. These findings suggest that incorporating an optimal proportion of licorice into the diet can effectively enhance the growth and health of Asian seabass in aquaculture.

Author Contributions: Conceptualization, Z.M. and R.Y.; methodology, W.Z. and Y.W.; software, J.H. and M.L.; validation, J.H.; formal analysis, Y.W. and W.Z.; investigation, M.L. and S.Z.; resources, S.Z. and Z.M.; data curation, W.Z. and Y.W.; writing—original draft preparation, R.Y. and W.Z.; writing—review and editing, Z.F. and Z.M.; visualization, S.Z. and R.Y.; supervision, Z.M.; project administration, R.Y. and J.H.; funding acquisition, Z.M. All authors have read and agreed to the published version of the manuscript.

Funding: This work was supported by the Central Public-Interest Scientific Institution Basal Research Fund, CAFS (grant number 2023TD58, 2023YJ03); the Central Public-Interest Scientific Institution Basal Research Fund, South China Sea Fisheries Research Institute, CAFS (grant number 2021SD09); the Hainan Provincial Natural Science Foundation of China (grant number 321QN094); the Hainan Province Science and Technology Special Fund (grant number ZDYF2022XDNY349); the Guangxi Science and technology planning project (grant number GUIKE AD21238026); the Hainan Province Natural Science Foundation enterprise talent project (grant number 324QY579).

Institutional Review Board Statement: This animal study was reviewed and approved by the Animal Care and Use Committee of South China Sea Fisheries Research Institute, the Chinese Academy of Fishery Sciences. The ethical code is 2020TD15, approved on 10 March 2020.

Informed Consent Statement: Not applicable.

Data Availability Statement: The data that support the findings of this study are available from the corresponding author upon reasonable request.

Conflicts of Interest: The authors declare no conflicts of interest.

References

1. Nandakumar, S.; Ambasankar, K.; Ali, S.S.R.; Syamadayal, J.; Vasagam, K. Replacement of fish meal with corn gluten meal in feeds for Asian seabass (*Lates calcarifer*). *Aquac. Int.* **2017**, *25*, 1495–1505. [CrossRef]
2. Zhao, W.; Hu, J.; Ma, Z.; Yu, G.; Yang, R.; Wang, L. Path analysis and growth curve fitting of morphological traits to body weight of juvenile *Lates calcarifer*. *J. South. Agric.* **2017**, *48*, 1700–1707.
3. Li, Z.; Yuan, F.; Lin, H.; Lu, X.; Yang, Q. Efects ofBacilhs lichen ifom is suppement on grwoth perform anceand digestive enzymes activities of the sea bass, *Lates calcarifer*. *J. Oceanogr. Taiwan Strait* **2011**, *30*, 43–48.
4. Lin, H.; Yuan, F.; Li, Z.; Lu, X. Effects of dietary photosyn thetic bacteria PS2 on growth performance, digestive enzymes and nonspecific immune enzymes. *South China Fish. Sci.* **2013**, *6*, 25–29.
5. Al-Khamees, A.M.A.S.M.E.-B.S. Influence of Dietary Garlic (Allium sativum) and/orAscorbic Acid on Performance, Feed Utilization, Body Composition and Hemato-Biochemical Parameters of Juvenile Asian Sea Bass (*Lates calcarifer*). *Animals* **2020**, *10*, 2396.
6. Chinnasamy, C.D.A. Dietary administration of natural immunostimulants on growth performance, haematological, biochemical parameters and disease resistance of Asian Sea bass *Lates calcarifer* (Bloch, 1790). *Aquac. Res.* **2017**, *48*, 1131–1145.
7. Talpur, A.D. *Mentha piperita* (Peppermint) as feed additive enhanced growth performance, survival, immune response and disease resistance of Asian seabass, *Lates calcarifer* (Bloch) against *Vibrio harveyi* infection. *Aquaculture* **2014**, *420–421*, 71–78. [CrossRef]
8. Razak, R.A.; Shariff, M.; Md Yusoff, F.; Safinar, I. Enhanced growth performance, haemato-biochemical and immune parameters of asian seabass, *lates calcarifer* (bloch, 1790) fed dietary supplementation with polygonum chinense. *Asian Fish. Sci.* **2019**, *32*, 19–28. [CrossRef]
9. Lu, X.; Lin, H.; Li, Z. Effect of dietary Chinese herbalm edicines on hematological and blood biochemical indices incultured seabass *Lates calarifer*. *J. Dalian Fish. Univ.* **2009**, *24*, 279–282.
10. Zhou, H.; Zhou, H.Q. Effects of Chinese herbal medicine on growth and immune function of *Litopenaeus vannamei*. *J. Hebei Fish.* **2008**, *9*, 44–48.
11. Shi, H.; Lou, B.; Hu, Z. Effects of Chinese herbal feed additives on growth and non-specific immunity of juvenile Bream (*Pagrosomus major*). *Prog. Mod. Biomed.* **2007**, *7*, 548–556.
12. Jin, E.; Xiong, X.; Su, Y. Effects of compound Chinese herbal preparation on immune function, antioxidant function and proliferative and apoptotic gene expression of bursa of Fabricius in green foot chicken. *J. Northeast Agric. Univ.* **2019**, *50*, 66–77.
13. Wang, Q.e.; Ren, H.; Zeng, X. Research and Utilization Statue of Licorice. *Chin. Agric. Sci. Bull.* **2011**, *27*, 290–295.
14. Zhao, S.; Nong, Z.; Zhong, Z. Experimental study on antitumor effect of the total flavonoids from radix glyeyrrhizae and its mechanisms. *Guangxi Med. J.* **2006**, *28*, 1496–1499.
15. Zhang, F.; Shen, Y. Advances in studies on cardioprotection of glycyrrhizie acid compound and flavonoids. *Drugs Clinie* **2012**, *27*, 429–434.
16. Wang, F.; Su, Y. Pharmacological action and clinical application of licorice. *Lishizhen Med. Mater. Medica Res.* **2002**, *13*, 303–304.
17. Zhu, Y.; Su, Y.; Chen, J. Inhibition of flavone from *Glycyrrhiza uralensis* on capsaicin-induced cough reflex in guinea pig. *Chin. Tradit. HerbalDrugs* **2006**, *37*, 1048–1051.

18. Zhang, F.; Shen, Y. Advances in studies on glyeyrrhizic acid and its derivatives in anti-inflammationand anti-allergy. *Drugs Climie* **2011**, *26*, 359–364.
19. Huang, Q.; Ma, Z. Pharmacological research progress of glyeyrrhizie acid. *Drug Eval. Res.* **2011**, *34*, 384–387.
20. Liu, Q. Summary of chemical composition and pharmacological action of licorice. *Chin. Med. Mod. Distance Educ. China* **2011**, *9*, 84.
21. Xie, H.; Du, X.; Yu, D. Research Progress in Pharmacological Effects of *Glycyrrhiza Polysaccharide*. *Chin. J. Vet. Drug* **2011**, *30*, 285.
22. Hu, J.; Ao, M.; Cui, Y. Anti-tumor Effect of Polysaccharide from Glycyrrhizaand Its Influence on immunological Function. *Nat. Prod. Res. Dev.* **2008**, *20*, 911–913, 938.
23. Zhu, L.; Liu, X.; Yin, S.; Wang, Q.; Ni, C.; Yang, F. Effects of Chinese herbal medicine on growth performance and antioxidant indexes of tilapia. *China Feed* **2014**, *23*, 38–41.
24. Tan, J.; Deng, F.; Cao, Y.; Yao, Y. Effects of Chinese herbal compound in feed on growth, musclecomposition and immune-related enzyme activities of juvenile grass carp (*Ctenopharyngodon idellus*). *Guangdong Agric. Sci.* **2015**, *10*, 109–113.
25. Wang, B.; Fang, P.; Lin, X. Effect of Liquorice Extracts on the Resistance of *Carassius auratus* to Stress and Pathogen Infection. *Freshw. Fish.* **2007**, *37*, 3–6.
26. Weng, Q.; Li, Z.; Lu, K. Effect of different levels of fermented licorice on the blood index and antioxidant ability of grouper under nitrite stress. *Feed Res.* **2019**, *5*, 24–27.
27. Chen, C.R.; Chen, X.H.; Chen, C.F. Effect of oral glycyrrhizin on anti-*Aeromonas hydrophila* infection of juvenile soft-shelled turtle, *Trionyx sinensis*. *J. Huazhong Agric. Unversity* **2000**, *19*, 577–580.
28. Adineh, H.; Naderi, M.; Yousefi, M.; Hamidi, M.K.; Ahmadifar, E.; Hoseini, S.M. Dietary licorice (*Glycyrrhiza glabra*) improves growth, lipid metabolism, antioxidant and immune responses, and resistance to crowding stress in common carp, *Cyprinus carpio*. *Aquac. Nutri.* 2021; 27, 417–426. [CrossRef]
29. Meng, X.; You, F.; Cao, H. Effects of dietary licorice (*Glycyrrhiza uralensis*) supplementation on growth performance, muscle quality, and immunity in the common carp (*Cyprinus carpio haematopterus*). *Aquac. Rep.* **2022**, *27*, 101331. [CrossRef]
30. Abdel-Tawwab, M.; El-Araby, D.A. Immune and antioxidative effects of dietary licorice (*Glycyrrhiza glabra* L.) on performance of Nile tilapia, *Oreochromis niloticus* (L.) and it susceptibility to *Aeromonas hydrophila* infection. *Aquaculture* **2021**, *530*, 735828. [CrossRef]
31. Zhang, H.; Xia, L.; Peng, X.; Zhao, M.; Lan, Y.; Tang, X. Study on growth and fatty liver repair of *Acipenser sinensis* by licorice and salvia miltiorrhiza. *Chin. Fish. Qual. Stand.* **2014**, *4*, 46–53.
32. Yang, R.; Han, M.; Fu, Z.; Wang, Y.; Zhao, W.; Yu, G.; Ma, Z. Immune Responses of Asian Seabass *Lates calcariferto* Dietary *Glycyrrhiza uralensis*. *Animals* **2020**, *10*, 1629. [CrossRef] [PubMed]
33. Ma, Z.; Yu, G.; Meng, X. *Culture Biology and Processing of Barramundi (Lates calcarifer)*; China Agriculture Press: Beijing, China, 2019.
34. Fu, Z.; Yang, R.; Chen, X. Dietary non-protein energy source regulates antioxidant status and immune response of barramundi (*Lates calcarifer*). *Fish Shellfish Immunol.* **2019**, *95*, 697–704. [CrossRef]
35. Picha, M.E.; Turano, M.J.; Beckman, B.R.; Borski, R.J. Endocrine Biomarkers of Growth and Applications to Aquaculture: A Minireview of Growth Hormone, Insulin-like Growth Factor (igf)-i, and Igf-binding Proteins as Potential Growth Indicators in Fish. *N. Am. J. Aquac.* **2008**, *70*, 196–211. [CrossRef]
36. Rolland, M.; Dalsgaard, J.; Holm, J.; Gómez-Requeni, P.; Skov, P.V. Dietary Methionine Level Affects Growth Performance and Hepatic Gene Expression of Gh–igf System and Protein Turnover Regulators in Rainbow Trout (*oncorhynchus Mykiss*) Fed Plant Protein-based Diets. *Comp. Biochem. Physiol. Part B Biochem. Mol. Biol.* **2015**, *181*, 33–41. [CrossRef]
37. Wan, Z. Studies on Juvenile Development and Digestive Physiology of *Takifugu rufifin*. Master's Thesis, Ocean University of China, Qingdao, China, 2005.
38. Figueiredo, M.; Lanes, C.; Almeida, D. The Effect of Gh Overexpression on Ghr and Igf-i Gene Regulation in Different Genotypes of Gh-transgenic Zebrafish. *Comp. Biochem. Physiol. Part D Genom. Proteom.* **2007**, *2*, 228–233.
39. Ma, X.; Zhang, Y.; Huang, W. cDNAs cloning of growth hormone, growth hormone receptor and the different expression between male and female Nile tilapia (*Oreochromis niloticus*). *Curr. Zool.* **2006**, *52*, 924–933.
40. Santis, C.; Gomes, G.; Jerry, D. Abundance of Myostatin Gene Transcripts and Their Correlation with Muscle Hypertrophy During the Development of Barramundi, *Lates Calcarifer*. *Comp. Biochem. Physiol. Part B Biochem. Mol. Biol.* **2012**, *163*, 101–107. [CrossRef]
41. Johnston, I. Muscle Development and Growth: Potential Implications for Flesh Quality in Fish. *Aquaculture* **1999**, *177*, 99–115. [CrossRef]
42. Gabillard, J.; Montserrat, B. Coordinated Regulation of the Gh/igf System Genes During Refeeding in Rainbow Trout (*oncorhynchus Mykiss*). *J. Endocrinol.* **2006**, *191*, 15–24. [CrossRef]
43. Nam, B.; Moon, J.; Kim, Y. Molecular and Functional Analyses of Growth Hormone-releasing Hormone (ghrh) from Olive Flounder (*paralichthys Olivaceus*). *Comp. Biochem. Physiol. Part B Biochem. Mol. Biol.* **2011**, *159*, 84–91. [CrossRef]
44. Canosa, L.; Chang, J.; Peter, R. Neuroendocrine Control of Growth Hormone in Fish. *Gen. Comp. Endocrinol.* **2007**, *151*, 1–26. [CrossRef]
45. Danhui, W.; Youbing, Y.; Ying, L. Effect of *Glycyrrhiza polysaccharide* in diets on growth performance, body size index and IGF-1 gene relative expression in broilers. *Mod. J. Anim. Husb. Vet. Med.* **2021**, *12*, 28–31.
46. Wenbin, C.; Xueying, W.; Cai, Z. Effect of *Glycyrrhiza polysaccharide* on growth performance and immunologic function of broilers. *Feed Res.* **2022**, *18*, 34–40.

47. Chen, W. Effects of *Glycyrrhiza polysaccharides* on Growth Performance, Immune Function and Intestinal Health of Broilers. Master's Thesis, Henan University of Science and Technology, Zhengzhou, China, 2022.
48. Youbing, Y.; Xinyue, L.; Ran, L. Effects of *Glycyrrhiza polysaccharides* on Growth Performance and Growth Gene Expression in Weaned Piglets. *Jiangsu Agric. Sci.* **2023**, *13*, 182–189.
49. Lou, R. Effects of *Glycyrrhiza polysaccharides* on Growth Performance and Expression of PRRSV-Related Genes in Piglets. Master's Thesis, Henan University of Science and Technology, Zhengzhou, China, 2021.
50. Li, J. Effects of *glycyrrhiza polysaccharides* on Growth Performance and Intestinal Health of Weaned Piglets. Ph.D. Thesis, Yangzhou University, Yangzhou, China, 2022.
51. Jianfang, L.; Rudu, W.; Pengli, Z. Effects of *Glycyrrhiza polysaccharides* on growth performance, blood routine, serum biochemical indexes and antioxidant properties of weaned piglets. *Feed. Ind.* **2022**, *43*, 32–37.

Journal of
Marine Science and Engineering

Article

An Experimental Study on Estimating the Quantity of Fish in Cages Based on Image Sonar

Guohao Zhu [1,2], Mingyang Li [1], Jiazhen Hu [1,2], Luyu Xu [1], Jialong Sun [1,3], Dazhang Li [4], Chao Dong [5], Xiaohua Huang [2,6,*] and Yu Hu [2,6,*]

1 School of Geomatics and Marine Information, Jiangsu Ocean University, Lianyungang 222001, China
2 Key Laboratory of South China Sea Fishery Resources Exploitation & Utilization, Ministry of Agriculture and Rural Affairs, South China Sea Fisheries Research Institute, Chinese Academy of Fishery Science, Guangzhou 510300, China
3 Jiangsu Marine Resources Development Research Institute, Lianyungang 222005, China
4 Zhejiang Provincial-Subordinate Architectural Design Institute, Hangzhou 310007, China
5 Key Laboratory of Marine Environmental Survey Technology and Application, Ministry of Natural Resources, Guangzhou 510300, China
6 Tropical Fisheries Research and Development Center, South China Sea Fisheries Research Institute, Chinese Academy of Fishery Science, Sanya 572018, China
* Correspondence: huangxhua@scsfri.ac.cn (X.H.); huyu@scsfri.ac.cn (Y.H.); Tel.: +86-020-3406-6940 (Y.H.)

Abstract: To address the highly demanding assessment of the quantity of fish in cages, a method for estimating the fish quantity in cages based on image sonar is proposed. In this method, forward-looking image sonar is employed for continuous detection in cages, and the YOLO target detection model with attention mechanism as well as a BP neural network are combined to achieve a real-time automatic estimation of fish quantity in cages. A quantitative experiment was conducted in the South China Sea to render a database for training the YOLO model and neural network. The experimental results show that the average detection accuracy mAP50 of the improved YOLOv8 is 3.81% higher than that of the original algorithm. The accuracy of the neural network in fitting the fish quantity reaches 84.63%, which is 0.72% better than cubic polynomial fitting. In conclusion, the accurate assessment of the fish quantity in cages contributes to the scientific and intelligent management of aquaculture and the rational formulation of feeding and fishing plans.

Keywords: cage fish; forward-looking image sonar; target recognition; quantity estimation

Citation: Zhu, G.; Li, M.; Hu, J.; Xu, L.; Sun, J.; Li, D.; Dong, C.; Huang, X.; Hu, Y. An Experimental Study on Estimating the Quantity of Fish in Cages Based on Image Sonar. *J. Mar. Sci. Eng.* **2024**, *12*, 1047. https://doi.org/10.3390/jmse12071047

Academic Editor: Sergei Chernyi

Received: 30 May 2024
Revised: 18 June 2024
Accepted: 20 June 2024
Published: 21 June 2024

1. Introduction

As a major agricultural country in the world, the development of China's agricultural economy is related to the development of the national economy [1]. As an important branch of aquaculture, fishery farming has always been an important pillar of China's agricultural economy. With the development of society, science, and technology, the level of agricultural modernization has rapidly improved, and the intelligent development of fish farming has accelerated. The monitoring and regulation of the breeding environment and the decision making of feed feeding have gradually shifted from completely relying on manual diagnosis, decision making, and adjustment to the mechanization and precision of monitoring equipment, and then to the digitalization and intelligence of the system [2].

At present, fish farming in China varies between pond and cage cultures. Among these, cage culture exhibits the highest level of intensification, with a myriad of issues arising during fish farming [3]. Fish quantity monitoring, as an important part of cage aquaculture production management, is of profound significance mainly in the following three aspects: 1. Intelligent management of aquaculture, allowing aquaculture managers to adjust the feeding amount and make fishery harvesting plans according to the fish production; 2. Early warning of the safety of the fishnet and the breakage of the fishnet in the case of abnormal fish quantity, to repair it in time to reduce losses; 3. Facilitation of the assessment of the financial assets of the catch, rendering necessary technical conditions for achieving financial assets of fishery harvesting [4].

Given the above requirements, experts and scholars at home and abroad put forward solutions based on different monitoring methods. Baumgartner et al. [5] observed fish in artificial ponds and calculated the fish quantity and body length by software, concluding that sonar was effective in observing fish activities and obtaining quantitative information. Ding et al. [6] collected 59h underwater data by using ARIS sonar and completed the automatic processing of a large number of acoustic data through an image processing algorithm, including target extraction and counting. A remote cage monitoring system that combines light and sound with motor rotation scanning was jointly developed by the Massachusetts Institute of Technology and Woods Hole Oceanographic Institution, which can identify individual fish well to achieve safe monitoring of fishnet [7]. However, its high cost and the prolonged acoustic imaging time required by motor rotation detection (compared to the standard imaging time of 3 min) cause the repeated detection of swimming fish in cages, resulting in a large error in fish quantity estimation. Domestically, the Fishery Machinery and Instrument Research Institute of the Chinese Academy of Fishery Sciences developed a multi-angle cage monitor by optical means [8]. Because of the turbid sea water in most coastal areas of our country, except Hainan, the instrument had been limited by effectively observing a range of underwater targets and higher power consumption. Given the limitation of the above optical monitoring technology in the actual condition of cages, most of the domestic research has prioritized acoustic monitoring methods. The Shanghai Acoustics Laboratory of the Chinese Academy of Science put forward the acoustic warning tape method and the remote-operated vehicle patrol method, which were mainly used for monitoring the size of netting and fish but were less able to obtain quantity data. Xiamen University has successively developed acoustic monitoring systems based on the vertical detection method and single-beam transducer motor-rotating horizontal scanning method. The circular multi-beam scanning detection method had high requirements for the estimation of fish swimming speed, and either the underestimation or overestimation will lead to partial fish missed detection or repeated detection [9–11]. Yihan Feng et al. [12] introduced an automated method for estimating fish abundance in sonar images based on the modified MCNN (multi-column convolutional neural network), named FS-MCNN. They also proposed the multi-dilation rate fusion loss, which improved the accuracy and robustness of the model. This method improved the impact of low pixels in sonar images and blurry edges of target objects in sonar images.

The target recognition technique was indispensable for locating and counting fish in acoustic images. Since the R-CNN (Region with CNN Features) was put forward in 2014, the target detection method based on deep learning has become the main technique, instead of the traditional method [13]. Initially, the two-stage method was adopted for target detection based on deep learning, that is, the detection process was explicitly divided into two stages: candidate region selection and target region judgment, with a high detection accuracy but slow detection speed. Later, in 2016, the one-stage target detection method represented by YOLOv1 came into being. Instead of extracting candidate regions in advance, the method directly predicted the category probability and position of the output target object, which attracted more attention by greatly reducing the consumption of computing resources and improving the detection speed [14]. The YOLO series of target detection methods, along with the development of single-stage object detection, has been regarded as a typical representative of the one-stage method. Ye Zhaobing et al. [15] proposed the YOLOv3-SPP (Spatial Pyramid Pooling) underwater target detection algorithm to solve the problem of missed detection and false detection caused by unclear images and the complex underwater environment in underwater target detection. Chen Yuliang et al. [16] put forward a method for detecting and identifying underwater biological targets in shallow water based on the YOLOv3 network, aiming to overcome the low detection accuracy of underwater biological targets in a shallow sea caused by color distortion, rough image, local overexposure, and large size difference in underwater images.

In response to the deficiencies of the aforementioned detection methods, this paper proposes a method for estimating the quantity of fish in net cage farming based on forward-looking imaging sonar. This method utilizes forward-looking sonar to generate acoustic images of aquaculture net cages, employs a YOLOv8 neural network model with an added attention mechanism to identify fish targets, and utilizes a BP neural network to invert feature data to estimate the overall quantity of fish. Quantitative detection experiments were conducted in constructed fish cages, with multiple sets of experimental results showing that the average accuracy of fish quantity assessment reached 84.63%, thereby validating the feasibility of this method. By using this method, fish farmers can gain real-time insights into the quantity of fish inside net cages during the farming process, enabling scientific aquaculture management and reducing farming risks.

2. Materials and Methods

2.1. Overall Process

On the whole, the adopted method is divided into three steps: Firstly, the image sonar is fixed on one side of the cage and observed for more than 10 min, recording sonar data and exporting it to video. Secondly, the improved YOLOv8 model is used to detect all the frames of the current video, and there is only one detection category, namely fish, Thirdly, the number of fish shoals detected in each frame of the video is sorted from the largest to the smallest, and the actual quantity of fish in the cage is estimated by using the trained neural network model according to the top 20 fish quantity. In the second step, the YOLOv8 model needs to be trained with fish sonar image data, and the neural network in the third step is trained by the mapping relationship between the previous observation data and the actual quantity.

2.2. Introduction to Image Sonar

The ARIS1800 (Adaptive Resolution Imaging Sonar) sonar used in the present study was introduced by Sound Metrics in 2012. When forward-looking sonar performs detection, the transducer at its top emits ultrasonic waves in the forward direction, and subsequently, the objects illuminated by these waves reflect them, forming echo signals. The sonar receives these signals to generate acoustic images. Typically, dividing the detection beam horizontally into multiple smaller fan-shaped beams can enhance imaging precision, with the vertical angle of each beam group remaining unchanged. Table 1 below lists the specific parameters of ARIS-1800 [17].

Table 1. Image sonar parameters of ARIS1800.

Item	Low-Frequency Mode	High-Frequency Mode
Operating frequency/MHz	1.1	1.8
Effective range/m	0.7–35	0.7–15
Resolution/mm	23	3
Maximum frame rate/second	3.5–15 frames	
Field of view (FOV)/(°)	28 × 14	
Size/cm	31 × 17 × 14	

Figure 1 is a physical diagram of ARIS1800 sonar. In the process of acoustic image generation, water reverberation, channel change, interference, and self-noise generated by target activity are usually accompanied. The non-sequential emission of the ARIS transducer elements can effectively reduce the influence of self-noise and crosstalk. As for how the ARIS system works, the transducer actively emits sound waves in the field of view according to the size of the reflected echo, thus forming acoustic images with different light and dark characteristics. The acoustic image includes a bright area corresponding to the bottom, a bright area representing the fish target, and a dark area corresponding to the water background, as shown in Figure 2 [18].

Figure 1. ARIS1800 sonar physical diagram.

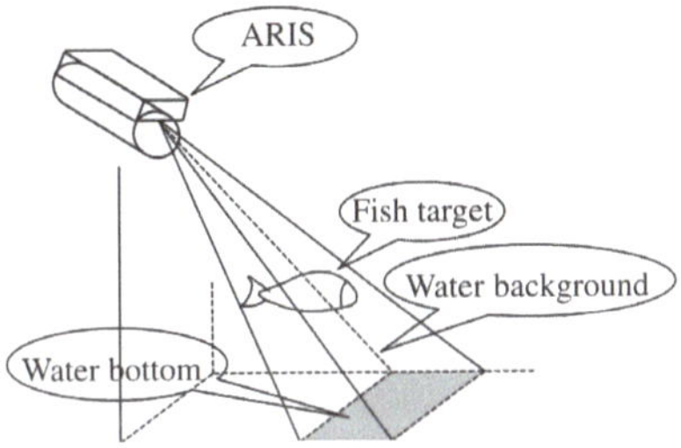

Figure 2. Schematic diagram of sonar fish detection.

2.3. *Sonar Data Acquisition of Fish Quantity in Cages*

The experimental data were measured in the sea area of Guishan Island, Zhuhai City, Guangdong Province, China, in March 2023 (latitude and longitude: 113.84473 and 22.12571, respectively). Figure 3 shows the satellite image of the experimental sea area, and Figure 4 shows the aerial image of the experimental base.

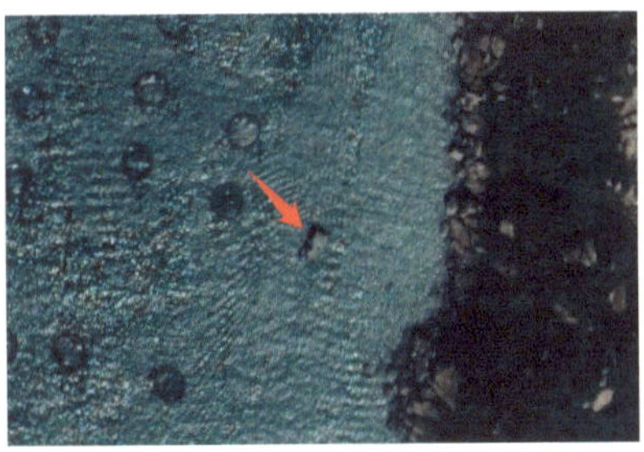

Figure 3. Satellite image of the experimental sea area.

Figure 4. Aerial image of the experimental base.

The cage used in this experiment is shown in Figure 5, with the width and height of the fishnet being 6 × 3 × 4 m, respectively. During the experiment, iron blocks were tied to the four corners of the fishnet as counterweights to open the netting. The object of sonar detection was a golden pomfret with a body length of about 15 cm, which was placed in the experimental cage.

Figure 5. Experimental cage.

It can be seen from Figure 6 that the ARIS sonar is tied to a lifebuoy and floating in the water, with the sonar probe placed at a depth ranging from 30 to 40 cm and a 45-degree angle inclined to the left. The sonar was placed in the middle of the short side of the netting, and the sonar signal covered as much water space as possible. Then, the sonar was connected to a laptop computer, and the supporting software ARISFish (v2.6.3) was used for data acquisition. The upper computer software ARISFish communicates with the sonar device to receive and process the sonar-collected data. It then displays the real-time processing results graphically. A high-frequency mode was used in the sonar, that is, the frequency was 1.8 MHz, and the detection distance was set to just observe the netting on the opposite side, which was about 4.6 m.

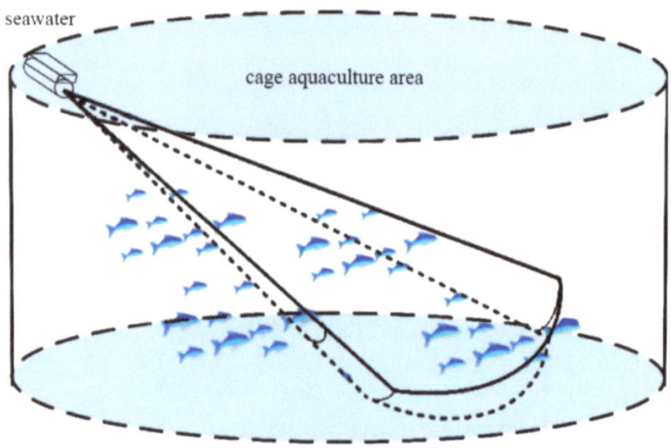

Figure 6. Schematic diagram of the sonar deployment.

The quantitative experiment was carried out with every 20 fish as the standard group, and 20, 40, 60, and 80 golden pomfrets were put into the experimental cage in turn. Each group of fish was continuously detected by sonar, and the data every 10 min were recorded as an ARIS source file, which was saved in the computer for subsequent processing.

The sonar images of different groups of fish are presented in Figure 7, from which the clear outlines of fish and netting were visible. The direction of the images was not the same because the waves were constantly beating the sonar, causing the sonar probe to swing left and right in a certain range. Meanwhile, only a 28° sonar opening angle made it impossible for the sonar detection waves to cover the entire cage, that is, not every fish was visible, which put forward higher requirements for the next estimation method.

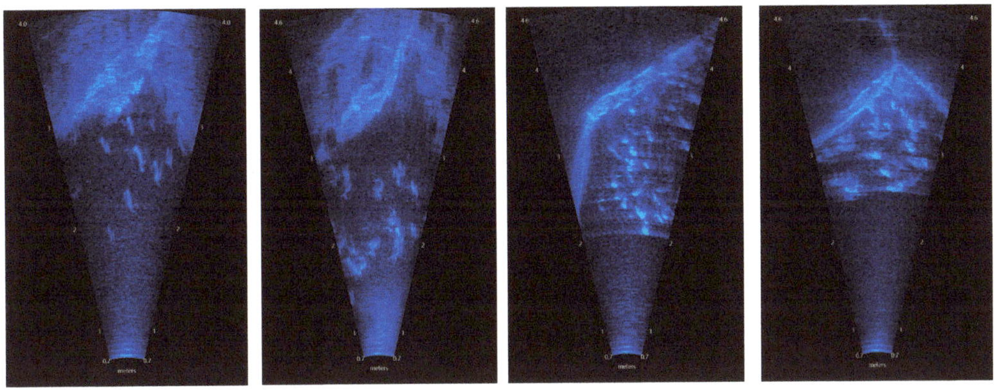

(**a**) (**b**) (**c**) (**d**)

Figure 7. Sonar images of different groups of fish. (**a**) Twenty fish; (**b**) forty fish; (**c**) sixty fish; and (**d**) eighty fish.

3. Recognition Algorithm

3.1. Introduction to YOLO Algorithm

YOLO is a two-step target detection model based on a neural network. Firstly, the input image is divided into S × S grids, and each grid generates B prediction frames, each of which is represented by a corresponding feature vector, generally taking S = 7 and B = 2. The feature vector is composed of: the coordinates of the center point of the corresponding prediction frame, the width and height of the prediction frame, and the confidence of the existence of the object, and each grid will generate a classification prediction feature vector. Finally, the prediction frame with high confidence and its classification are returned to the original input image [19].

By adding the feature fusion method to the feature extraction network, the algorithm adopts the backbone network of Darknet-53. The feature extraction network structure of the YOLO model is shown in Figure 8. The network is a full convolution network, which is trained and tested on the COCO dataset, and finally outputs a feature map of size 13 × 13 × 255. After the feature map is input to the target detection layer 1, position regression and classification regression are performed. Moreover, the feature map of the last layer and the feature map of the middle layer are fused by the above sampling method and input into the target detection layer 3 and the target detection layer 2, respectively, to achieve position regression and classified regression on the feature maps of multiple sizes [20].

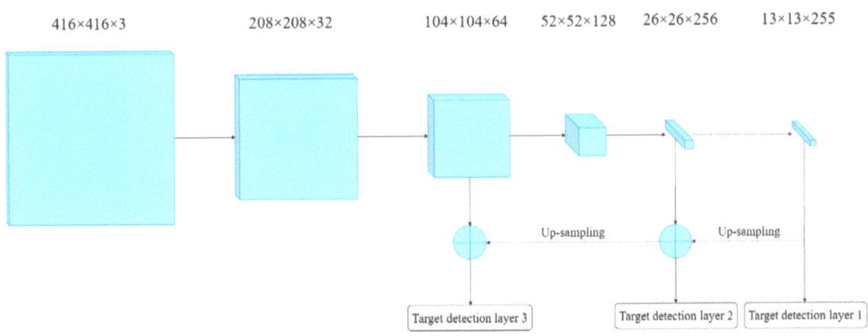

Figure 8. YOLO feature extraction network.

Considering the performance and stability of the model comprehensively, the YOLOv8 model was used for fish target detection in this study. YOLOv8 directly transforms the problem of fish detection into a regression problem. After a regression, not only the position coordinates of each fish group are generated, but also the probability of each candidate region belonging to the category is obtained.

3.2. YOLO Algorithm Improvement

On the premise of satisfying real-time performance and high detection accuracy, a target detection model based on an improved YOLOv8 algorithm is proposed. Considering that every fish is a small target in a sonar image, the core idea of the improved algorithm is to improve the network's perception ability of small target feature information [21]. Firstly, the CBAM (Convolutional Block Attention Module) [22] is improved by using the attention mechanism, and the channel-space attention module CSAM is proposed, which is lighter and can focus on the dimensional features of a small target space. The CSAM is embedded after each convolution of the backbone network to extract features. Then, a 4-fold down-sampling process is added to the YOLOv8 backbone network using 4-scale detection. After the input image is down-sampled by 4 times, a large shallow feature map is obtained. Because of the small receptive field, the feature map contains rich position information to improve the detection effect of small targets [23].

CAM is the channel attention module in CBAM. It consists of two fully connected layers to capture non-linear cross-channel interaction. However, the introduction of the fully connected layer causes a large amount of computation. Even if the channel characteristics are compressed, the parameter quantity is still proportional to the square of the number of channels [24]. For a reduced computational burden, a one-dimensional convolution with convolution kernel length k is used to achieve local cross-channel interaction by referring to the idea of ECANet, aiming to extract the dependency between channels [25]. L-CAM represents the improved lightweight channel attention module, and the convolution kernel length k is calculated by Formula (1):

$$k = \psi(C) = \left| \frac{\mathrm{lb}C}{\gamma} + \frac{b}{\gamma} \right|_{\mathrm{odd}} \tag{1}$$

where C is the number of channels of the input characteristic map, and γ and b are set to 2 and 1, respectively. "lb" means log-based binary.

SAM stands for the spatial attention module in CBAM. In this study, a new channel-spatial attention structure CSAM was constructed by using the improved L-CAM and SAM modules, as shown in Figure 9. Firstly, L-CAM and SAM were used to obtain the channel attention weight Mc and spatial attention weight M_s, respectively. Then, the map of attention M_c and M_s was extended to the size of $\mathrm{R}^{W\times H\times C}$; W and H represent the width and height of the image, respectively; and C represents the number of channels. The sum of elements and sigmoid normalization were carried out to obtain the attention weight matrix M_{cs} based on the space and channel. The weight reflects the attention distribution in the feature map so that the model can obtain more effective features in the more accurate attention area, as shown in Formula (2):

$$M_{cs} = \mathrm{sigmoid}(M_c + M_s) \tag{2}$$

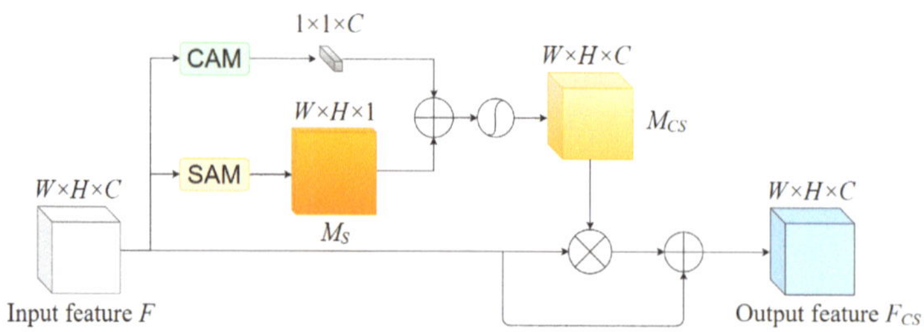

Figure 9. CSAM module.

Finally, the mixed attention weight matrix M_{cs} was multiplied with the input feature map F element by element and added to the original input feature map to obtain a refined feature map F_{cs}, which was calculated as shown in Formula (3):

$$F_{cs} = F + F \otimes M_{cs} \tag{3}$$

The attention mechanism tells the model where to concentrate more calculations and improve the expressive force of the region of interest [26]. The idea of CSAM was to obtain attention weight matrices M_c and M_s from the input feature map F along the spatial dimension and the channel dimension, respectively, to improve the effective flow of feature information in the network. This module emphasizes paying attention to meaningful features, focusing on important features and suppressing invalid features in the two dimensions of channel and space. For small targets, a single feature region gains more weight and contain more effective targets. The model will place a much higher premium on learning the features of this region to extract features better with limited computing resources.

3.3. Experimental Analysis

3.3.1. Dataset Making

One hundred sonar images of fish schools were intercepted from experimental data and processed by the MakeSense online data labeling website. There was only one labeling category, namely fish. After the completion of all labeling, the label file was exported, and each sonar image corresponded to a text file with the same name to record the labeling results. The labeled datasets were divided into two categories by random numbers, with 80 images as the training sets and 20 as the test sets.

3.3.2. Experimental Environment

The Windows 10 system was used in the experiment, with NVIDIA GeForce RTX 3070 (8 GB) as the GPU and Intel i9-12900H as the processor. The experimental environment was python3.9.13, pytorch1.13.1, and cuda11.7.

3.3.3. Evaluation Indicators

For the detection performance, the average precision (*mAP*), parameter quantity (Params), calculation quantity (GFLOPs), and speed (FPS) were used as evaluation indexes [27]. In the process of calculating *mAP*, it was necessary to calculate the average accuracy (*AP*) first, which represents the average accuracy of a category in the dataset. The calculation process is shown in Formula (4). Then, the *AP* values of different categories were averaged to obtain a *mAP*, and the calculation process was shown in Formula (5):

$$AP = \int_0^1 p(r)dr \tag{4}$$

$$mAP = \frac{1}{N}\sum_{i=1}^{N} AP_i \tag{5}$$

where P represents the precision ratio, that is, the ratio of the correct result of model recognition among all the recognized results; r represents the recall ratio, that is, the ratio of the correct results of model recognition to the results that need to be recognized in the dataset; N represents the number of categories of samples, and $N = 1$ in this study.

3.3.4. Training Process

When training the detection network model, the number of iterations was set to 300, the weight attenuation coefficient to 0.0005, the initial learning rate to 0.01, the learning rate momentum to 0.937, and the batch size to 16. As shown in Figures 10 and 11, the model triggered "Early Stopping" to stop training after 120 iterations, at which time the loss decreased to 0.6 and the mAP50 reached 73.02%.

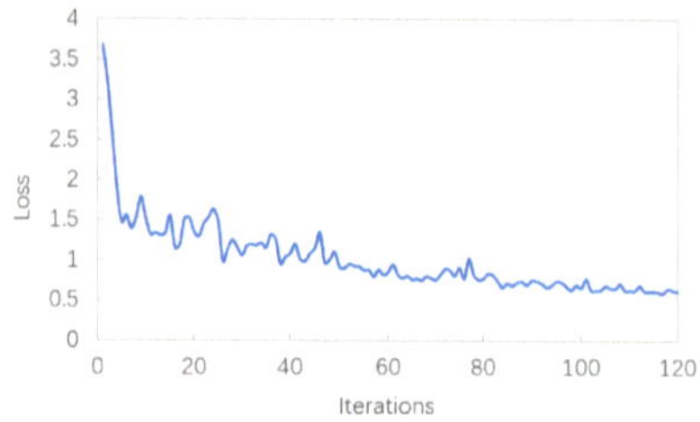

Figure 10. Training process (loss).

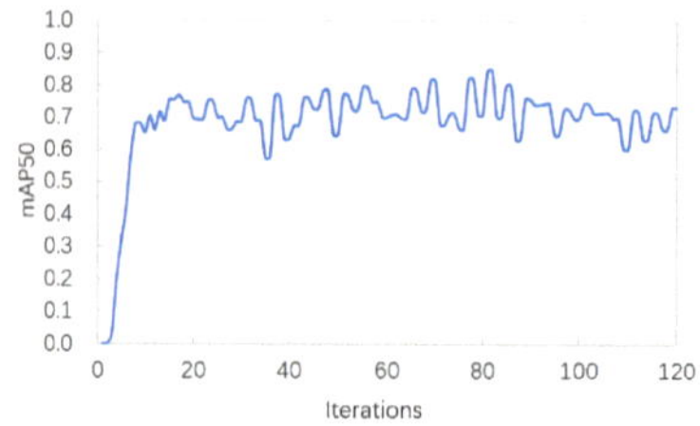

Figure 11. Training process (mAP50).

3.3.5. Ablation Experiments

To verify the effectiveness of the channel-space attention mechanism CSAM proposed in this paper, different modules were added to the YOLOv8 detection algorithm under the same experimental conditions, and the influence of each module on the performance of the detection algorithm was evaluated. The results are shown in Table 2. In the added attention module, CSAM improved the accuracy of the detection algorithm the most, which was 3.81 percentage points, while CSAM also ensured fewer parameters, less computation, and the real-time performance of the algorithm.

Table 2. Comparative results of the ablation experiments.

Models	Params/10^6	FLOPs/10^9	mAP50/%	FPS
YOLOv8	25.90	78.9	69.21	18.87
YOLOv8+CBAM	32.07	104.6	71.92	6.58
YOLOv8+CSAM(Ours)	27.20	96.5	73.02	9.72

3.3.6. Comparative Test

To verify the superiority of the improved detection algorithm, three mainstream detection algorithms were selected for comparative experiments, as shown in Table 3. When the input sizes were all set to 640×640 pixels, the detection accuracy of the improved detection algorithm in this paper was better than other algorithms based on ensuring real-time detection. Compared to the Faster RCNN, mAP50 increased by 18.06 percentage points, while Params and FLOPs decreased by 1.59×10^8 and 8.56×10^{10}, respectively. Compared to YOLOv5, the mAP50 of this algorithm increased by 8.24%. On the whole, the improved detection algorithm added the attention module CSAM to the backbone network, which improved the feature extraction ability of small targets and made the model better in detecting fish sonar images.

Table 3. Comparative experimental results of the different detection algorithms.

Models	Size/Pixel	Params/10^6	FLOPs/10^9	mAP50/%	FPS
Faster RCNN	640×640	186.3	182.1	54.96	2.00
SDD	640×640	23.8	188.0	53.00	2.86
YOLOv5	640×640	7.2	16.5	64.78	18.01
YOLOv8+CSAM(Ours)	640×640	27.2	96.5	73.02	9.72

3.3.7. Comparison of the Detection Images

A comparison between the algorithm in this paper and YOLOv8 in detecting fish sonar images without an attention mechanism is presented in Figure 12. Figure 12a–c show our algorithm and Figure 12d–f show YOLOv8 in this paper. The upper and lower parts correspond to the same frame image. It can be seen that the model can distinguish the fish from the netting, and the detected fish was selected by the red identification box. By comparing Figure 12a and Figure 12d, it can be observed that the algorithm in this paper has detected the leftmost small fish, but YOLOv8 has not, and instead mislabeled the rightmost blackfish. Comparing with Figure 12b and Figure 12e, it can also be observed that the algorithm in this paper recognized one more small fish than the original algorithm. Figure 12c turns on the label and confidence display, and it can be seen that the average confidence of fish identification was higher than 80%, which shows that the neural network model can identify fish well.

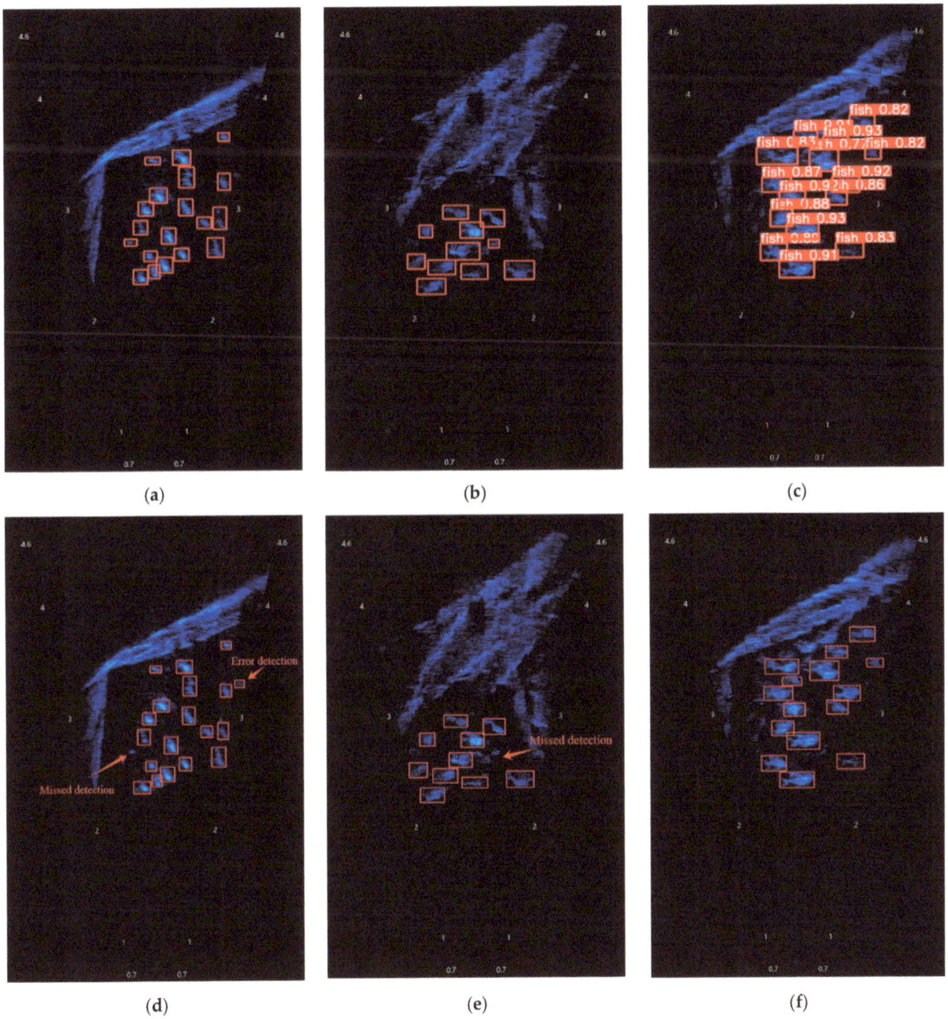

Figure 12. Comparison of detection images. **(a–c)** our algorithm; **(d–f)** YOLOv8.

4. Data Fitting

4.1. Introduction to the BP Neural Network

In this study, the estimation of the detected fish quantity to the actual quantity was a nonlinear mapping problem. Neural networks boast strong applicability in dealing with nonlinear mapping and are considered an effective method of data fitting and widely used [28].

The BP (back propagation) neural network is a widely used algorithm at present. The training steps are: initializing the weights and thresholds of each layer, inputting sample data in the input layer, and finally outputting the results in the output layer after calculation in the hidden layer. In the process of the forward transmission of each layer, the current layer only affects the adjacent next layer. If the results of the output layer do not meet the expected output value, the error with the expected value will be propagated back to the network, so that the error function will decrease along the negative gradient direction [29].

The BP neural network includes one input layer, one or more hidden layers, and one output layer. The basic topological structure of the BP neural network (taking one hidden layer as an example) is shown in Figure 13.

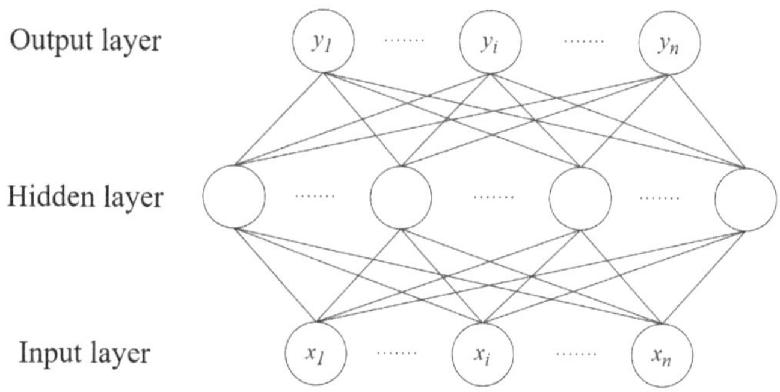

Figure 13. Topology of the neural network.

In neurons, the input acts on another function after a series of weighted summations, and this function is the activation function here. The function of the activation function in a neural network is to transform multiple linear inputs into nonlinear relationships, to achieve the mapping function from linear to nonlinear. The definition of a sigmoid function is shown in Formula (6) [30].

$$Sigmoid(x) = \frac{1}{1+e^{-x}} \tag{6}$$

4.2. Experimental Analysis

4.2.1. Training Data

To automatically obtain the fish quantity in the cage, human subjective factors and manual intervention should be minimized. In this paper, all the images collected by sonar were selected for target recognition and detection. Sonar data were divided into four groups: 20 fish, 40 fish, 60 fish, and 80 fish, and each group had 10 continuous detection videos with a frame rate of 15 frames per second. Seven videos from each group were randomly selected as the fitting data, and the remaining three were used as detection data. These 40 sonar videos were detected by this algorithm, and the identification data of each frame was saved as a text file.

Each 10-minute video had nearly 10,000 images. If all such vast data were used for fitting, it would not only be a vast amount of calculation but also make it difficult for the algorithm to learn the key features of the data. Considering that the goal is to obtain the fish quantity in the cage, and there was a certain mapping relationship between the quantity of fish in the detection image and the actual quantity, the amount of fish detected in a single frame in each video was sorted from large to small in this paper, taking the top 30 fish quantity detected. The statistical results are shown in Figure 14.

Serial No.	1	2	3	4	5	6	7	8	9	10	11	12	13	14	15	16	17	18	19	20	21	22	23	24	25	26	27	28
Actual quantity	20	20	20	20	20	20	20	40	40	40	40	40	40	40	60	60	60	60	60	60	60	80	80	80	80	80	80	80
Quantity detected 1	11	13	13	15	14	12	14	18	19	18	20	19	18	20	16	17	20	24	23	27	20	23	24	28	25	24	27	26
Quantity detected 2	11	12	13	13	14	12	13	18	18	18	20	19	18	20	15	17	20	20	22	27	19	22	23	27	25	23	26	26
Quantity detected 3	11	12	12	12	13	11	12	17	18	18	20	18	18	19	15	16	19	20	22	25	19	22	23	27	24	22	25	25
Quantity detected 4	10	11	12	11	12	11	12	17	17	18	19	18	18	19	15	16	17	19	22	24	19	22	23	26	24	22	25	24
Quantity detected 5	10	11	11	11	12	11	12	17	17	18	19	18	18	19	15	16	17	19	20	24	19	22	22	24	24	22	25	23
Quantity detected 6	10	11	11	11	11	11	12	17	17	18	19	17	18	19	15	16	17	19	20	24	18	22	22	24	24	22	25	23
Quantity detected 7	10	11	11	11	11	11	11	17	17	18	18	17	18	19	15	16	17	19	19	24	18	21	22	24	24	22	24	23
Quantity detected 8	10	11	11	11	11	10	11	16	17	18	18	17	18	19	15	16	17	18	19	24	18	21	22	23	24	22	24	23
Quantity detected 9	10	10	11	11	11	10	11	16	17	18	18	17	17	19	14	16	17	18	19	24	18	20	21	23	23	21	24	23
Quantity detected 10	10	10	11	11	11	10	11	16	17	18	18	17	17	18	14	16	17	18	19	24	18	20	20	23	23	21	24	23
Quantity detected 11	10	10	11	11	11	10	11	16	16	17	18	17	17	18	14	16	16	18	19	23	17	20	20	22	23	21	24	23
Quantity detected 12	10	10	11	11	11	10	11	16	16	17	18	17	17	18	14	15	16	17	19	23	17	20	20	22	23	21	24	23
Quantity detected 13	10	10	10	11	11	9	11	16	16	17	18	17	17	18	14	15	16	17	19	23	17	20	20	21	23	21	24	22
Quantity detected 14	10	10	10	11	11	9	10	16	16	17	18	16	17	18	14	15	16	17	18	23	17	19	20	21	23	21	24	22
Quantity detected 15	10	10	10	11	11	9	10	16	16	17	18	16	16	18	14	15	16	17	18	23	17	19	20	21	22	21	24	22
Quantity detected 16	10	9	10	11	11	9	10	16	16	17	18	16	16	17	14	15	16	17	18	23	17	19	19	21	22	21	24	22
Quantity detected 17	10	9	10	11	11	9	10	16	16	17	18	16	16	17	14	15	16	17	18	23	17	19	19	21	22	21	24	22
Quantity detected 18	10	9	10	11	11	9	10	16	16	17	18	16	16	17	13	15	16	17	18	23	16	19	19	21	22	21	24	22
Quantity detected 19	10	9	10	10	11	9	10	16	16	17	18	16	16	17	13	15	16	17	18	22	16	19	19	20	22	21	21	22
Quantity detected 20	10	9	10	10	10	9	10	15	16	17	17	16	16	17	13	15	16	17	18	22	16	18	19	20	22	21	21	22
Quantity detected 21	9	9	10	10	10	9	10	15	16	17	17	16	16	17	13	15	16	17	18	22	16	18	19	20	21	21	21	22
Quantity detected 22	9	9	10	10	10	9	9	15	16	17	17	16	16	17	13	15	16	17	18	22	16	18	18	20	21	21	21	20
Quantity detected 23	9	9	10	10	10	9	9	15	16	17	17	16	16	17	13	15	16	17	18	22	16	18	18	20	21	21	21	19
Quantity detected 24	9	9	10	10	10	9	9	15	16	16	17	16	16	17	13	15	15	17	18	22	16	18	18	20	21	21	21	19
Quantity detected 25	9	9	10	10	10	9	9	15	16	16	17	16	15	17	13	15	15	17	18	22	16	18	18	20	21	20	21	19
Quantity detected 26	9	9	10	10	10	9	9	15	16	16	17	16	15	17	13	15	15	17	17	22	16	18	18	20	20	20	21	19
Quantity detected 27	9	9	10	10	10	9	9	15	16	16	17	16	15	17	13	15	15	17	17	22	16	18	18	20	20	20	21	19
Quantity detected 28	9	9	10	10	10	9	9	15	16	16	17	16	15	17	13	15	15	17	17	22	15	18	18	20	20	20	21	19
Quantity detected 29	9	9	10	10	10	9	9	15	15	16	17	16	15	17	13	15	15	17	17	22	15	18	18	19	20	20	21	19
Quantity detected 30	9	9	10	10	10	9	9	15	15	16	17	15	15	17	13	15	15	17	17	22	15	18	18	19	20	20	21	19

Figure 14. Statistical diagram of the maximum quantity detected.

4.2.2. Evaluation Indicators

In the process of neural network training, the error between the predicted or fitted data and the measured data can be expressed by the *MSE* (mean square error), as shown in Formula (7):

$$MSE = \frac{1}{n}\sum_{i=1}^{n}(Y_i - \hat{Y}_i)^2 \tag{7}$$

In Equation (7), "n" represents the data quantity, "Y_i" represents the measured data, and "$\hat{Y}_i$" represents the predicted or fitted data based on the neural network model.

4.2.3. Training Process

The top 10, top 20, and top 30 fish abundance detected were input into the network for training, and the fitting target was the corresponding actual quantity. After comparative experiments, the best effect parameters were the top 20 fish quantity detected in fitting, and the best number of neurons in the hidden layer was 30. Bayesian regularization was used for training. There were 28 groups of data, 85% of which were randomly selected as training data and the remaining 15% as test data.

Based on the above parameters, the neural network was trained, and the training results are shown in Figures 15 and 16. Figure 15 shows the change in the sample mean square error. After 45 training operations, the MSE of the training group produced the best result, with a value of 85.1716. Figure 16 shows the prediction errors of the training group and the test group, in which the vast majority of sample errors were between −12 and 12, with positive numbers indicating that the prediction was greater than the actual quantity and negative numbers indicating that the prediction was lower than the actual quantity.

Figure 15. The sample mean square error.

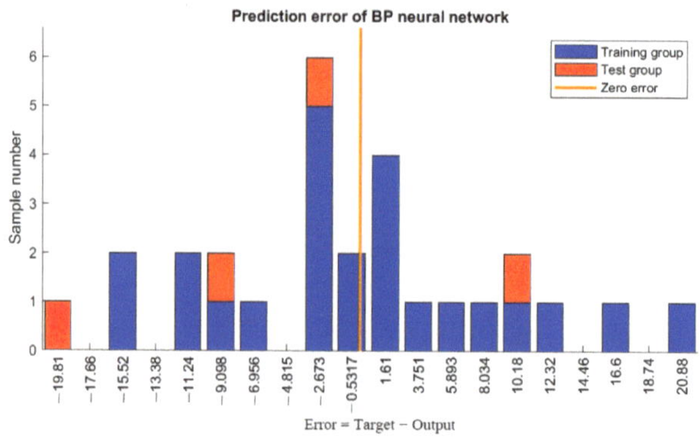

Figure 16. Prediction error of the training group and the test group.

The learning results of the BP neural network are shown in Figure 17. The regression results of the training group, the test group, and all data, that is, the fitting degree between the output value and the target value, are shown in these three small graphs. As it can be seen from the figure, most of the data are concentrated near the diagonal, and some data are far away, and the fitting results are all above 0.82, indicating that the fitting effect is relatively good.

Figure 17. Regression results of the BP neural network model.

4.2.4. Fitting Test

The BP neural network was used to estimate the top 20 fish quantity of the three tests in each group of test data, and the fitting results are shown in Table 4. Error number = total fitting quantity-actual quantity, error percentage = absolute value of error quantity/actual quantity; the average error was the average of all error percentages and average accuracy = 1 − average error.

Table 4. The statistical results of the method in this paper on the test dataset.

Serial No.	Actual Quantity	Maximum Quantity Detected	Fitting Total Quantity	Error Quantity	Error Percentage/%	Precision Percentage/%
1		12	17.68	−2.32	11.60	88.4
2	20	14	26.63	6.63	33.13	66.87
3		13	24.48	4.48	22.40	77.6
4		19	44.91	4.91	12.27	87.73
5	40	16	35.59	−4.41	11.02	88.98
6		17	37.87	−2.13	5.34	94.66
7		22	52.56	−7.44	12.40	87.6
8	60	23	68.46	8.49	14.10	85.9
9		22	60.99	0.99	1.66	98.34
10		21	74.42	−5.78	6.97	93.03
11	80	13	58.00	−22.00	27.50	72.5
12		18	59.19	−20.81	26.01	73.99
				Average	15.37	84.63

It can be seen from Table 4 that the algorithm in this paper had a high accuracy in fitting the sonar image data of 20, 40, and 60 groups and achieved a single-digit error. However, when fitting the sonar data of 80 fish, the error was large, and the quantity sequence detected was small, resulting in a large error of about 27%. The manual inspection of the detection videos with serial numbers 11 and 12 showed that there were few fish in the sonar images. It was speculated that the sonar probe shook badly during this period due to heavy sea waves, and the swimming trajectory of the fish was different from the usual one; so, the data detected by the sonar did not reflect the real situation in the cage. The solution can be to observe in multiple periods, obtain multiple groups of sonar image data and carry out target recognition and detection, eliminate detection sequences with too large data differences, and then estimate by a neural network. The obtained data were more objective and more realistic after averaging.

4.2.5. Data Fitting and Comparison

The commonly used data fitting methods are linear fitting and polynomial fitting. Because they can only deal with one-to-one mapping relationships, it is necessary to extract key data from the detection sequence [31]. In this paper, the quantity of fish detected in a single image in each video was sorted from large to small, and the maximum quantity of fish detected, the average of the top 10 fish quantity, and the average of the top 20 fish quantity were statistically analyzed.

The training data and neural network fitting were the same. Firstly, linear fitting and cubic polynomial fitting were carried out for these three statistical data, and the results are shown in Figure 18. The upper left corner of each small graph shows the fitting formula and fitting coefficient R^2, which reflect the overall accuracy of the model, that is, the fitting degree. The closer its value is to 1 shows that the model accurately reflects the changes in the observed data, and the better the reliability of the data. It can be seen from the figure that the R^2 of cubic polynomial fitting is greater than the corresponding linear fitting, and the fitting results of the average of the top 10 fish quantity in the two fitting methods are better than those of the maximum quantity of detected and the average of the top 20 fish quantity.

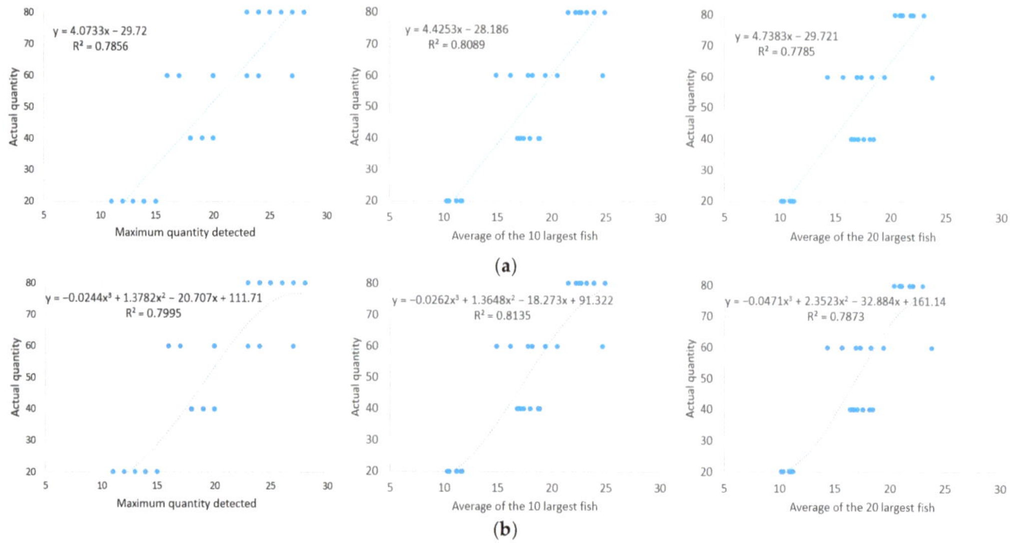

Figure 18. Comparison of the data fitting results. (**a**) Linear fitting; (**b**) cubic polynomial fitting.

When the polynomial fitting was performed on the average of the top 10 fish quantity, the fitting results of the quadratic, cubic, and quartic polynomials are compared as shown in Figure 19. It can be seen that the best fitting result of the quartic polynomial was $R^2 = 0.8387$, but the highest term was too high, which leads to a better effect on sample data, but the effect of test data will decline, that is, there will be over-fitting. Generally, the highest term was not higher than three times when the polynomial fitting was used.

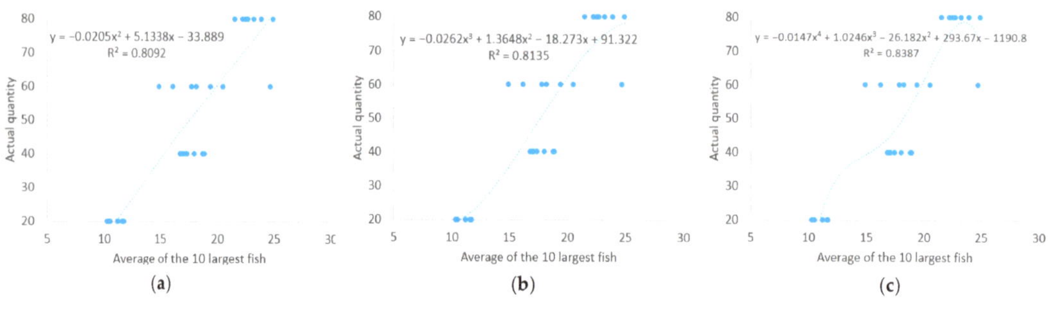

Figure 19. Comparison of the fitting results of higher order polynomials. (**a**) Quadratic polynomial; (**b**) cubic polynomial; and (**c**) quartic polynomial.

To sum up, the cubic polynomial was selected to fit the fish quantity in the comparison test, and the equation is shown in Formula (8):

$$y = -0.0262x^3 + 1.3648x^2 - 18.273x + 91.322 \quad (8)$$

where x was the average of the top 10 fish abundance detected, y was the estimated fish quantity in the cage, and R^2 of the equation was 0.8135, which can be understood as the theoretical accuracy of data fitting as about 81.35%.

Formula (8) was used to estimate the average of the top 10 fish quantity of the three tests in each group of test data, and the fitting results are shown in Table 5. It can be seen that the average accuracy is 83.58%, which is lower than the 84.63% of neural network fitting.

Table 5. The statistical results of high-order polynomial fitting on the test dataset.

Serial No.	Actual Quantity	Average of the Top 10 Fish Quantity Detected	Fitting Total Quantity	Error Quantity	Error Percentage/%	Precision Percentage/%
1	20	12.6	25.35	5.35	26.74	73.26
2		10.5	19.59	−0.41	2.03	97.97
3		11.3	21.30	1.30	6.52	93.48
4	40	15.3	37.39	−2.61	6.52	93.48
5		14.5	33.44	−6.56	16.40	83.6
6		16.7	44.77	4.77	11.92	88.08
7	60	19	57.12	−2.88	4.80	95.2
8		21.8	70.14	10.14	16.90	83.1
9		18.2	52.88	−7.12	11.86	88.14
10	80	20.7	65.49	−14.51	18.14	81.86
11		17.3	48.01	−31.99	39.98	60.02
12		18	51.80	−28.20	35.24	64.76
				Average	16.42	83.58

5. Discussion

5.1. Comparison of the Fish Quantity Estimation Methods

The traditional methods to obtain the fish abundance in cages are the mark–recapture method, fish finder measurement, annular underwater acoustic multi-beam detection, and others [32]. Due to the impossibility of conducting comparison experiments in the same environment, the instruments and equipment used in various methods vary. At present, there are few reports on the estimation algorithm and estimation accuracy of fish abundance in cages. In this paper, a comparison table of the different estimation methods was made based on previous studies by scholars, which is shown in Table 6.

Table 6. Comparison of the different estimation methods.

Methods	Equipment Used	Precision	Advantages	Disadvantages
Mark–recapture method [33]	Fishing net, stain	Large discrete interval	No electronic equipment is needed	Low precision, time-consuming, and laborious, affecting the growth of fish
Fish finder measurement [34]	Fish detector	About 50%	Low equipment cost	Low accuracy, fish density, sometimes vast errors
Annular underwater acoustic multi-beam detection [35]	Annular multi-beam detector	60%–70%	Wide detection angle, high precision	Expensive equipment, difficult layout
The method in this study	Image sonar	About 84%	High precision, automatic measurement, simple layout	Expensive equipment

It can be seen from Table 6 that the forward-looking image sonar used in the method presented in this paper is more expensive than the equipment used in previous methods, and the average purchase unit price is USD 30,000, but the layout is relatively simple. After the sonar is installed, the data can be obtained and processed automatically, and the estimated fish abundance in the cage can be obtained without additional manual intervention. Compared to the traditional methods, the accuracy of this method is significantly improved, reaching about 84%.

As a high-definition image sonar, ARIS1800 is widely used in fishery. Both at home and abroad, image sonar is mainly used in the study of fish behavior, rather than in the assessment of fish quantity. This paper makes a very meaningful attempt to evaluate the quantity of fish in cages by using the imaging characteristics of ARIS1800, based on fixed detection and prediction methods. ARIS1800 can display the size, shape, and position of fish in the cage with high-definition images. It eliminates the limitation of traditional fish finders only being able to assess fish quantities by target strength, achieving a higher credibility.

5.2. Error Analysis

The estimation of fish quantity in cages is a major challenge in fish acoustics research, which is influenced by various factors: complicated and changeable ocean factors, such as wind and waves and tidal currents in aquaculture areas; feeding, sailing, and other interferences; transducer reverberation blind area and strong sea-floor reflection; obscuration of beam detection by fish in dense schools; some fish swim close to the wall, which make the fish echo and the net echo overlap and difficult to distinguish; and repeated detection caused by swimming fish [36]. These uncontrollable factors cause the data collected by sonar at different times to be inconsistent, and in turn, the estimated neural network model has inevitable errors, affecting the final estimated quantity of fish.

Figure 20 is a histogram of the estimation and error of the fitting test in Table 4. The error of the neural network estimation in groups 1–10 is relatively small, and the prediction results in groups 11 and 12 are affected by the large wave fluctuation. By observing the cages in different periods, it was possible to estimate the average value of multiple groups of data to reduce the error.

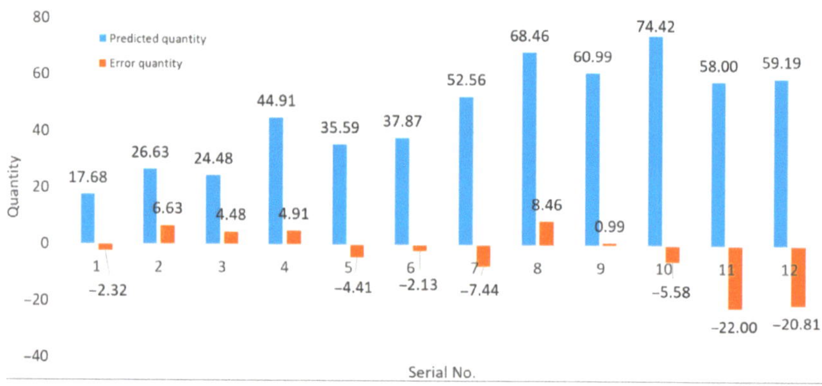

Figure 20. Estimation and error bar chart of the fitting test.

Given the measurement results of 80 fish, three additional observation data at different times were selected in this paper, and a new test dataset was formed together with the three data in the previous fitting test. The fish quantity was estimated by using this method, and the results are shown in Table 7.

Table 7. The statistical results of the method in this paper across multiple time periods of datasets.

Serial No.	Actual Quantity	Testing Time	Fitting Total Quantity	Error Quantity	Error Percentage/%	Precision Percentage/%
1	80	9:00–9:15	74.42	−5.78	6.97	93.03
2		9:30–9:45	58.00	−22.00	27.50	72.5
3		10:00–10:15	59.19	−20.81	26.01	73.99
4		13:00–13:15	82.32	2.32	2.9	97.1
5		15:00–15:15	72.45	−7.55	9.44	90.56
6		17:00–17:15	69.45	−10.55	13.19	86.81
		Average	69.31	−10.69	14.33	85.67

Table 7 reveals the fluctuations in the data measured and predicted in different periods, but only the two groups of data with serial numbers 2 and 3 have a deviation of 20, with the other groups having an error of less than 10. The average value of six groups of data prediction was 69.31, the error was −10.69, and the average accuracy was 85.67%, which was significantly improved by 5.83 percentage points compared to the average accuracy of 79.84% of data only using the same period.

6. Conclusions

This paper proposes a method for estimating the quantity of fish in net cages based on forward-looking imaging sonar. The method first investigates the YOLO neural network model and makes improvements for underwater fish identification tasks. An attention mechanism is introduced into the YOLO model construction, allocating more computing power to focus on small targets, thereby enhancing the performance of the region of interest, especially for small targets. Through ablation experiments, the addition of the CSAM module is shown to improve the accuracy of the detection algorithm by 3.81 percentage points, and compared to the YOLOv5, the improved algorithm in this paper increases the mAP50 by 8.24 percentage points. Subsequently, quantitative detection experiments for 80 oval damselfish are conducted in the constructed fish cages. Due to the limited visual angle of the sonar, the experiments are conducted by deploying the sonar on one side of the net cage and continuously observing to obtain video images. The improved and trained YOLOv8 model is used to detect fish shoals in sonar images. The detection quantity results are sorted from large to small, and the quantity of fish in the net cage is estimated based on the top 20 maximum counts using a trained BP neural network. Multiple experimental results show that the average accuracy of fish quantity assessment reaches 84.63%, validating the feasibility of this method.

Through research on detection methods, target identification, and quantity inversion of fish in net cages, a new method for estimating the quantity of fish in net cage farming based on imaging sonar has been developed. This method achieves a high-precision assessment of fish quantity in net cage farming, providing technical support for the development of intelligent equipment for net cages in China.

Nevertheless, there are still the following problems in this research method, which need to be improved in future research:

1. In the part of fish target recognition, the background of the image is not removed in advance, and the netting in the background fluctuates with the waves. In some cases, fish will swim against the netting, and the two are mixed in the sonar image, which will affect the fish recognition effect of the YOLO model and make the recognition quantity fluctuate [37];
2. The YOLO target detection model and neural network prediction model used in this method are highly dependent on training data. For this reason, quantitative fish data collection should be carried out under the condition that the cage size and sonar layout are consistent before practical application. The above two models can only be applied to the fish quantity prediction after learning the collected data. As for the simplification of the model training process and the production of general datasets, further in-depth research is needed;
3. The quantitative experiment in this paper was carried out in a small fishing raft, and it is planned to be applied to a large deep-sea cage in the future. With the increase in the cage scale and the quantity of fish, the density of fish will increase obviously, and more fish will overlap and block each other. In theory, when detecting training data, the situation of fish occlusion is roughly the same as that when estimating the quantity, and the neural network will be relatively accurate when fitting the total quantity. However, as to whether the actual prediction effect can meet the precision of a small-scale quantitative experiment, it still needs to be tested.

Author Contributions: Conceptualization, G.Z. and Y.H.; methodology, M.L.; software, J.H.; validation, L.X.; investigation, D.L.; resources, Y.H.; data curation, G.Z.; writing—original draft preparation, G.Z.; writing—review and editing, C.D., X.H., Y.H. and J.S.; supervision, Y.H.; project administration, X.H.; funding acquisition, X.H. and Y.H. All authors have read and agreed to the published version of the manuscript.

Funding: This research received funding from the Major Science and Technology Plan of Hainan Province (Grant No. ZDKJ2021013), Hainan Province Science and Technology Special Fund (Grant No. ZDYF2021XDNY305, ZDYF2023XDNY066), Central Public-interest Scientific Institution Basal Research Fund, CAFS (Grant No. 2023TD97), Central Public-interest Scientific Institution Basal Research Fund, South China Sea Fisheries Research Institute, CAFS (No. 2022TS06), and Project supported by Southern Marine Science and Engineering Guangdong Laboratory (Zhuhai) (No. SML2023SP204).

Institutional Review Board Statement: Not applicable.

Informed Consent Statement: Not applicable.

Data Availability Statement: Data are contained within the article.

Conflicts of Interest: The authors declare no conflicts of interest.

References

1. Garcia Serge, M.; Rosenberg Andrew, A. Food security and marine capture fisheries: Characteristics, trends, drivers and future perspectives. *Philos. Trans. R. Soc. B* **2010**, *365*, 2869–2880. [CrossRef] [PubMed]
2. Yu, J.; Yan, T. Analyzing Industrialization of Deep-Sea Cage Mariculture in China: Review and Performance. *Rev. Fish. Sci. Aquac.* **2023**, *31*, 483–496. [CrossRef]
3. Huan, X.; Shan, J.; Han, L.; Song, H. Research on the efficacy and effect assessment of deep-sea aquaculture policies in China: Quantitative analysis of policy texts based on the period 2004–2022. *Mar. Policy* **2024**, *160*, 105963. [CrossRef]
4. Kleih, U.; Linton, J.; Marr, A.; Mactaggart, M.; Naziri, D.; Orchard, J.E. Financial services for small and medium-scale aquaculture and fisheries producers. *Mar. Policy* **2013**, *37*, 106–114. [CrossRef]
5. Baumgartner, L.J.; Reynoldson, N.; Cameron, L.; Stanger, J. Assessment of a Dual-Frequency Identification Sonar (DIDSON) for Application in Fish Migration Studies. *Fish. Final Rep.* **2006**, *84*, 1449–1484.
6. Shahrestani, S.; Bi, H.; Lyubchich, V.; Boswell, K.M. Detecting a nearshore fish parade using the adaptive resolution imaging sonar (ARIS): An automated procedure for data analysis. *Fish. Res.* **2017**, *191*, 190–199. [CrossRef]
7. Guan, M.; Cheng, Y.; Li, Q.; Wang, C.; Fang, X.; Yu, J. An Effective Method for Submarine Buried Pipeline Detection via Multi-sensor Data Fusion. *IEEE Access* **2019**, *7*, 125300–125309. [CrossRef]
8. Qiu, Z.W.; Jiao, M.L.; Jiang, T.C.; Zhou, L. Dam Structure Deformation Monitoring by GB-InSAR Approach. *IEEE Access* **2020**, *8*, 123287–123296. [CrossRef]
9. Liu, Y.; Wang, R.; Gao, J.; Zhu, P. The Impact of Different Mapping Function Models and Meteorological Parameter Calculation Methods on the Calculation Results of Single-Frequency Precise Point Positioning with Increased Tropospheric Gradient. *Math. Probl. Eng.* **2020**, *35*, 9730129. [CrossRef]
10. Sun, P.; Zhang, K.; Wu, S.; Wang, R.; Wan, M. An investigation into real-time GPS/GLONASS single-frequency precise point positioning and its atmospheric mitigation strategies. *Meas. Sci. Technol.* **2021**, *32*, 115018. [CrossRef]
11. Cai, J.; Zhang, Y.; Li, Y.; Liang, X.S.; Jiang, T. Analyzing the Characteristics of Soil Moisture Using GLDAS Data: A Case Study in Eastern China. *Appl. Sci.* **2017**, *7*, 566. [CrossRef]
12. Feng, Y.; Wei, Y.; Sun, S.; Liu, J.; An, D.; Wang, J. Fish abundance estimation from multi-beam sonar by improved MCNN. *Aquat. Ecol.* **2023**, *57*, 895–911. [CrossRef]
13. Viswanatha, V.; Chandana, R.K.; Ramachandra, A.C. Real-Time Object Detection System with YOLO and CNN Models: A Review. *arXiv* **2022**, arXiv:2208.00773.
14. He, S.; Lu, X.; Gu, J.; Tang, H.; Yu, Q.; Liu, K.; Ding, H.; Chang, C.; Wang, N. RSI-Net: Two-Stream Deep Neural Network for Remote Sensing Imagesbased Semantic Segmentation. *IEEE Access* **2022**, *10*, 34858–34871. [CrossRef]
15. Ye, Z.B.; Duan, X.H.; Zhao, C. Research on Underwater Target Detection by Improved YOLOv3-SPP. *Comput. Eng. Appl.* **2023**, *59*, 231–240.
16. Chen, Y.L.; Dong, S.J.; Zhu, S.K. Detection of underwater biological targets in shallow water based on improved YOLOv3. *Comput. Eng. Appl.* **2023**, *59*, 190–197.
17. Guo, H.; Li, R.; Xu, F.; Liu, L. Review of research on sonar imaging technology in China. *Chin. J. Oceanol. Limnol.* **2013**, *31*, 1341–1349. [CrossRef]
18. Shen, W.; Peng, Z.; Zhang, J. Identification and counting of fish targets using adaptive resolution imaging sonar. *J. Fish Biol.* **2023**, *104*, 422–432. [CrossRef] [PubMed]
19. Kang, C.H.; Kim, S.Y. Real-time object detection and segmentation technology: An analysis of the YOLO algorithm. *JMST Adv.* **2023**, *5*, 69–76. [CrossRef]

20. Wang, Z.; Zhou, D.; Guo, C.; Zhou, R. Yolo-global: A real-time target detector for mineral particles. *J. Real-Time Image Process.* **2024**, *21*, 85. [CrossRef]
21. Lü, H.; Xie, J.; Xu, J.; Chen, Z.; Liu, T.; Cai, S. Force and torque exerted by internal solitary waves in background parabolic current on cylindrical tendon leg by numerical simulation. *Ocean Eng.* **2016**, *114*, 250–258. [CrossRef]
22. Woo, S.; Park, J.; Lee, J.Y.; Kweon, I.S. Cbam: Convolutional block attention module. In Proceedings of the European Conference on Computer Vision (ECCV), Munich, Germany, 8–14 September 2018; pp. 3–19.
23. Selvaraju, R.R.; Cogswell, M.; Das, A.; Vedantam, R.; Parikh, D.; Batra, D. Grad-cam: Visual explanations from deep networks via gradient-based localization. In Proceedings of the IEEE International Conference on Computer Vision, Venice, Italy, 22–29 October 2017; pp. 618–626.
24. Wei, X.; Wang, Z. TCN-attention-HAR: Human activity recognition based on attention mechanism time convolutional network. *Sci. Rep.* **2024**, *14*, 7414. [CrossRef] [PubMed]
25. Cui, Z.; Wang, N.; Su, Y.; Zhang, W.; Lan, Y.; Li, A. ECANet: Enhanced context aggregation network for single image dehazing. *Signal Image Video Process.* **2023**, *17*, 471–479. [CrossRef]
26. Guo, M.H.; Lu, C.Z.; Liu, Z.N.; Cheng, M.M.; Hu, S.M. Visual attention network. *Comput. Vis. Media* **2023**, *9*, 733–752. [CrossRef]
27. Zhu, G.; Shen, Z.; Liu, L.; Zhao, S.; Ji, F.; Ju, Z.; Sun, J. AUV dynamic obstacle avoidance method based on improved PPO algorithm. *IEEE Access* **2022**, *10*, 121340–121351. [CrossRef]
28. Liu, J.; Yu, L.; Sun, L.; Tong, Y.; Wu, M.; Li, W. Fitting objects with implicit polynomials by deep neural network. *Optoelectron. Lett.* **2023**, *19*, 60–64. [CrossRef]
29. Zhang, J.; He, X. Earthquake magnitude prediction using a VMD-BP neural network model. *Nat. Hazards* **2023**, *117*, 189–205. [CrossRef]
30. Nabizadeh, E.; Parghi, A. Artificial neural network and machine learning models for predicting the lateral cyclic response of post-tensioned base rocking steel bridge piers. *Asian J. Civ. Eng.* **2024**, *25*, 511–523. [CrossRef]
31. Guan, M.; Li, Q.; Zhu, J.; Wang, C.; Zhou, L.; Huang, C.; Ding, K. A method of establishing an instantaneous water level model for tide correction. *Ocean Eng.* **2019**, *171*, 324–331. [CrossRef]
32. Zhang, X.; Xu, X.; Peng, Y.; Hong, H. Centralized Remote Monitoring System for Bred Fish in Offshore Aquaculture Cages. *Trans. Chin. Soc. Agric. Mach.* **2012**, *43*, 178–182+187.
33. Lin, W.Z.; Chen, Z.X.; Zeng, C.; Karczmarski, L.; Wu, Y. Mark-recapture technique for demographic studies of Chinese white dolphins—Applications and suggestions. *Acta Theriol. Sin.* **2018**, *38*, 586–596.
34. Garg, R.; Phadke, A.C. Enhancing Underwater Fauna Monitoring: A Comparative Study on YOLOv4 and YOLOv8 for Real-Time Fish Detection and Tracking. In *Artificial Intelligence and Sustainable Computing*; Pandit, M., Gaur, M.K., Kumar, S., Eds.; ICSISCET 2023. Algorithms for Intelligent Systems; Springer: Singapore, 2024.
35. Connolly, R.M.; Jinks, K.I.; Shand, A.; Taylor, M.D.; Gaston, T.F.; Becker, A.; Jinks, E.L. Out of the shadows: Automatic fish detection from acoustic cameras. *Aquat. Ecol.* **2023**, *57*, 833–844. [CrossRef]
36. Li, D.; Du, L. Recent advances of deep learning algorithms for aquacultural machine vision systems with emphasis on fish. *Artif. Intell. Rev.* **2022**, *55*, 4077–4116. [CrossRef]
37. Maki, T.; Horimoto, H.; Ishihara, T.; Kofuji, K. Tracking a Sea Turtle by an AUV with a Multibeam Imaging Sonar: Toward Robotic Observation of Marine Life. *Int. J. Control. Autom. Syst.* **2020**, *18*, 597–604. [CrossRef]

Journal of
Marine Science and Engineering

MDPI

Article

The Effect of Corner Structure on the Optimisation of Fishable Flow Field in Aquaculture Tanks

Fan Zhang [1,2], Mingchao Cui [1,*], Huang Liu [1] and Chen Zhang [1,2]

1 Fishery Machinery and Instrument Research Institute, Chinese Academy of Fishery Sciences, Shanghai 200092, China; zinfanity@126.com (F.Z.)
2 School of Navigation and Naval Architecture Engineering, Dalian Ocean University, Dalian 116023, China
* Correspondence: cuimingchao@126.com

Abstract: As coastal waters face constraints such as the deterioration of the aquaculture environment and limitations on the scale of operation, aquaculture will move towards the deep and distant sea. Large-scale aquaculture vessels are a new method of deep-sea aquaculture, and improving the utilisation efficiency of aquaculture tanks to ensure the best growth conditions for fish inside while ensuring the efficient discharge of particulate matter in these tanks will affect the productivity of aquaculture and the profitability of aquaculture vessels. This study investigated the effects of the tank structure ratio on the flow field characteristics and particulate removal efficiency in the aquaculture tanks of an aquaculture vessel. Numerical simulations of the flow field characteristics in the aquaculture tanks of an 8000 t-class aquaculture vessel at anchor were conducted using the FLOW-3D software to quantitatively evaluate the effects of the corner ratio on the fishability of aquaculture tanks and the efficiency of particulate emission using the parameters related to flow velocity, turbulence intensity, capacity utilisation rate, and particulate removal efficiency. The simulation results show that the tanks with corner structures have better flow field characteristics, which include a higher flow velocity, turbulence intensity, and discharge effect. When the corner length is more than 1/3 of the tank length, increasing the corner distance does not significantly enhance the optimisation of the flow field characteristics in the tank. Overall, this study's results provide a reference basis for the structural design and optimisation of aquaculture tanks in aquaculture vessels.

Keywords: aquaculture vessel; corner ratio; fishability; effluent effect; eddy strength

Citation: Zhang, F.; Cui, M.; Liu, H.; Zhang, C. The Effect of Corner Structure on the Optimisation of Fishable Flow Field in Aquaculture Tanks. *J. Mar. Sci. Eng.* **2024**, *12*, 1185. https://doi.org/10.3390/jmse12071185

Academic Editor: Muk Chen Ong

Received: 18 June 2024
Revised: 10 July 2024
Accepted: 11 July 2024
Published: 15 July 2024

1. Introduction

According to a report by the Food and Agriculture Organization of the United Nations (FAO), driven by aquaculture expansion and the recovery of capture fisheries, aquaculture production exceeded capture fisheries production for the first time in 2022, with 94.4 million tonnes of aquaculture production, accounting for 51 percent of the world's total aquatic animal production [1]. In search of more space and better water quality, aquaculture is moving into the deep sea [2,3]. The safety and fishability of aquaculture equipment are key to expanding the scale of deep-sea aquaculture, and many scholars have conducted experiments on the hydrodynamic response of deep-sea cages [4,5]. Deep-sea cage aquaculture is often exposed to deep waters, where poor sea conditions can lead to problems such as fish escape and structural safety [6]. Shipboard tank technology has provided a new approach to deep-sea aquaculture. Industrial aquaculture at sea was first proposed in the late 1970s [7] with the systematic study of aquaculture vessels by Xu and Cai et al. [3,8]. Cui et al. provided specific ideas for the construction of large-scale aquaculture vessels [9] and, together with Guo and Li et al., proposed the problem of the fishability of tanks and investigated the effects of external incentives, water entry modes, and other factors on the flow pattern [10–13]. The size of the flow rate and the uniformity of the flow in aquaculture tanks have a critical impact on the removal of particulate matter, such as feed, bait, and faeces, from the environment. An aquaculture water body with an appropriate flow rate not only enables the uniform distribution of dissolved oxygen in the limited aquaculture space but also facilitates the removal of particulate matter consisting of bait residues, faeces, and secondary metabolites from the tanks. Therefore, utilising a larger culture water body area and ensuring the comprehensive performance of the flow field are contradictory points. Determining the corner ratio that combines the comprehensive performance of the tank with the aquaculture vessel area can improve the vessel's economic efficiency and provide a reference design basis.

Currently, fishability studies in confined environments have mostly been based on recirculating aquaculture systems (RASs), and rectangular and circular tanks are the most widely used among fully or semi-closed aquaculture tanks. Despite these problems, rectangular tanks are still widely used in aquaculture because they are easier to construct than other shapes and have a higher overall space efficiency. Aquaculture tanks that tend to be circular have the advantages of good hydrodynamic characteristics and high self-purification capacity but have a low space utilisation rate [14–16]. In 2004, Oca et al. compared the distribution of flow fields in rectangular aquaculture ponds with different intake structures. The results showed that horizontal tangential inlets resulted in higher and more uniform flow rates, thus preventing the deposition of solid particles [17]. In 2007, Oca et al. analysed the simplest inlet and outlet configurations to create uniform cyclonic pools in rectangular aquaculture ponds. They combined the advantages of rectangular and circular ponds, suggesting that higher flow rates can impact the self-cleaning performance of aquaculture ponds [18]. In 2013, Lee et al. modelled the effect of inlet structures on the flow field in flow tanks for tangential and normal flow in culture tanks [19]. In 2015, Davidson et al. also experimentally comparatively analysed the effect of the inlet structure of Cornell dual-channel aquaculture tanks on the hydraulic mixing performance and the movement of solid particles, with appropriately increased flow velocities at the walls of the tanks tending to favour the discharge of particulate matter [20]. In 2017, Liu et al. modelled the hydrodynamics of octagonal aquaculture ponds in a recirculation system and demonstrated 90% particle removal in octagonal aquaculture ponds [21]. The above study shows that the numerical model of spatial hydrodynamics of octagonal aquaculture can be further designed and optimised.

The optimisation of the tank structure has attracted much attention in the research field of ship and marine engineering. In marine engineering, the research on the ship's tank focuses on the impact pressure of the liquid on the bulkhead and the safety problems it creates. An aquaculture vessel is a special vessel used for aquaculture with the characteristics of both the vessel and the aquaculture vessel. The vessel tank aquaculture technology features the newly emerging aquaculture mode, with fewer research results and a lack of practical application. The traditional land-based factory farming model and the marine and shipbuilding industries cannot be used to determine the effect of tank structure on the fishability of the flow field in the tank. As the construction of large-scale aquaculture fishing vessels is advancing, Cui and Guo studied the effect of changing external incentive conditions on the fishability of the flow field in tanks [10–12,22–24]. Xiong et al. investigated the effects of the number of intake pipes and the bottom water discharge rate on the exclusion rate of solid particles such as faeces and residual feed in aquaculture tanks. The results showed that the bottom water discharge rate did not have a significant effect on the discharge of solid particles [25]. Xue et al. investigated the effects of the tank structure on the fluid characteristics and solid particulate exclusion rate of the tank. They showed that the fluid characteristics of the rectangular tank were significantly improved after chamfering the two right-angled edges of the tank [26].

In the early stage of aquaculture vessel design, it is often necessary to consider the problem of the flow field in aquaculture tanks. The tank type is particularly important for improving the fishability of the flow field in these tanks. If the tank type is unreasonable, even if the inlet and outlet arrangement and flow rate are changed, an aquaculture tank cannot obtain a high fishability of the flow field. Therefore, it is particularly important to simulate the flow pattern in different tanks. As mentioned before, aquaculture tanks with square-shaped tanks have a higher aquaculture water body area, but the poor flow regime and low energy utilisation efficiency of the tanks lead to their poor economy, while circular tanks have an excellent tank flow field, but the decline in their space utilisation also affects the number of fish cultured, and the area of the aquaculture water body and the flow regime of the tanks become a pair of contradictions. The aim of this study was to find the corner ratio with higher comprehensive performance and to improve the aquaculture vessel's overall economic efficiency by increasing the aquaculture water body area as much as possible while ensuring that the flow pattern in the tank is suitable for fishing.

Based on computational fluid dynamics (CFD) technology, this study used the FLOW-3D software to numerically simulate the flow pattern of aquaculture tanks in an 8000 t-class aquaculture vessel and analysed the influence of the proportion of the corner cut of the aquaculture tanks on the flow pattern of these tanks using mathematical and statistical methods. Numerical simulation was used to analyse the effect of the tank corner ratio on the tank culture flow state and provide theoretical guidance for the shipboard tank culture technology. In this study, statistical methods were used to evaluate the flow characteristics in tanks with different corner ratios. This study is organised as follows: Section 2 describes the data-processing methodology, the physical model of the cultured tank, and the setup in the CFD software. Numerical simulations are carried out in Section 3, and the results are analysed. The effect of the corner type on the flow velocity, energy utilisation efficiency, vortex strength, and particulate emission efficiency is described in detail. In Section 4, based on the numerical simulation, the proposed corner cut ratio is presented, and the findings of this study are summarised.

2. Research Object and Model Construction

2.1. Study Objects and Numerical Model Setup

The 8000 t-class aquaculture vessel studied is shown in Figure 1, with a length of 64.5 m, a beam of 17.0 m, a depth of 10.2 m, a draught of 7.7 m, a displacement of 8654.0 tonnes, a longitudinal centre of gravity of 32.3 m, a transverse centre of gravity of 0 m, a vertical centre of gravity of 4.7 m, a rolling radius of 5.7 m, a pitching radius of 20.6 m, and a yawing radius of 20.9 m. Each tank was 12.0 m long, 12.0 m wide, and 9.75 m high, and the volume of the water body in the tank at the highest liquid level was 1190 m^3. The bottom surface of the tank was about 1.5 m from the ship's baseline, the working liquid level was about 10.5 m high compared with the ship's baseline, and the maximum depth of the water in the tank was 9 m. Each tank was equipped with 16 inlet ports and 1 outlet port, with the inlet ports located at the corners of the wall surface of the aquaculture tank and the outlet ports located in the central position of the tank bottom. The inlet was located at the corner of the wall of the culture tank, and the outlet was located at the centre of the bottom. The inlets and outlets of the tanks were flow-controlled, but different tank types inevitably lead to changes in the volume of the water in the tanks. In order to ensure the one-way analysis of the corner ratio, the water in the tanks was always changed 16 times per day in this study.

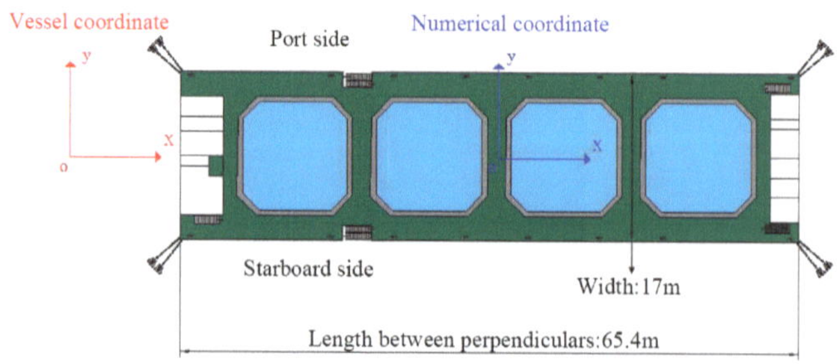

Figure 1. Sketch and marks of the aquaculture fish vessel.

2.2. Model Construction

In this study, the water body in the aquaculture vessel was a three-dimensional incompressible fluid, and the continuity equation of the control equation is

$$\frac{\partial u_i}{\partial x_i} = 0 \quad (1)$$

The momentum equation is

$$\begin{cases} \frac{\partial u}{\partial t} + \frac{1}{V_F}\left(uA_x\frac{\partial u}{\partial x} + vA_y\frac{\partial u}{\partial y} + wA_z\frac{\partial u}{\partial z}\right) = -\frac{1}{\rho}\frac{\partial p}{\partial x} + G_x + f_x - \frac{R_{SOR}}{\rho V_F}(u - u_s) \\ \frac{\partial v}{\partial t} + \frac{1}{V_F}\left(uA_x\frac{\partial v}{\partial x} + vA_y\frac{\partial v}{\partial y} + wA_z\frac{\partial v}{\partial z}\right) = -\frac{1}{\rho}\frac{\partial p}{\partial y} + G_y + f_y - \frac{R_{SOR}}{\rho V_F}(v - v_s) \\ \frac{\partial w}{\partial t} + \frac{1}{V_F}\left(uA_x\frac{\partial w}{\partial x} + vA_y\frac{\partial w}{\partial y} + wA_z\frac{\partial w}{\partial z}\right) = -\frac{1}{\rho}\frac{\partial p}{\partial z} + G_z + f_z - \frac{R_{SOR}}{\rho V_F}(w - w_s) \end{cases} \quad (2)$$

where (u, v, w) are the velocity components in the coordinate directions (x,y,z), respectively; (A_x, A_y, A_z) are the fractional area open to flow in each coordinate direction; (G_x, G_y, G_z) and (f_x, f_y, f_z) are the body accelerations and the vicious accelerations, respectively; (u_s, v_s, w_s) are the velocity components of the fluid from the surface of a source, where the shape and the normal direction of such surface can be arbitrary, and each velocity component is estimated according to the normal direction of the source surface; V_F is the fractional volume open to flow; R_{SOR} is the mass momentum source; ρ is the fluid density; and p is the pressure.

The turbulence model was the RNG K-ε turbulence model, which has a wider range of applicability and is better able to simulate rotating flows [27].

$$\frac{\partial k_T}{\partial t} + \frac{1}{V_F}\left(uA_x\frac{\partial k_T}{\partial x} + vA_y\frac{\partial k_T}{\partial y} + wA_z\frac{\partial k_T}{\partial z}\right) = P_T + D_{k_T} - \varepsilon_T \quad (3)$$

$$\frac{\partial \varepsilon_T}{\partial t} + \frac{1}{V_F}\left(uA_x\frac{\partial \varepsilon_T}{\partial x} + vA_y\frac{\partial \varepsilon_T}{\partial y} + wA_z\frac{\partial \varepsilon_T}{\partial z}\right) = c_1\frac{P_T\varepsilon_T}{k_T} + D_{\varepsilon_T} - c_2\frac{{\varepsilon_T}^2}{k_T} \quad (4)$$

where K_T and P_T are the turbulent kinetic energy and its generating term; ε_T is the turbulent dissipation rate; D_{k_T} and D_{ε_T} are the turbulent kinetic energy diffusion term and the turbulent dissipation diffusion term, respectively; V_F is the fractional volume; t is the time; and C_1 and C_2 are the coefficient constants, with C_1 = 1.44 and C_2 = 1.92 in the classical K-ε turbulence model.

To ensure the validity of the numerical model, the results obtained from the numerical model were compared with the experimental data of Liu [28]. Comparing the numerical simulation results with Liu's experimental results, the two sets of results have good consistency, indicating that the numerical model developed in this study is correct, and its application to the simulation of the flow field in the aquaculture vessel's tanks is feasible. The results of the numerical model validation are shown in Figure 2. In order to verify the influence of the number of model grids on the computational accuracy, a 7 s transverse rocking period and a 5° transverse rocking angle were used to verify the convergence of the grids. A uniform grid was used, and the grid was encrypted at the inlet and outlet of the ship, with sizes of 0.1 m, 0.15 m, and 0.4 m, respectively. The total number of grids was 28,000, 430,000, and 1,420,000, respectively, and the computational length was 1500 s. The velocity amplitude of the 0.1 m grid was calculated for the 0.1 s grid size. Taking the velocity amplitude calculated for the 0.1 m grid size as the reference, the velocity amplitudes for the 0.15 m and 0.4 m grids were 99.8% and 96.0%, respectively. Taking 98% as the convergence criterion and considering the calculation accuracy and efficiency, the following calculations were performed with a 0.15 m grid resolution.

The numerical simulation in this study was completed using the CFD software FLOW-3D, the solver version is 11.2.0.16. In the FLOW-3D software, the computational domain of the water body was realised by the Geometry module, and the model type was "Complement". The mesh structure adopted a 0.15 m uniform mesh, and a 0.05 m mesh was encrypted at the inlet and outlet. The grid structure was a 0.15 m uniform grid, and a 0.05 m grid was encrypted at the inlet and outlet. The fluid was 20 °C water. The boundary condition of the solid–liquid coupling interface of the tank was set as a no-slip boundary (wall). The boundary conditions of the gas–liquid coupling interface were set as the atmospheric pressure with a fluid fraction of 0. The inlet and outlet water flow of the tank was realised using the inlet source of the mass momentum source; the generation of particles was set as the generation of particle sources, with the particle sources at the corners of the tank. The z-axis height of the particle sources was higher than that of the water surface by 0.5 m. The rate of the generation of each particle source was 250 per second. The z-axis height of the particle sources was 0.5 m above the water surface. The rate of generation of each particle source was 250 particles per second, and the four particle sources generated 5000 particles in 5 s. The numerical model was based on the Cartesian grid coordinate system, and the area and volume porosity functions were used to analytically calculate the area boundary and type, called the FAVOR method (fractional area–volume obstacle representation method) [29].

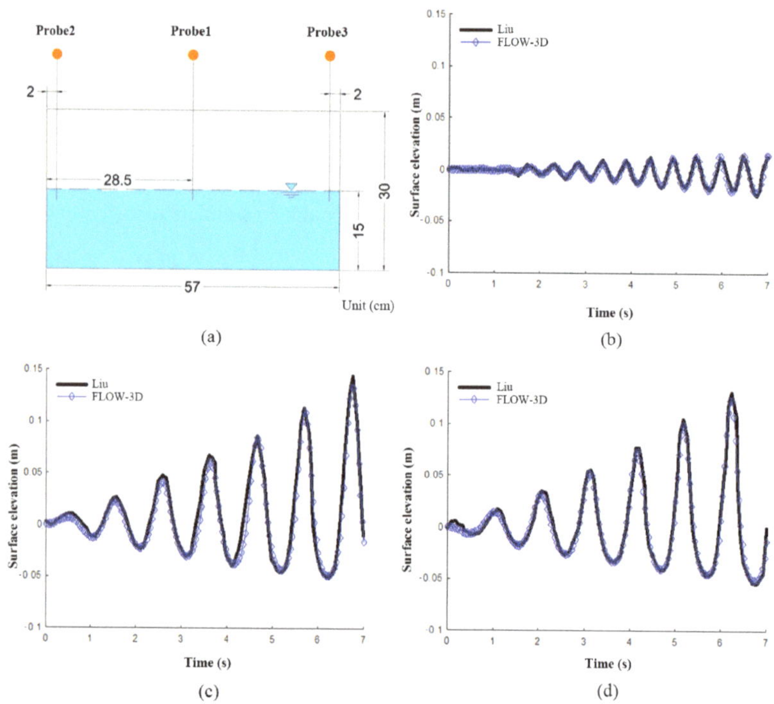

Figure 2. (**a**) Rectangular tank geometry model for numerical model validation; (**b**–**d**) show the variation curves of the flow velocity in Probes 1, 2, and 3, respectively.

2.3. Experimental Group

In this study, based on the model scale setting of the culture tank, the number of daily water changes was fixed, and numerical simulation calculations were conducted for different working conditions to analyse the influence of the corner ratio on the characteristics of the flow field in the tank. Figure 3 shows the schematic diagrams of different corner ratios. The selection of corner ratios was based on the width of the tank of 12 m, and the ratio of the corner distance of the aquaculture tank to the half-length of the tank was defined as the corner ratio C. The grouping of the numerical conditions of the aquaculture tank is shown in Table 1.

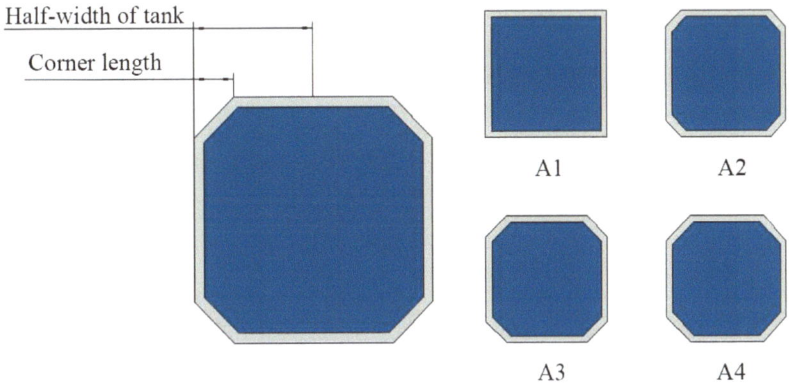

Figure 3. Situation groups.

Table 1. Simulation working condition groups.

	Corner Ratio	Corner Distance (m)	Water Inflow (m^3/s)	Water Outflow (m^3/s)
A1	0:1	0	43.43	694.81
A2	0.26:1	1.56	42.00	672.04
A3	0.33:1	2.0	41.06	656.88
A4	0.40:1	2.4	40.02	640.34

2.4. Data Processing

2.4.1. Drag Coefficient C_t

Oca et al. [18] defined the drag coefficient C_t as the energy consumed by a tank to overcome the drag that is equal to the energy supplied by the inlet impact force for a tank system operating in a steady-state flow field condition. The formula is as follows:

$$C_t = \frac{2Q(V_{in} - V_{avg})}{AV_{avg}^2} \tag{5}$$

where Q is the inlet flow rate, V_{in} is the inlet velocity, V_{avg} is the average velocity in the pool, and A is the wet week.

2.4.2. Energy Use Efficiency η_e

The energy of the water-body circulation motion is mainly provided by the inlet jet, and the energy loss mainly comes from overcoming the resistance of the breeding pool and the viscous resistance of the relative motion between the water-body mass points. Evaluating the energy conversion in the tank at different corner ratios, the change in the energy use efficiency η_e was analysed [30], and the energy use efficiency η_e was calculated as follows:

$$\eta_e = DU_{50}\left(\frac{V_{avg}}{V_{in}}\right)^2 \tag{6}$$

where DU_{50} [31] is calculated by the formula

$$DU_{50} = \frac{V_{50}}{V_{avg}} \tag{7}$$

where V_{50} is the average of 50 percent of the lower velocities in the breeding tank, V_{avg} is the average velocity in the tank, and V_{in} is the inlet velocity.

2.4.3. Percentage of Areas with Suitable Flow Rates

The size of the flow velocity in the aquaculture water environment has an important effect on the growth of fish and the discharge of particles, such as residual feed and faeces, to the environment [22,32,33]. The movement forced by the flow velocity can improve the muscle tone of cultured fish, which, in turn, improves the survival rate of the fish species after release [14]. However, too low a flow rate can lead to a significant decrease in the growth performance of cultured fish [33], while too high a flow rate can cause cultured fish to lose their swimming ability and even lead to fish death [34]. Many scholars have studied the appropriate flow rate for fish and found that the fishable flow rate is related to its body length (hereafter, "bl") [34], which is usually the distance from the end of the muzzle to the base of the caudal fin. Wang et al. investigated the growth performance of juvenile Hsu's flatfish at different flow rates and found that the growth performance of Hsu's flatfish was significantly improved in the 1.5 bl/s flow rate group compared with the 0 bl/s and 0.5 bl/s flow rate groups [33]. Timmerhaus et al. studied the effects of different flow rates (0.5, 1.0, 1.8, and 2.5 bl/s) on the growth and muscle development of juvenile Atlantic salmon (Salmo salar) and found that the growth rate increased with an increase in the flow rate, but when the flow rate was greater than 1.8 bl/s, the number of inflamed muscle fibres, gill lesions, and fin lesions increased significantly. The upper limit of the fishable flow rate was related to the bl of cultured fish and varied between species [34].

At present, in aquaculture vessel production, we generally select fry with a body length of about 0.1 m to start aquaculture and salvage and sell the fish when they grow to about 0.5 m in body length. We took the flow velocity of the tank for 1 bl fish as a reference, i.e., when the flow velocity V is controlled at 0.1 m/s–0.5 m/s, the effect of the flow velocity on the fish is relatively small; therefore, in this study, we set the fishery-suitable flow velocity at $0.1\ \mathrm{m/s} < V < 0.5\ \mathrm{m/s}$, the flow velocity V in the range of 0.1 m/s, and the proportion S of the fish-suitable flow velocity in the outboard chamber to 0.5 m/s ($V < 0.5\ \mathrm{m/s}$). The ratio of the number of grids in the range of 0.1 m/s and 0.5 m/s to the overall number of grids was taken as the proportion of the fishable flow velocity in the tank S. The formula for calculating the proportion of the fishable flow velocity S is as follows:

$$S = \frac{\sum n_{(0.1-0.5)}}{\sum n_t} \tag{8}$$

where n_t is the number of grids in the tank, and $n_{(0.1-0.5)}$ is the number of tanks with flow velocities V in the range of 0.1 m/s–0.5 m/s.

The flow field characteristics extracted by the RAS system are usually limited to a certain point and cross-section velocity. Compared with the RAS system, the size of an aquaculture tank in an aquaculture vessel is large, and the variation in the flow velocity data at individual points, lines, or planes in the flow field of the aquaculture tank cannot completely reflect the characteristics of the whole flow field. Therefore, in order to analyse the distribution characteristics of the flow velocity (V) in the tank more comprehensively, probability density (PD) and cumulative probability density (CPD) were used for statistical analysis in this study. By plotting CPD curves with variations in the flow velocity (V), the distribution of the flow velocity in the whole flow field of the tank can be more comprehensively investigated. In order to avoid the influence of the large flow velocity at the inlet and outlet on the overall data in the tank, the 99% quantile of the flow velocity data was used as the representative value of the maximum flow velocity. The maximum flow velocity in the tank was defined as the 99% quantile of the flow velocity data, denoted as $V_{99\%}$. In this study, the processing of data was performed using MATLAB R2021a.

3. Results and Discussion

3.1. Effect of Corner Ratio on the Percentage of Areas with Fishable Flow Rates

In order to analyse the effect of the ratio of the corner on the flow field in the tank, the flow field in the tank was numerically calculated for four working conditions, namely,

C = 0, C = 0.26, C = 0.33, and C = 0.40. Figure 4 shows the distribution of the $V_{99\%}$ flow velocity under different proportions of corners. As shown in Figure 4, the proportion of the corner has a significant effect on the distribution of the flow velocity in the tank. The flow velocity in the tank increases with the increase in the corner of the tank, and the average velocity of the four conditions shows a significant upward trend. For conditions A1 and A2, i.e., when the diameter-to-depth ratios are C = 0 and C = 0.26, the low-velocity area of the flow velocity in the tank (the area where the flow velocity is less than 0.1 m/s) is higher, the quality of the flow field in the tank is relatively low, and the flow velocity area that can satisfy the requirements of fish culture is 50% and 45%, respectively. Compared with conditions A1 and A2, the low-velocity area of the flow velocity in the tank is significantly reduced in conditions A3 and A4, i.e., when the diameter-to-depth ratio C > 0.33. Compared with the A1 and A2 conditions, the A3 and A4 conditions, i.e., when the diameter-to-depth ratio C > 0.33, the low-velocity region of the flow velocity in the tank is significantly reduced, and the quality of the flow field in the tank is higher at this time. The percentage of the flow velocity region that can satisfy the requirements of fish culture (0.1 m/s < V < 0.5 m/s) is 72% and 75%, respectively. The maximum flow velocities, $V_{90\%}$ and $V_{50\%}$, in the tanks with a corner design (e.g., with a corner ratio of 0.33:1) were significantly increased, which was consistent with the findings of Milad et al. in their CFD simulation of land-based tanks, and proved that the corner structure had a significant effect on the velocity of the flow field in the tanks [35].

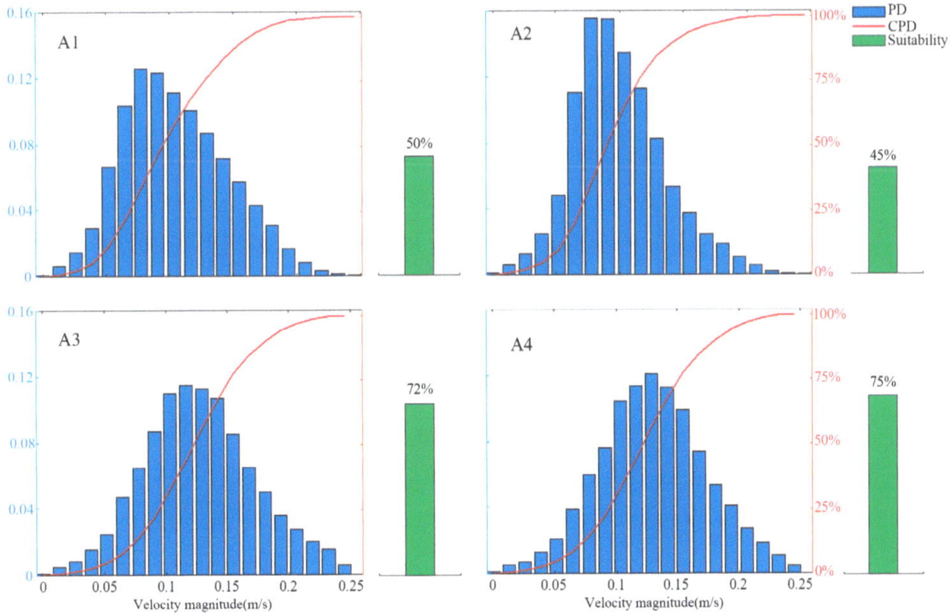

Figure 4. Probability density distribution of the streamflow and proportion of fishable area at each corner ratio.

3.2. Energy Transfer Impact

Figure 5 shows the variation in the flow velocity, residence time, drag coefficient, and energy conversion efficiency for different corner ratios. As shown in Figure 5a, the characteristic velocities obtained through statistical analysis are positively correlated with the corner ratio. An increase in the corner ratio leads to a decrease in the probability of a low flow velocity in the tank, and the velocity $V_{90\%}$ increases from 0.16 m/s to 0.19 m/s, while the velocity $V_{50\%}$ increases from 0.1 m/s to 0.13 m/s. In contrast, the water residence time is not sensitive to the corner ratio, as shown in Figure 5a. The residence time $T_{90\%}$

increases slightly, the median residence time changes less, the residence time $T_{90\%}$ increases from 2727 s to 2811 s, and the residence time $V_{50\%}$ always fluctuates above and below 2100 s. As shown in Figure 5b, the drag coefficient C_t and the energy conversion efficiency in the tank increase and then decrease and finally increase with an increase in the corner ratio C of the tank. When the ratio C > 0.26 in the corner of the tank, the value of the drag coefficient C_t in the tank shows a significant decline, and the efficiency of the energy utilisation shows a significant growth trend.

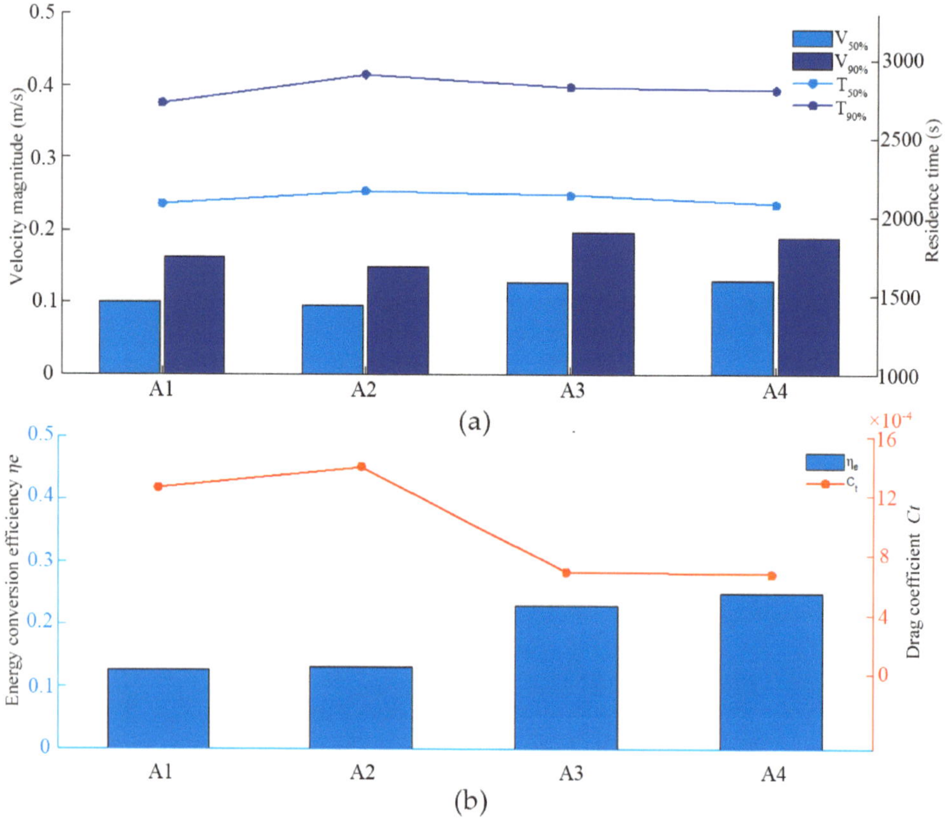

Figure 5. (**a**) The variation in the statistical flow rate versus the variation in the residence time. (**b**) The variation in the drag coefficient C_t versus the capacity utilisation efficiency in the tank.

In the A3 and A4 conditions, the wall surface of the culture tank with a tangent angle smoothly guides the incident water movement trajectory to steer, which slows down the impact of the incident water flow. The energy loss caused by the water flow-steering process is changed into energy to maintain the movement of the water body, and the flow velocity in the tank is increased while the discharge capacity is enhanced. When the corner ratio C > 0.33, the effect of the corner ratio on the flow velocity and the proportion of the fishable area in the tank are small.

3.3. Local Vectors at Corners

Figure 6 shows the flow velocity distribution characteristics of the surface layer (Z = 7.5 m), middle layer (Z = 4.5 m), and bottom layer (Z = 0.5 m) profiles with four corner ratios. Overall, the water in the tank forms a counterclockwise-rotating flow under the inlet push current, and there is a low-flow velocity area in the centre of the tank. The magnitude of the flow velocity gradually increases from the vertical centre to the wall of the tank, and with the decrease in the liquid height, the flow velocity shows a decreasing trend. In the case of the square tanks, the corners of the tanks show a low flow velocity in the stagnant water area, and in the other three cases, no low-velocity area is found in the corners.

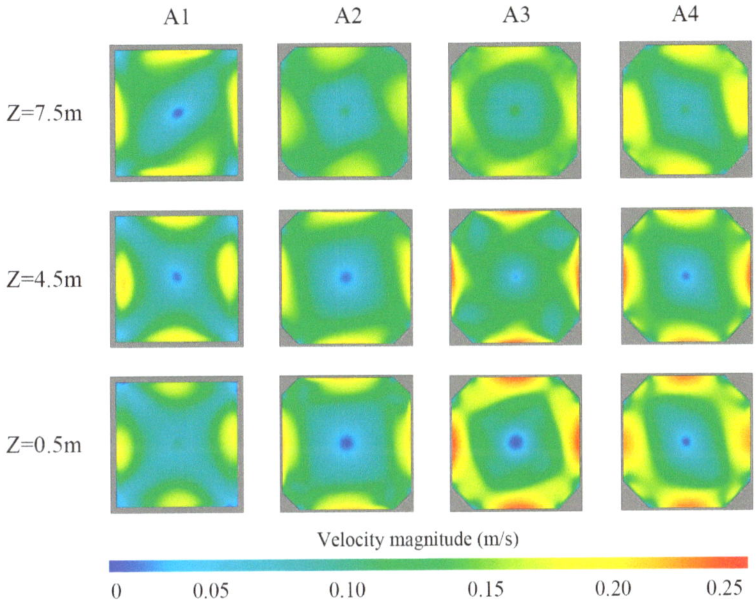

Figure 6. Flow velocity cloud at Z = 7.5 m, Z = 4.5 m, and Z = 0.5 m profiles for different corner scales.

Figure 7 shows the flow velocity vector map at the corner. When the square culture tank without a corner or corner distance is small, the water body mass point movement regularity is poor, and the flow pattern is turbulent. With the increase in the corner distance of the square culture tank, the corner of the water flow guidance strengthens the role of the water body to reduce the water body mass point and the wall between the collision. The irregular collision between the water body mass point is also relatively reduced. The trajectory is gradually changed to the centre of the tank around the spinning flow, and the flow pattern is more stable. Its trajectory gradually changes to a rotating flow around the centre of the culture chamber, and the flow pattern is more stable. This proves that the corner structure can reduce the proportion of the low-speed zone in the tank, and the larger the distance between the corners of the square tank and the larger the distance between the corners, i.e., where the tank tends to be rounded, the better the tank's velocity distribution.

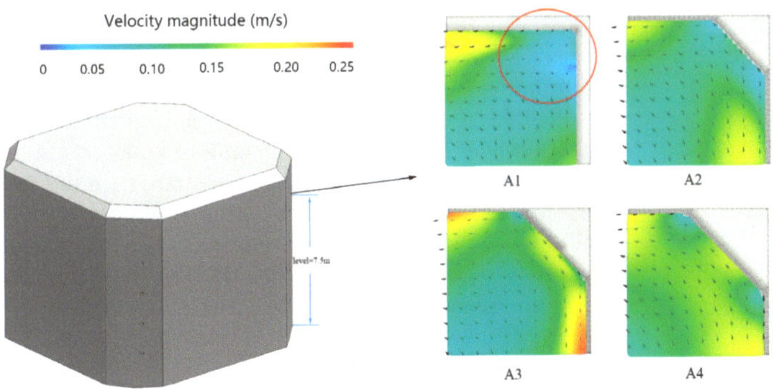

Figure 7. Vector diagram of the flow field at the corners.

3.4. Impact of Effluent Efficiency

Solid particles produced by organisms in tanks need to be quickly removed as their hydrolysis, leading to decomposition, reduces dissolved oxygen levels and releases organic molecules, finely suspended solids, and ammonia, which can cause water quality problems. Another danger of using seawater in aquaculture vessels travelling for long periods of time in the deep sea is that toxic H2S can also accumulate in the tanks, and this can be harmful to fish if the solids are not rinsed away quickly [36,37]. To eliminate particle build-up, the design of the tanks needs to be self-cleaning so that fish growth and welfare are not negatively affected. By monitoring the mean kinetic energy and turbulent kinetic energy of the fluid in the computational grid in FLOW-3D, it was found that the computation converged to a steady state after 1500 s. At 1800 s, the particles started to drop, with a total of 5000 particles with a diameter of 1 cm, and the particles were released at the free surface with an initial downward velocity of 1 cm/s. Figure 8 shows the combined information on the particle dispersion in the tank.

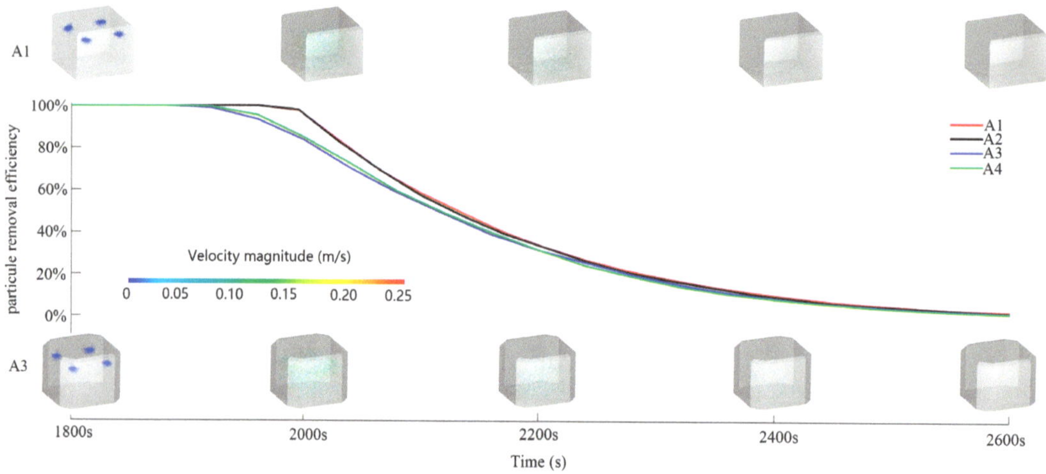

Figure 8. Efficiency of particulate exclusion in tanks.

As shown in Figure 8, the solid particle exclusion effect of the A1 and A2 conditions was always in the lead in the 1000 s after particle placement, and the particle visualisation also shows that there were more low-flow particles in the culture tanks in the A1 and A2 conditions compared with the A3 and A4 conditions. After 60% of the particles were excluded, the particle removal efficiencies of the four conditions gradually approached each other. As mentioned before, in condition A1, the overall flow velocity in the tank was low, and there were obvious low-flow velocity zones in the corners and centre of the tank. The existence of these low-flow velocity zones made it easy for impurities such as faeces and feed residues to be deposited and difficult to remove, which led to the accumulation of waste and pollution at the bottom water, which increases the risk of disease in the aquaculture and might even trigger the spread of parasites and diseases. In actual operation, the organisms in culture tanks continuously produce faeces; therefore, tanks with a corner-cutting design can strengthen the solid particle exclusion efficiency, which makes these tanks obtain better fishability and also ensures good water quality conditions. When the corner-cutting ratio $C > 0.33$, the corner-cutting ratio has a small effect on the particle exclusion efficiency, and the increase in the corner-cutting ratio yields limited improvement to the particle exclusion efficiency.

3.5. Vortex Structural Characteristics

In addition to flow velocities, vortices in the flow field are also important factors affecting the water environment in tanks. The combination of tangential and axial movements of fluids in tanks generates vortices and plays an important role in the transfer of momentum in the water column of the tanks [38,39]. In recent decades, many vortex identification methods have been developed to describe the vortex structure of fluid flow as a result of in-depth research by scholars. The q-method is one of the most commonly used vortex structure visualisation methods, and in practical applications, the Q-value is usually used to identify vortex structures in fluids [13]. When the Q-value is greater than zero, it indicates that the rotational rate is greater than the shear rate, which helps to identify potential vortex regions. Therefore, the Q-value is often used to visualise vortexes in fluids, especially in the study of complex flow phenomena (e.g., turbulence and vortex motions). An excellent and stable flow field forms a flow regime that generates rotation at the centre of the breeding tank. When monitoring the mean kinetic energy and mean turbulent kinetic energy of the fluid in the computational grid in FLOW-3D, the computation tended to exhibit a steady state after 1500 s.

Figure 9 shows the vortex intensity at 2500 s for each condition, and according to the Q criterion, the vortex maps at Q = 0.05 were selected. The vortex rings around the inlet and outlet are the two main vortex areas, which are closely related to the transfer of turbulent energy in the tank. An increase in the vortex intensity can strengthen the efficiency of tank discharge, but the torque generated by excessive vortex can cause fish to capsize and lose balance. Fish use their pectoral fins to restore balance and sharply increase their body's hydrodynamic resistance and spatial equilibrium energy consumption to resist capsizing caused by vortex filaments of different scales, which make them swim slower or even lose their swimming ability [40,41]. As shown in Figure 9, when the corner ratio is less than 0.33, the large-scale vortex structure breaks down into many close-to-small-scale structures, vortex filament structures, and irregular vortex distributions. The colouring of the flow velocity also shows that the flow velocity is higher near the tank wall. When the ratio of the angle of tangency is greater than 0.33, the vortex column in the centre of the tank is gradually stabilised, and the water moves regularly inside the tank, gradually appearing as a large vortex ring.

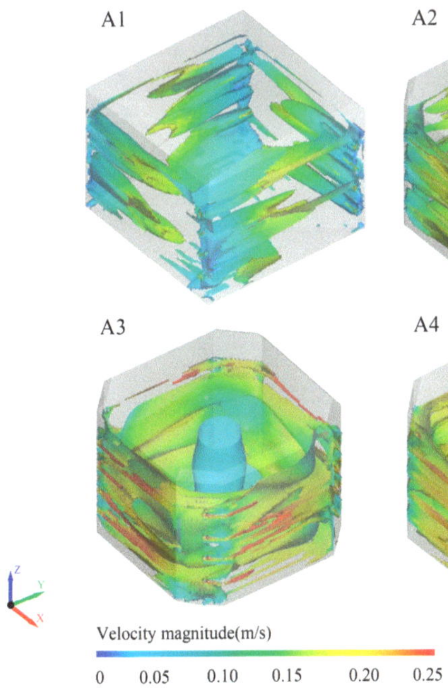

Figure 9. Characteristics of vortex strength for different corner ratios.

4. Conclusions

In this study, the flow field in the aquaculture tank of an 8000 t-class aquaculture vessel was analysed using computational fluid dynamics (CFD) methods. The effect of the proportion of tank corners on the flow field of the aquaculture tank was calculated, and the efficiency of particle removal in the flow field was analysed. Based on the calculation results, a reasonable scheme was given, which provides the necessary parameters for the development and design of aquaculture tanks in aquaculture vessels and offers important reference engineering value.

(1) From the perspective of the fishability of tanks, for the same culture volume, tanks with a larger corner ratio C have better hydrodynamic characteristics, including a more uniform velocity distribution, a smaller share of the low-speed zone, and a high intensity of eddies, which also contribute to the mixing of dissolved oxygen and the aggregation and discharge of solid particles.

(2) From the point of view of circulating water utilisation efficiency, when the corner ratio C is small, the average speed of the aquaculture vessel is low, the input energy utilisation efficiency of the jet is low, and the jet speed has to be increased in order to maintain the appropriate speed range, which, in turn, generates more wastewater and reduces the overall circulating water utilisation and economic efficiency of the aquaculture vessel tanks.

(3) From the point of view of the space utilisation of culture tanks, the utilisation of the culture space is reduced when the proportion of corners in square tanks is large, thereby increasing the bioculture density, which is not conducive to fish welfare.

When the corner structure is not used, the overall flow pattern in the tank is poor, and problems such as low energy utilisation efficiency, poor particulate exclusion effect, and low vortex strength are present. This tank type is not recommended. When the corner ratio C = 0.26, the tank has a high volume of cultured water, but the flow field in the tank is weak. An excellent flow field can be maintained by changing the position of the inlet and outlet ports or by increasing the number of daily water exchanges and other measures. When the corner ratio C = 0.33, the square-cut aquaculture vessel has a higher comprehensive performance, including a higher fishable flow rate, energy use efficiency, and particulate removal efficiency, as well as higher flow field characteristics and aquaculture volume. This helps to improve the overall economic efficiency of the aquaculture vessel and should be considered a priority in the construction of aquaculture vessels. When the corner ratio C = 0.4, the tank has an excellent flow field performance and can reduce the number of daily water changes to save energy loss under the condition of meeting the requirements of fishability. When the corner ratio C = 0.4, the flow field in the chamber is excellent, and the number of daily water changes can be appropriately reduced to save energy losses, provided that the requirements for fishability are met.

When the aquaculture vessel is cruising, the large transverse rocking of the vessel will lead to a large increase in the flow velocity in the tanks. Hence, the above conclusions are only applicable to the case of vessels at anchor or operating in small winds and waves. This study's results can still be used as a reference for the design and construction of aquaculture vessels, which are mostly at anchor, except when there is an impending typhoon or unsuitable water temperature.

During the actual operations of aquaculture vessels, both fish density and tail-swinging characteristics can impact the flow field in the tanks. Future work extending this study will consider the effect of fish characteristics on the flow field characteristics in the tanks, where flow velocities can be significantly reduced when fish densities are too high and where fish wagging can result in localised increases in flow velocities within the flow field. Fish characteristics will inevitably create complex flow dynamics within the tanks, so flow velocities and effluent efficiencies, for example, become more difficult to predict. In subsequent research, we plan to use the multiple reference frame (MRF) technique to model the cultured fish body. The MRF method can be used to complement and improve this study. The results of this study will be considered in future computational studies of tanks.

Author Contributions: Methodology, F.Z. and M.C.; validation, F.Z. and C.Z.; writing—original draft preparation, F.Z.; writing—review and editing, M.C. and H.L. All authors have read and agreed to the published version of the manuscript.

Funding: This study was supported by the Key R&D Program of Shandong Province, China (no. 2021SFGC0701), the National Key Research and Development Program of China (no. 2022YFD2401101), and the Central Public-Interest Scientific Institution Basal Research Fund (CAFS) (no. 2023TD84).

Institutional Review Board Statement: Not applicable.

Informed Consent Statement: Not applicable.

Data Availability Statement: The data supporting the reported results can be made available upon reasonable request.

Conflicts of Interest: This research was conducted in the absence of any commercial or financial relationships that could be construed as a potential conflict of interest.

References

1. FAO. *The State of World Fisheries and Aquaculture 2024*; Blue Transformation in Action; FAO: Rome, Italy, 2024.
2. Zhang, Y.; Ni, Q.; Liu, H.; Zhang, C. Status quo of industrialized aquaculture of Atlantic salmon in Norway and its implications for China. *Trans. Chin. Soc. Agric. Eng.* **2020**, *36*, 310–315.
3. Xu, H.; Chen, Z.; Cai, J.; Huang, Y.; Liu, H. Research on the development of deep sea aquaculture engineering equipment in China. *Fish. Mod.* **2016**, *43*, 1–6.
4. Guo, Y.C.; Mohapatra, S.C.; Guedes, S.C. Review of developments in porous membranes and net-type structures for breakwaters and fish cages. *Ocean Eng.* **2020**, *200*, 107027. [CrossRef]
5. Xu, Z.; Qin, H. Fluid-structure interactions of cage based aquaculture: From structures to organisms. *Ocean Eng.* **2020**, *217*, 107961. [CrossRef]
6. Huang, X.; Pang, G.; Yuan, T.; Hu, Y.; Wang, S.; Guo, G.; Tao, Q.Y. Review of engineering and equipment technologies for deep-sea cage aquaculture in China. *Prog. Fish. Sci.* **2022**, *43*, 121–131.
7. Xu, H.; Jiang, T. Development strategy of offshore aquaculture engineering in China. *Fish. Mod.* **2012**, *39*, 1–7.
8. Cai, J.; Zhang, Y.; Li, J. General technology research of 100 thousand ton deep sea aquaculture platform. *Ship Eng.* **2017**, *39*, 198–203.
9. Cui, M.; Jin, J.; Huang, W. Discussion about system construction and general technology of aquaculture platform. *Fish. Mod.* **2019**, *46*, 61–66.
10. Cui, M.; Wang, Q.; Zhang, B. Numerical Analysis on Water Flow Field Characteristics of Aquaculture Vessel under Pitch Motion. *Ship Eng.* **2020**, *10*, 56–60.
11. Tao, Y.; Zhu, R.; Gu, J.; Li, Z.; Zhang, Z.; Xu, X. Experimental and numerical investigation of the hydrodynamic response of an aquaculture vessel. *Ocean Eng.* **2023**, *279*, 114505. [CrossRef]
12. Wang, W.; Li, M.; Fan, G.; Zhang, K.; Huang, Y. Engineering. Analysis of Fluid Field in Fish Tank of Breeding Vessel with Perforated Broadsides under Wave Conditions. *J. Mar. Sci. Eng.* **2024**, *12*, 367. [CrossRef]
13. Wang, Y.; Li, Z.; Guo, X.; Cui, M. Numerical investigation on the flow field in prototype aquaculture tanks. *Fish. Mod.* **2022**, *49*, 24–34.
14. Burrows, R.E.; Chenoweth, H.H. The Rectangular Circulating Rearing Pond. *Progress. Fish-Cult.* **1970**, *32*, 67–80. [CrossRef]
15. Zhang, J.; Wang, M.H.; Jia, G.C. Effect of Structures on Hydrodynamic Characteristics of Recirculating Aquaculture Pond. *Trans. Chin. Soc. Agric. Mach.* **2022**, *53*, 311–320.
16. Summerfelt, S.T.; Davidson, J.W.; Waldrop, T.B.; Tsukuda, S.M.; Bebak-Williams, J. A partial-reuse system for coldwater aquaculture. *Aquac. Eng.* **2004**, *31*, 157–181. [CrossRef]
17. Oca, J.; Masaló, I.; Reig, L. Comparative analysis of flow patterns in aquaculture rectangular tanks with different water inlet characteristics. *Aquac. Eng.* **2004**, *31*, 221–236. [CrossRef]
18. Oca, J.; Masaló, I. Design criteria for rotating flow cells in rectangular aquaculture tanks. *Aquac. Eng.* **2007**, *36*, 36–44. [CrossRef]
19. Lee, J.-V.; Loo, L.; Chuah, Y.; Tang, P.; Tan, Y.; Wong, C. The design of a culture tank in an automated recirculating aquaculture system. *Int. J. Eng. Appl. Sci.* **2013**, *2*, 67–77.
20. Plew, D.R.; Klebert, P.; Rosten, T.W.; Aspaas, S.; Birkevold, J. Changes to flow and turbulence caused by different concentrations of fish in a circular tank. *J. Hydraul. Res.* **2015**, *53*, 364–383. [CrossRef]
21. Liu, Y.; Liu, B.L.; Lei, J.L.; Guan, C.T.; Huang, B. Numerical simulation of the hydrodynamics within octagonal tanks in recirculating aquaculture systems. *Chin. J. Oceanol. Limnol.* **2017**, *35*, 912–920. [CrossRef]

22. Guo, X.; Li, Z.; Cui, M.; Wang, B. Numerical investigation on flow characteristics of water in the fish tank on a force-rolling aquaculture platform. *Ocean Eng.* **2020**, *217*, 107936. [CrossRef]
23. Li, Z.; Guo, X.; Cui, M. Numerical investigation of flow characteristics in a rearing tank aboard an aquaculture vessel. *Aquac. Eng.* **2022**, *98*, 102272. [CrossRef]
24. Cui, M.; Li, Z.; Zhang, C.; Guo, X. Statistical investigation into the flow field of closed aquaculture tanks aboard a platform under periodic oscillation. *Ocean Eng.* **2022**, *248*, 110677. [CrossRef]
25. Xiong, Z.X.; He, M.X.; Zhu, W.Y.; Sun, Y.; Hou, X.R. Analysis of Flow Field Characteristics of Aquaculture Cabin of Aquaculture Ship. *J. Mar. Sci. Eng.* **2023**, *11*, 390. [CrossRef]
26. Xue, B.R.; Zhao, Y.P.; Bi, C.W.; Cheng, Y.; Ren, X.Z.; Liu, Y. Investigation of flow field and pollutant particle distribution in the aquaculture tank for fish farming based on computational fluid dynamics. *Comput. Electron. Agric.* **2022**, *200*, 107243. [CrossRef]
27. Pope, S.B. Turbulent flows. *J. Meas. Sci. Technol.* **2001**, *12*, 2020–2021. [CrossRef]
28. Liu, D.; Lin, P. Three-dimensional liquid sloshing in a tank with baffles. *Ocean Eng.* **2009**, *36*, 202–212. [CrossRef]
29. Hirt, C.; Sicilian, J. A porosity technique for the definition of obstacles in rectangular cell meshes. In Proceedings of the International Conference on Numerical Ship Hydrodynamics, 4th, Washington, DC, USA, 24–27 September 1985.
30. Xue, B.-r.; Yu, L.-p.; Zhang, Q.; Ren, X.-z.; Bi, C.-w. A numerical study of the effect of relative inflow distance on hydrodynamic characteristics in the single-drain rectangular aquaculture tank with arc angles. *J. Fish. China* **2021**, *45*, 444–452.
31. Masalo, I.; Oca, J. Influence of fish swimming on the flow pattern of circular tanks. *Aquac. Eng.* **2016**, *74*, 84–95. [CrossRef]
32. Gorle, J.M.R.; Terjesen, B.F.; Summerfelt, S.T. Hydrodynamics of Atlantic salmon culture tank: Effect of inlet nozzle angle on the velocity field. *Comput. Electron. Agric.* **2019**, *158*, 79–91. [CrossRef]
33. Wang, J.; Zhang, J.; Zhang, X.; Li, H.-X.; Hu, Y.; Ma, Z. Flow velocity on growth and behavior in black rockfish (*Sebastes schlegelii*). *Acta Hydrobiol. Sin.* **2023**, *47*, 973–981.
34. Timmerhaus, G.; Lazado, C.C.; Cabillon, N.A.R.; Reiten, B.K.M.; Johansen, L.-H. The optimum velocity for Atlantic salmon post-smolts in RAS is a compromise between muscle growth and fish welfare. *Aquaculture* **2021**, *532*, 736076. [CrossRef]
35. Moghadam, M.M.; Islami, H.R.; Ezam, M.; Mousavi, S.A. Optimizing flow uniformity and velocity fields in Aquaculture Tanks by modifying water inlets and nozzles arrangement: A Computational Fluid Dynamics Study. *Aquac. Eng.* **2024**, *106*, 102431. [CrossRef]
36. Letelier-Gordo, C.O.; Aalto, S.L.; Suurnäkki, S.; Pedersen, P.B. Increased sulfate availability in saline water promotes hydrogen sulfide production in fish organic waste. *Aquac. Eng.* **2020**, *89*, 102062. [CrossRef]
37. Gorle, J.M.R.; Terjesen, B.F.; Summerfelt, S.T. Influence of inlet and outlet placement on the hydrodynamics of culture tanks for Atlantic salmon. *Int. J. Mech. Sci.* **2020**, *188*, 105944. [CrossRef]
38. Lingfei, Z.; Boru, X.; Yunpeng, Z. The influence of The bend ratios on The hydrodynamic characteristics in The tank of aquaculture vessel. *Fish. Mod.* **2024**, *51*, 22–31.
39. Schlatter, P.; Orlu, R. Quantifying the interaction between large and small scales in wall-bounded turbulent flows: A note of caution. *Phys. Fluids* **2010**, *22*, 051704. [CrossRef]
40. Lupandin, A.I. Effect of flow turbulence on swimming speed of fish. *Biol. Bull.* **2005**, *32*, 461–466. [CrossRef]
41. Hockley, F.A.; Wilson, C.; Brew, A.; Cable, J. Fish responses to flow velocity and turbulence in relation to size, sex and parasite load. *J. R. Soc. Interface* **2014**, *11*, 20130814. [CrossRef]

Journal of
Marine Science and Engineering

MDPI

Article

A Method for Estimating the Distribution of *Trachinotus ovatus* in Marine Cages Based on Omnidirectional Scanning Sonar

Yu Hu [1,2], Jiazhen Hu [1,3], Pengqi Sun [1,3], Guohao Zhu [1,3], Jialong Sun [3,4], Qiyou Tao [1,5], Taiping Yuan [1,2], Gen Li [1,2], Guoliang Pang [1,2,5] and Xiaohua Huang [1,*]

1 Key Laboratory of South China Sea Fishery Resources Exploitation & Utilization, Ministry of Agriculture and Rural Affairs, South China Sea Fisheries Research Institute, Chinese Academy of Fishery Sciences, Guangzhou 510300, China; huyu@scsfri.ac.cn (Y.H.)
2 Tropical Fisheries Research and Development Center, South China Sea Fisheries Research Institute, Chinese Academy of Fishery Sciences, Sanya 572018, China
3 School of Geomatics and Marine Information, Jiangsu Ocean University, Lianyungang 222005, China
4 Jiangsu Marine Resources Development Research Institute, Lianyungang 222005, China
5 Southern Marine Science and Engineering Guangdong Laboratory (Zhuhai), Zhuhai 519000, China
* Correspondence: huangxhua@scsfri.ac.cn; Tel.: +86-20-3406-6940

Abstract: In order to accurately estimate the distribution of *Trachinotus ovatus* in marine cages, a novel method was developed using omnidirectional scanning sonar and deep-learning techniques. This method involved differentiating water layers and clustering data layer by layer to achieve precise location estimation. The approach comprised two main components: fish identification and fish clustering. Firstly, omnidirectional scanning sonar was employed to perform spiral detection within marine cages, capturing fish image data. These images were then labeled to construct a training dataset for an enhanced CS-YOLOv8s model. After training, the CS-YOLOv8s model was used to identify and locate fish within the images. Secondly, the cages were divided into water layers with depth intervals of 40 cm. The identification coordinate data for each water layer were clustered using the DBSCAN method to generate location coordinates for the fish in each layer. Finally, the coordinate data from all water layers were consolidated to determine the overall distribution of fish within the cage. This method was shown, through multiple experimental results, to effectively estimate the distribution of *Trachinotus ovatus* in marine cages, closely matching the distributions detected manually.

Keywords: marine cage; distribution of fish; omnidirectional scanning sonar; density clustering; *Trachinotus ovatus*

Citation: Hu, Y.; Hu, J.; Sun, P.; Zhu, G.; Sun, J.; Tao, Q.; Yuan, T.; Li, G.; Pang, G.; Huang, X. A Method for Estimating the Distribution of *Trachinotus ovatus* in Marine Cages Based on Omnidirectional Scanning Sonar. *J. Mar. Sci. Eng.* **2024**, *12*, 1571. https://doi.org/

Academic Editors: Zhenhua Ma, Jianguang Qin and Angelo Rubino

Received: 13 July 2024
Revised: 31 August 2024
Accepted: 4 September 2024
Published: 6 September 2024

1. Introduction

Cage aquaculture is the main aquaculture mode for the development of modern marine pastoralism [1], playing an indispensable role in the conservation and utilization of marine resources. To ensure culture efficiency and fish health, simply estimating the distribution of fish in the marine cages is insufficient for a comprehensive understanding of the state of aquaculture [2,3]. The distribution of fish in marine cage, especially at different time periods and in different areas, is important for optimizing feeding management, preventing disease outbreaks, and improving culture efficiency [4]. Firstly, the distribution of fish in the marine cage at different time periods is affected by a variety of factors that are not, or are only partially, controllable in open sea cages by aquaculture managers; these include daylight, temperature [5], feed placement time, waves [6], and other environmental conditions. Understanding these distribution patterns is useful for aquaculture managers to determine when fish reach satiation during feeding, so that they can adjust feeding strategies, such as optimizing the timing and location of feed placement, thereby improving feed utilization and reducing waste. Secondly, understanding the spatial distribution of fish helps to identify normal and abnormal behavior, serving as an indicator of sub-optimal rearing conditions or good management practices. Dense fish populations are more likely to lead to increased stress, skin or fin damage, and the rapid spread of diseases. By understanding the spatial distribution of fish and adjusting high-density areas, the risk of disease is effectively reduced [7]. Therefore, accurately estimating the distribution of fish in marine cage culture is an important challenge [8].

The estimation methods for the distribution of fish in cages mainly include three categories: acoustic, optical, and acousto-optical methods [9]. However, acousto-optical devices are too expensive, and the implementation conditions and steps are too cumbersome. Therefore, the current mainstream methods are acoustic and optical methods. Optical methods have advantages such as non-invasiveness and high resolution [10]. Some scholars have proposed methods of monitoring and estimation with different optical devices and have achieved good results [11–13]. Under seawater, optical devices have low visibility; the close-range resolution is high, but the long-range resolution is low. The resolution is also very low when the seawater is turbid. Therefore, it is difficult to fully exploit the advantages of optical methods. By contrast, the detection range of acoustic methods is larger, and the resolution is more constant. Data can be acquired with various sonar devices to generate image information and obtain data on the fish in marine cages [14]. In recent years, the utilization of sonar devices has achieved some positive results in the monitoring of fishery resources [15–19]. Feng et al. [20] proposed an automatic estimation method for fish abundance in sonar images. This method was based on an improved multi-column convolutional neural network (MCNN) and supplemented by multi-expansion rate fusion loss, improving the accuracy of the model and enhancing the level of information technology in aquaculture. Duffy et al. [21] used side-scan sonar and underwater video to draw sediment maps of reefs near the northern coast of Illinois in Lake Michigan to describe and quantify the distribution of lake trout and the eggs laid. Jing et al. [22] used DIDSON imaging sonar to obtain water data, calculate the average density of objects, and estimate the quantity of fish in the water in combination with the water area. This method also had high statistical accuracy. Chevallay et al. [23] combined small sonar tags with active acoustic and high-resolution GPS, pressure, motion, and optical sensors to explore the oceanographic parameters and distribution of seal-aimed pelagic fish and the behaviors of seals. Cui et al. [24] proposed a fish quantity estimation method based on imaging sonar. In this method, denoising, edge extraction, object segmentation, and other treatments were performed on the data acquired by the sonar to determine the image features of fish. Afterward, the measurement area of the sonar was counted to calculate the average volume density and realize the estimation of fish quantity in combination with the area of the culture area. The above studies have shown that monitoring fish with sonar is efficient and convenient. However, the opening angle problem of the sonic wave of the sonar during operation results in a limited scanning area and the existence of visual blind spots in the detected data, leading to double counting or the missed counting of fish. This has a significant influence on the estimation of fish distribution. Therefore, solving the visual blind spots during sonar measurement needs to be further explored to estimate fish distribution more accurately.

In summary, this study proposes a method for fish distribution estimation in marine cage aquaculture based on omnidirectional scanning sonar. This method mainly involves fish identification and counting as well as fish clustering. The method achieves distribution estimation through water layer differencing and column-by-column clustering, which not only estimates the number of fish in aquaculture cages but also provides detailed information about fish distribution. This detailed information could offer effective data support for improving aquaculture efficiency and management levels.

2. Materials and Methods

2.1. Experimental Materials and Collection

2.1.1. Experimental Site and Biological Samples

The experimental data were measured and obtained in the adjacent waters of Guishan Island, Zhuhai, Guangdong province (113.84473° E, 22.12571° N, salinity 21.5‰ and visibility of approximately 1.5 m) on 7 January, 2024. The satellite map of the experimental area and the aerial view of the experimental site are shown in Figures 1 and 2, respectively.

Figure 1. Satellite image map of the experimental site.

Figure 2. Aerial view of the experimental site.

The cage used in this experiment is shown in Figure 3. The dimensions of the cage are 4 m × 4 m × 4 m. The four corners of the cage had additional weights to ensure that it remained stable underwater. However, affected by the buoyancy of seawater, the depth of the cage was approximately 2.9 m. The fish used in this experiment were *Trachinotus ovatus* under normal culture conditions. The body length of the fish was approximately 24 cm (Figure 4).

Figure 3. Cage used in the experiment.

(a) (b)

Figure 4. *Trachinotus ovatus* used in the experiment. (**a**) Body length; (**b**) Body height.

2.1.2. Device and Operating Mode

The device used in this paper was the omnidirectional scanning imaging sonar Scanfish-II (A2), launched by T-SEA Marine Technology Co., Ltd. (Zhangjiagang, China) This small imaging sonar was designed for underwater detection. Its specific technical indexes are shown in Table 1.

Table 1. Technical indexes of high-frequency horizontal mechanical scanning sonar.

Item	Technical Index
Operating frequency	667 kHz
Range of action	3 m
Beam opening angle(Vertical × Horizontal)	$(7.5 \pm 1.5)^\circ \times (2.6 \pm 0.3)^\circ$
Distance pixel/precision	5 cm
Scanning area (polar coordinates)	360°
Maximum working depth	3 m
Pan tilt rotation frequency	2.5 s/circle
Cylinder size	Φ57 mm × 99 mm

The working principle of the sonar is shown in Figure 5a. The range of the sonar is denoted as R, with a beam opening angle of 2.6° horizontally and 7.5° vertically. The sonar emits ultrasonic waves, and the beam is rotated sequentially through the gimbal to achieve 360° horizontal measurement.

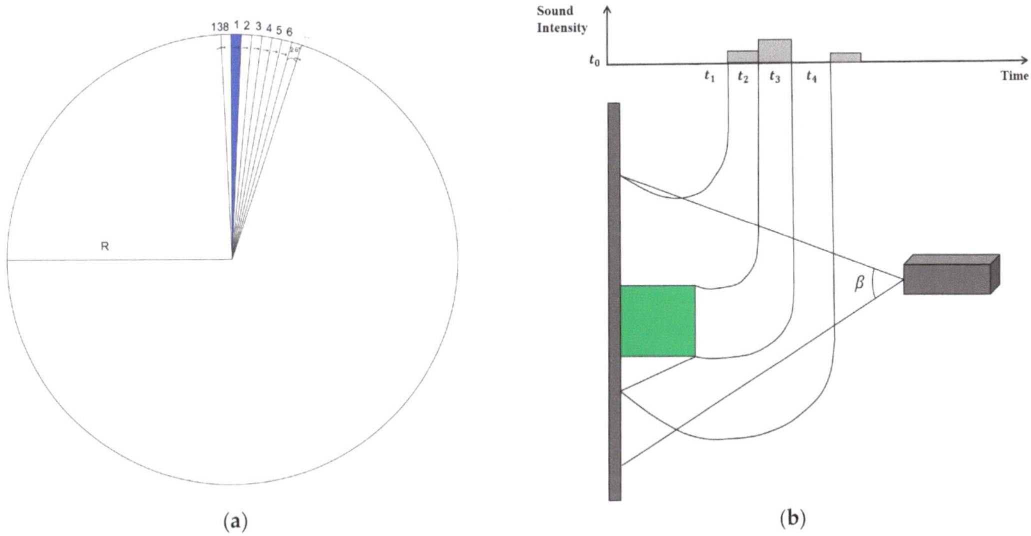

Figure 5. Working principle diagram of omnidirectional scanning sonar. (**a**) Schematic diagram of sonar scanning; (**b**) Work diagram.

In the measurement of each detection angle, the sonar emits a sound signal forward, and the echo signal reflected from the target object within the beam's irradiation area is received by the sonar for imaging. The detection beam emitted by the sonar is horizontally divided into several fan-shaped beams with vertical opening angles. Each beam illuminates the target, as shown in Figure 5b, forming a set of distance and intensity information. The echo intensity information of all beams is arranged based on position relationship to create the sonar image. The intensity of the echo signal reflects the acoustic reflection characteristics of different target objects, while the resulting time delay is used to determine the actual distance of the target object relative to the sonar position.

2.1.3. Data Acquisition

This experiment was a quantitative and temperature study that compared different quantities and culture temperatures of *Trachinotus ovatus*, with a difference of 50 fish in each group. Two hundred of the tens of thousands of fish cultured in the same batch were caught for the experiment, so there was no great disparity between the individual fish.

The sonar is shown in Figure 6, and it was designed to be connected to the end of the silver electric telescopic rod depicted in Figure 7. The telescopic rod drove the sonar to achieve unidirectional, uniform, vertical movement. Both the telescopic rod and the sonar were connected to 24 V batteries for their power supply. The maximum telescopic speed of the rod was 2.8 cm·s^{-1}, with a maximum range (fully extended) of 6.41 m (allowing access to the bottom of the cage), a fixed length (not extended) of 1.5 m, and a telescopic capacity of 4.9 m. The rod was supported by an orange float to keep it suspended above the water surface. The omnidirectional scanning sonar was used for underwater measurement, and the data were transmitted back to the computer. The sonar operated in low-frequency mode, and the range was set to 3 m, based on the maximum radius of the cage (approximately 2.8 m).

Figure 6. Omnidirectional scanning sonar.

Figure 7. Sonar Assembly.

The steps for a single data collection were as follows: Suspend the telescopic rod with the sonar at the center of the water surface of the cage, ensuring the sonar is in contact with the water surface. Activate both the sonar and the electric telescopic rod. The telescopic rod drives the sonar to move underwater at a constant speed of 2.4 cm/s for measurement. The telescopic rod should stop working when its extension length (from the water surface to the bottom of the sonar) reaches 2.9 m, which is the maximum depth affected by buoyancy in the cage. This measurement depth is recorded as experimental data. The sonar data are recorded at 160 frames per second, with each recording lasting approximately 2 min. After the data acquisition, retrieve the sonar to the water surface. Note: As shown in Figure 8, the sonar was positioned on the center axis of the cage. During the scanning operation, the telescopic rod was extended so that the sonar moved at a uniform speed along the center axis in a unidirectional manner. Consequently, the scanning area was not limited to the horizontal plane but formed a spiral measurement pattern.

In order to explore the effects of temperature on fish distribution in sea cages, we designed an experimental scheme. Firstly, at 8:00 (approximately 16 degrees Celsius), 100 fish were manually removed from the right side of the experimental cages, as shown in Figure 3, and placed in the cage for the experiment. The previous collection steps were repeated and the experimental data were recorded. Next, another 50 fish were manually placed into the cage, bringing the total fish to 150. The previous collection procedure was then repeated to collect the experimental data. Finally, another 50 fish were added to the cage to bring the total fish to 200. The collection procedure was repeated to collect the experimental data. We waited until 13:00 (approximately19 degrees Celsius) and repeated the collection procedure. Data were collected for three different numbers of fish at this time. At 18:00 (approximately 18 degrees Celsius), the collection procedure was repeated. Data were collected for three different numbers of fish at this time. A total of 19 sets of experimental data were collected.

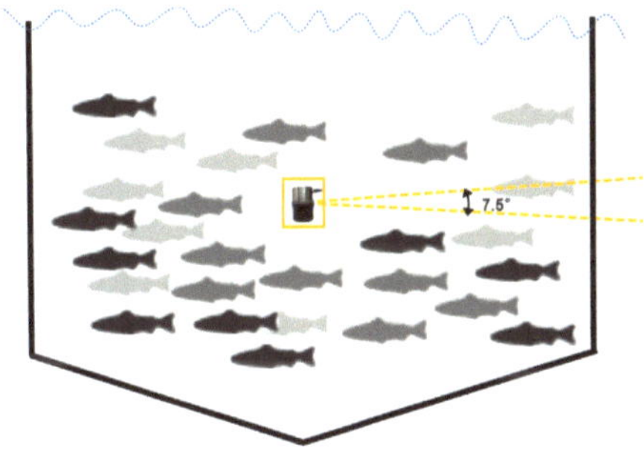

Figure 8. Side view of the sea cage. The sonar in the yellow box is located on the center axis of the net cage, and the yellow dotted lines shows the range covered by the sonar.

The 19 groups of data were numbered; 7 groups of data with 200 tails were numbered as 13, 14, 15, 16, 17, 18, 19; 6 groups of data with 100 tails were numbered as 1, 2, 3, 4, 5, 6; and 6 groups of data with 150 tails were numbered as 7, 8, 9, 10, 11, 12. Of these, numbers 1, 2, 7, 8, 13, and 14 are 8:00 data; numbers 3, 4, 9, 10, 15, and 16 are 13:00 data; and numbers 5, 6, 11, 12, 17, 18, and 19 are 18:00 data.

At each time period, the number of fish in each group was measured two times, with one randomly selected as the training group for the target recognition model and the other as the test group.

2.2. Data Saving Method

The image data, collected 19 times by the sonar, were replayed and processed to obtain video format image data for subsequent target recognition using a trained recognition model. There was a total of 19 sets of video data. Each set of data corresponded to the time per second of the sonar during spiral measurement, with a duration of approximately 2 min and a resolution of 804 pixels in width by 604 pixels in height.

2.3. Object Identification Module

2.3.1. Identification Model CS-YOLOv8s

The YOLOv8 model by Ultralytics is a SOTA model based on previous YOLO versions [25]. On the premise of maintaining real-time performance and high detection accuracy, this paper proposes an improved object detection model, CS-YOLOv8s, based on the YOLOv8s model.

Considering that each fish in sonar images is a small object, the core idea of the improved algorithm was to enhance the network's ability to perceive the feature information of small objects. CS-YOLO focuses on optimizing the structure of the original model and introduces the CoT attention mechanism [26] to improve small object detection. Additionally, redundant information is processed using the thin neck module of Slim-Neck [27], and the GSConv [27] model is integrated to enhance the model's identification performance while maintaining a lightweight design.

2.3.2. Dataset Creation

The sonar image data obtained were recorded as video data, with each second corresponding to each second during the spiral measurement. From this image data, 200 fish sonar images were randomly captured and annotated with MakeSense (category—fish). The annotated datasets were then randomly divided into two categories: 160 images were used as training sets, and 40 images were used as test sets.

2.3.3. Evaluation Indexes

Evaluation indexes are essential for assessing model performance. To evaluate the effectiveness of the proposed method, this study focused on four measures: Params, GFLOPs, mAP, and Size. GFLOPs indicate the amount of computation required by the model, reflecting its complexity. Params describe the number of parameters contained in the model.

The specific type of Average Precision (AP), referred to as mAP (mean Average Precision), was obtained by sorting the predicted values and the recall values of the model [25]. It involved calculating the area enclosed by the lines generated by the vertical axis and the recall values represented on the horizontal axis in the Cartesian coordinate system. This area provided a measure of the model's precision and recall performance across different thresholds.

$$\mathrm{AP} = \sum_{n} (\mathrm{R}_{n+1} - \mathrm{R}_n)\cdot \mathrm{P}_n \quad (1)$$

where R is Recall, the recall represented by the horizontal axis; P is Precision, the precision represented by the vertical axis.

MAP@0.5 refers to the value of Average Precision (AP) when the Intersection over Union (IoU) threshold is set at 0.5 and MAP@0.5:0.95 represents the mean Average Precision across IoU thresholds ranging from 0.5 to 0.95 in increments of 0.05 [25]. There is a positive correlation between the precision of the model and the mAP [25].

$$\mathrm{mAP@0.5:0.95\%} = \frac{\mathrm{AP}_{\mathrm{IoU}=0.5} + \mathrm{AP}_{\mathrm{IoU}=0.55} + \ldots + \mathrm{AP}_{\mathrm{IoU}=0.95}}{\mathrm{n}} \quad (2)$$

where the value of n was 10.

2.3.4. Training Process

When training the detection network model, the following parameters were set: the number of iterations was 300, the weight decay coefficient was 0.0005, the initial learning rate was 0.01, the learning rate momentum was 0.937, and the batch size was 16. As shown in Figures 9 and 10, the model stopped training after 300 iterations. At this point, the loss had decreased to 0.69 and the mAP@0.5% had reached 87.4%.

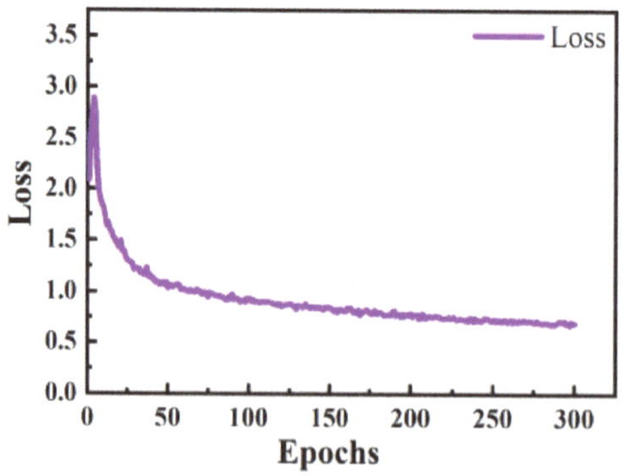

Figure 9. Training process (Loss).

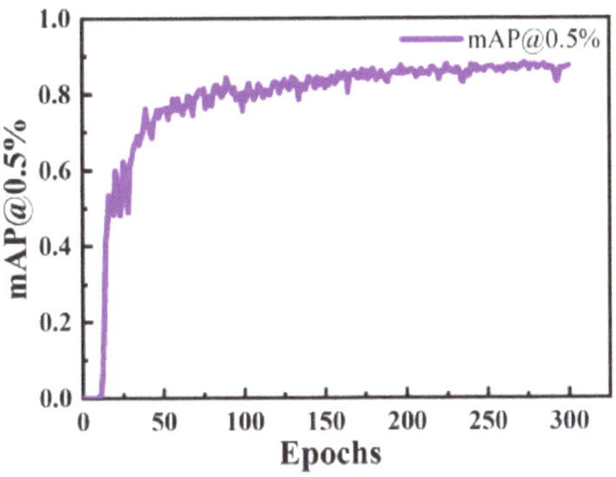

Figure 10. Training process (mAP@0.5%).

2.4. DBSCAN Clustering Module

DBSCAN (Density-Based Spatial Clustering of Applications with Noise) is a clustering algorithm that focuses on the density of samples [28]. The core principle of DBSCAN is density-based classification, where samples within a certain density range are considered to belong to the same cluster. The clustering effect is achieved by classifying samples based on their density. To illustrate the DBSCAN process in detail, we will use the measurement data of 150 fish numbered 10 as an example.

2.4.1. Pre-Processing of Clustering Data

After processing with the improved CS-YOLOv8s identification algorithm, the coordinate positions and quantity information of the fish identified in each frame of the video data were obtained. The quantity of fish detected in each frame and their corresponding coordinate information were recorded. Subsequently, the two-dimensional coordinate data were converted from the normalized image coordinates to the actual coordinates in the real-world context.

$$\begin{aligned} X &= (x-n)*(2r/h)*w \\ Y &= (y-n)*(2r) \end{aligned} \tag{3}$$

where (X, Y) represent the point p' coordinates in Figure 11a, (x, y) represent the point p coordinate in Figure 11b, n was set to 1/2, r is the sonar range, which was 300 cm, h is the image height in pixels, and w is the image width in pixels.

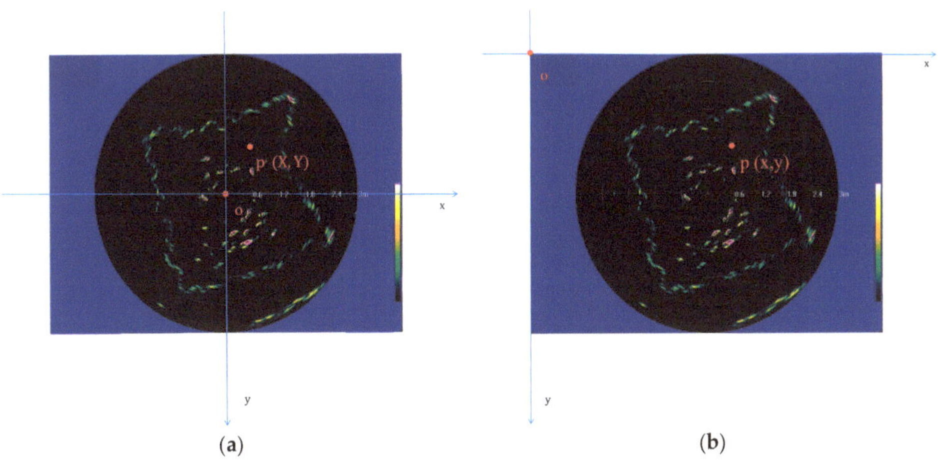

Figure 11. Coordinate transformation. (**a**) A coordinate system with sonar as the origin; (**b**) A coordinate system with the top left corner as the origin.

To convert the two-dimensional coordinate data into three-dimensional coordinates, depth information calculated based on the time of the video data was added. The depth calculation formula can be expressed as follows:

$$d = t * v \tag{4}$$

where d is the water depth, t is the frame-by-frame time of the video, and v is the descending motion speed of the sonar.

In Figure 12, the purple and blue boxes represent different water layers. The blue and green hollow circles denote the radius of clustering. Solid circles represent fish. The purple solid circle indicates the core object of a cluster, while the red solid circle represents the data of the same fish observed in different water layers. The summarized three-dimensional coordinate data were divided into 11 water layers, each with a depth interval of 40 cm. This spacing was determined using a trigonometric function based on the sonar beam opening angle and the scanned radius of the cage. The data were segmented into 11 water layers, with an overlap of 15 cm between adjacent layers, based on the body length and height of the fish used in the experiment. The overlap between water layers was intentional to account for the possibility that fish might move between layers during measurement. This overlap helped minimize the impact of repeated scanning data on the coordinates of fish after clustering, ensuring a more accurate representation of fish locations.

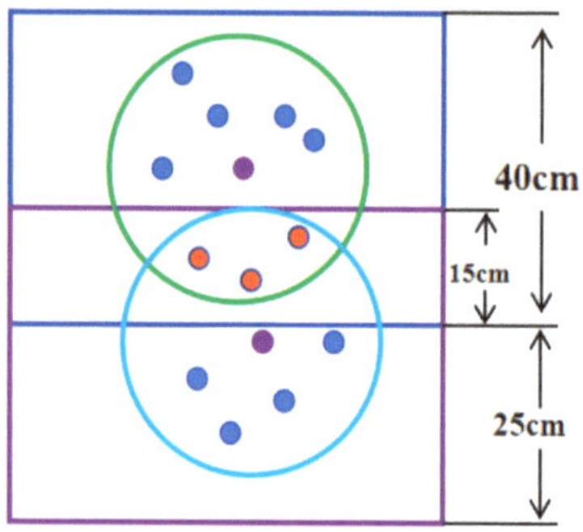

Figure 12. Water layer division.

2.4.2. DBSCAN Clustering

The 3D data for each water layer were clustered using DBSCAN clustering analysis, with the following parameter settings: epsilon = 20.0, min_samples = 1. Considering the body length of the fish (approximately 24 cm) and their body height (approximately 12 cm), as well as the tendency for fish to school, the radius of clustering was set to 20.0 cm. This radius was chosen because fish typically do not stay very close to each other for extended periods. In this clustering setup, a hypersphere with a center at the coordinate of each fish and a radius of 20 cm was used. If additional coordinate points fell within this hypersphere, they were considered to belong to the same fish, since multiple fish are unlikely to be positioned very close to each other for long periods. Thus, the DBSCAN parameters were chosen to ensure that individual fish could be distinguished even if they were close together, reflecting the experimental conditions and the natural behavior of the fish.

Taking the data from the 100–140 cm water layer as an example, the spatial distribution of all coordinate points within this water layer is illustrated in Figure 13a. After traversing all points in the water layer, points that fell within the hypersphere with an epsilon radius of 20 cm were classified into the same cluster. This resulted in a total of 5384 coordinate points in this water layer. After applying DBSCAN clustering, these points were divided into 16 clusters, including noise clusters. The clusters were identified by different colors, as shown in Figure 13b.

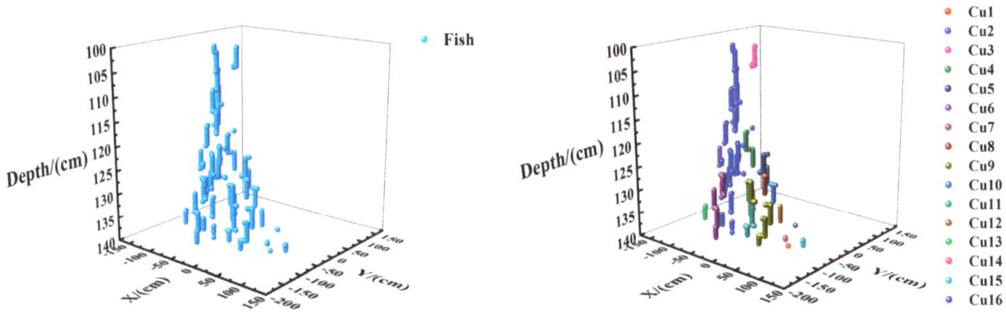

(a) (b)

Figure 13. Clustering in the 100–140 cm water layer. (**a**) Fish distribution; (**b**) Cluster effect diagram.

2.4.3. Reduction and Display of Noise Point

During clustering, attention should be focused on two types of data: core objects and noise points. Core objects are the coordinate points that meet the clustering parameters set (e.g., within the epsilon radius) and are considered the main components of a cluster. Core objects are crucial as they form the basis of the clusters and represent grouped fish. Noise points are coordinate points that do not meet the clustering parameters and are classified as noise during the clustering process. From a practical perspective, noise points indicate locations where there is only one fish within a certain range without interference from other fish. Each noise point represents a fish that should not be disregarded. In subsequent data processing, it is essential to comprehensively consider both core objects and noise points to ensure accurate fish count and distribution analysis.

As shown in Figure 14, the data of 5384 points in the 100–140 cm water layer were selected from the sonar images for experimental analysis. All data points of the water layer are shown in Figure 14a, and experimental results after clustering parameters (epsilon = 20.0, min_samples = 1) are displayed in Figure 14b. In Figure 14b, the red "★" (Cu1) represents the noise point identified by clustering, totaling one noise point. The remaining clusters, composed of core objects, are also shown, with a total of 15 clusters identified, excluding noise points. In the statistical process, clusters are counted as single fish. Noise points are counted individually as each represents an independent fish. The "Noise" cluster, which consists solely of noise points, should not be counted as a fish. This method ensures that both core objects and noise points are accurately accounted for in the fish distribution analysis.

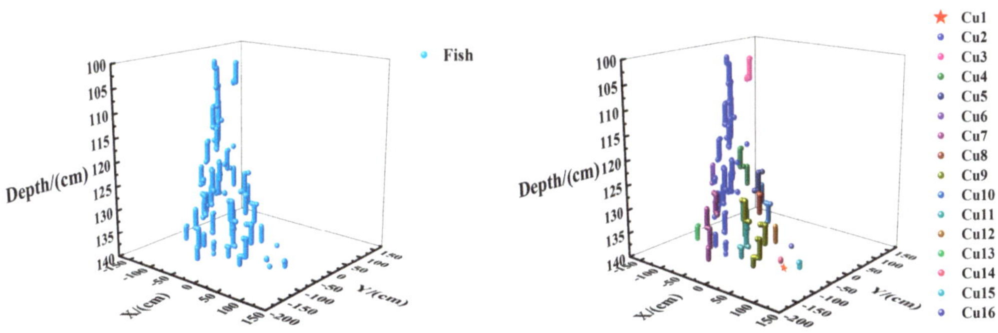

(a) (b)

Figure 14. Water layer data and the data after clustering. (**a**) Fish distribution map before clustering; (**b**) Cluster diagram of noisy points.

2.4.4. Reduction and Experiment Display of Center

To display clusters and centers more intuitively, the study used data from 5384 points in the 100–140 cm water layer, selected from the previous sonar images. The clustering parameters used were epsilon = 20.0 and min_samples = 1. After clustering, clusters representing coordinate points identified as the same fish were obtained. Given the difficulty in counting all coordinate points for spatial distribution analysis, each cluster was centralized to select a representative center point. This approach facilitated the study of fish distribution. Note that the "Noise" cluster, consisting solely of noise points, was not included in this centralization process and was counted point by point. The centers of each cluster were calculated, and their coordinates are provided in Tables 2 and 3.

Table 2. Coordinates of cluster centers in the 100–140 cm water layer.

Cluster Center	X/cm	Y/cm	Depth/cm
1	−35.4	2.0	22.8
2	16.2	16.3	8.8
3	−133.6	72.8	17.1
4	−157.0	42.1	17.2
5	−68.4	132.8	19.6
6	−107.9	106.8	22.9
7	−106.3	−2.6	18.8
8	−12.6	42.6	33.0
9	−38.8	99.6	34.0
10	−151.1	99.0	29.5
11	−86.9	84.3	29.5
12	37.1	−21.9	35.1
13	−69.4	104.5	36.3
14	−28.4	−113.4	38.8

Table 3. Coordinates of cluster noise point in the 100–140 cm water layer.

Noise	X/cm	Y/cm	Depth/cm
1	16.4	40.4	11.9

The calculated center coordinates and noise points of each cluster are displayed in Figure 15. In Figure 15a, the red "★" markers indicate noise points, while cyan "★" markers indicate the centers of each cluster. The color shading of the icons reflects the distance of coordinate points within each cluster along with the center of each cluster. The clustering effect is demonstrated to be effective.

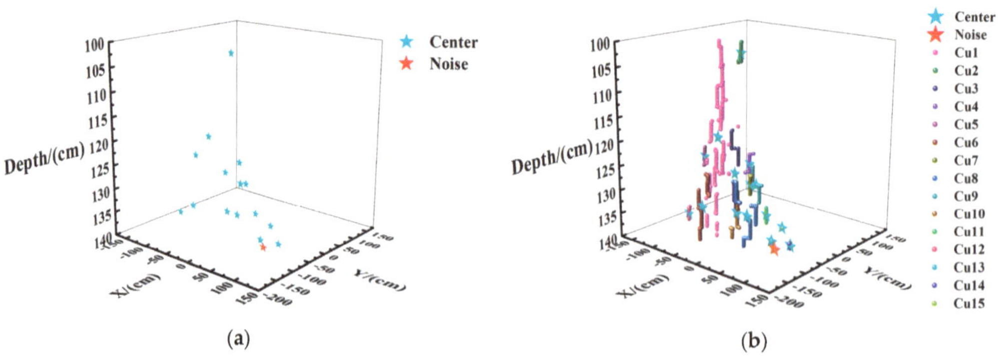

Figure 15. Center and noise point of core object. (**a**) Distribution map of center point and noise point; (**b**) Centralization rendering.

2.4.5. Data Statistics of Water Layer

The quantity of fish in each water layer was determined by clustering the data for each water layer using the parameters epsilon = 20.0 and min_samples = 1. The results are summarized in Table 4.

Table 4. Statistics of clustering results.

Water Layer/cm	Total Number of Coordinate Points/Piece	Number of Clusters (Including Noise Clusters)/Piece	Number of Noise Points/Piece	Number of Fish in This Layer/Piece
0~40	439	2	0	2
25~65	0	0	0	0
50~90	792	4	0	4
75~115	1794	5	0	5
100~140	5384	16	1	16
125~165	11,242	31	1	31
150~190	15,273	19	1	19
175~215	15,431	18	2	19
200~240	11,861	40	2	41
225~265	4432	26	0	26
250~290	253	2	0	2
Total	66,901	162	7	164

As shown in Table 4, clustering of the data from all water layers, totaling 66,901 points, resulted in 162 clusters, including 7 noise point clusters. The quantity of fish was estimated based on these clusters, with a total of 164 fish detected in the entire cage. Given that there were originally 150 fish in the target cage, the estimated quantity of 164 fish represents an error of 14 fish. This results in an accuracy rate of 90.7% and an error rate of 9.3%. Despite the slight overestimation, the high accuracy rate indicates that the clustering method was effective. Additionally, the method provided valuable information on the number of fish in each water layer, which is useful for analyzing the vertical distribution of fish in marine cages.

3. Results

3.1. Result of Cage Data Acquisition

The sonar images depicting varying quantities of fish at a given moment are shown in Figure 16. The figure highlights several key observations:

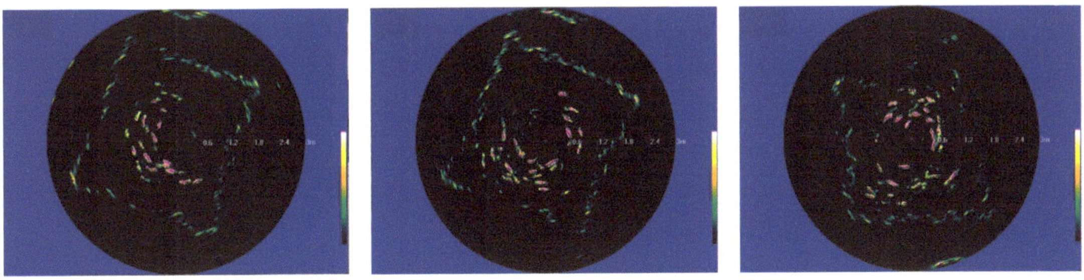

(**a**) (**b**) (**c**)

Figure 16. Sonar images of different numbers of fish. (**a**) 100 fish; (**b**) 150 fish; (**c**) 200 fish.

Outer Contour: The outermost square in each image represents the contour of the netting within the cage.

Fish Contours: The fish-like shapes within the central portion of the netting indicate the presence of fish.

Additional Points: Smaller image points throughout the net garment represent fish at different locations and some noise points resulting from sea wave interference.

It is noted that the netting contours across the images are not identical, with missing segments visible in the middle and right images. These discrepancies were attributed to the changes of the net garment caused by the movement of sea waves.

3.2. Data Processing

This paper concentrated on the analysis of the cage area to obtain detailed insights into the quantity and spatial distribution of fish across different water layers. The clustering analysis provided the quantity and coordinates of fish in each layer, allowing for an in-depth examination of their distribution patterns. This information was crucial for understanding fish behavior and enhancing feeding efficiency.

To ensure the fairness and effectiveness of model training, data processing was performed on each dataset of 200, 150, and 100 fish, respectively. The following evaluation metrics were used to assess the performance of the proposed method: accuracy, error rate, and number of errors. Sets of data from the 100 fish, 150 fish, and 200 fish datasets were selected (numbered 4, 10, and 16) as an example to illustrate the following in this section.

Number of Errors: This was calculated as the absolute value of the difference between the actual number of fish and the estimated number.

Error Rate: The error rate was determined as the percentage of the number of errors relative to the actual number of fish.

Average Error: This was computed as the mean of all error rates.

Average Precision: This was calculated as one minus the average error rate.

These metrics were used to comprehensively evaluate the accuracy and reliability of the proposed method.

The sonar image data, recorded as video data, were processed using the trained improved CS-YOLOv8s model for object identification (Figure 17). This model analyzed the sonar video data to extract information on the identified fish.

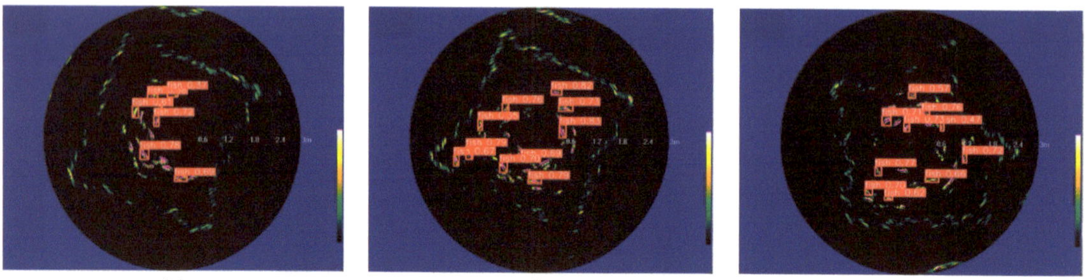

(a) (b) (c)

Figure 17. Fish for object identification. (**a**) 100 fish; (**b**) 150 fish; (**c**) 200 fish.

The image coordinates were converted into actual coordinate data using the ratio of image pixels to the actual cage size. Depth data, based on the sonar's motion rate, were added to the coordinate information of the fish to form 3D coordinate data. The cage was then divided into 11 water layers, each with a depth interval of 40 cm. The coordinate data for fish identified in each water layer were obtained and clustered using the DBSCAN algorithm to estimate the fish coordinates within each layer.

The estimated distribution positions of the fish in the cage were summarized based on the estimated coordinate data. Additionally, a comparative analysis of fish distribution was conducted to evaluate the spatial arrangement and density of fish in different layers of the cage.

3.3. Result Analysis

3.3.1. Effects of Quantity on Fish Distribution

Sets of data from the 100 fish, 150 fish, and 200 fish datasets were selected (numbered 4, 10, and 16) as an example to illustrate the following in this section. Based on the data on fish quantities in each layer obtained from clustering, the trend of fish numbers with depth was plotted, as shown in Figure 18. From Figure 18, it can be observed that the number of fish initially increased with depth before gradually decreasing. A significant concentration of fish was found within the depth range of 1.25 to 2.45 m, with the highest density at a depth of 2.25 m. This distribution pattern suggested that fish primarily preferred this depth range. Considering the impact of human factors [29,30], it was noted that the vertical distribution of fish might have been influenced by activities on the fishing rafts. Fish tended to seek refuge below 1.25 m in response to human presence, indicating behavioral adjustments to avoid disturbances.

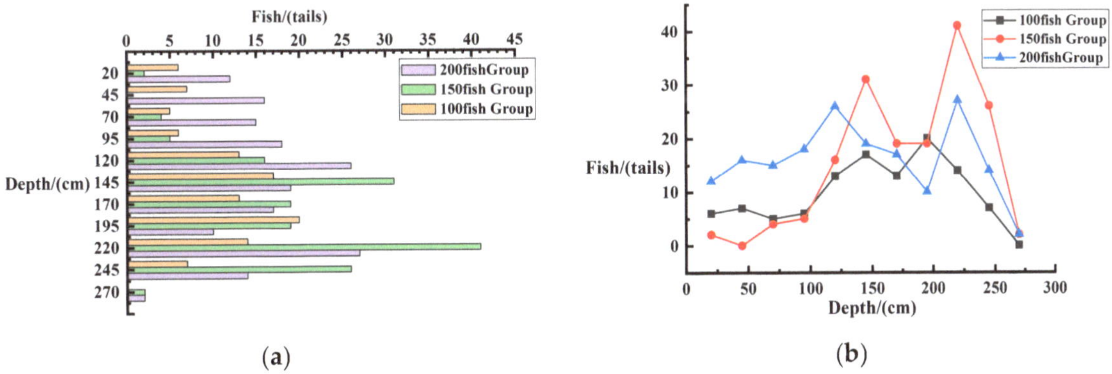

Figure 18. Trend of number of fish with depth. (**a**) Comparison bar chart of depth distribution; (**b**) Comparison line chart of fish depth distribution.

Horizontal Spatial Distribution of Fish

According to the coordinates of fish in each layer obtained from clustering, the horizontal distribution maps of fish for different experimental groups were plotted, as shown in Figure 19. In Figure 19, the origin of the coordinates (0, 0) denotes the position of the sonar. It can be observed from Figure 19 that as the number of fish increased, the horizontal distribution range expanded from a radius of approximately 1 m to a radius of 1.5 m. Additionally, Figure 19a shows that a majority of fish were located around the sonar, suggesting that the moving sonar had some level of attraction for the fish. The summary of horizontal fish distribution across the three experimental groups is presented in Figure 20. The distribution was mainly concentrated within a radius of 1.5 m from the sonar, with a fairly even spread in all directions and only a few fish located near the edges. Considering the effects of fish domestication [31], the even distribution can be attributed to the variability in daily feeding positions, as fish were fed at various points in the cage. Consequently, no significant aggregation in a corner was observed, indicating a uniform horizontal distribution throughout the cage.

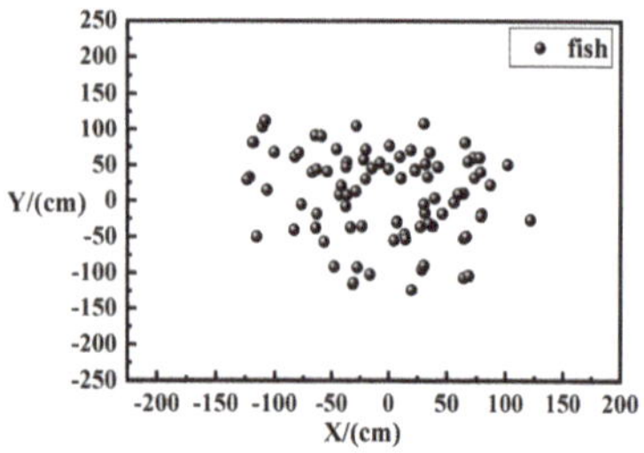

(a)

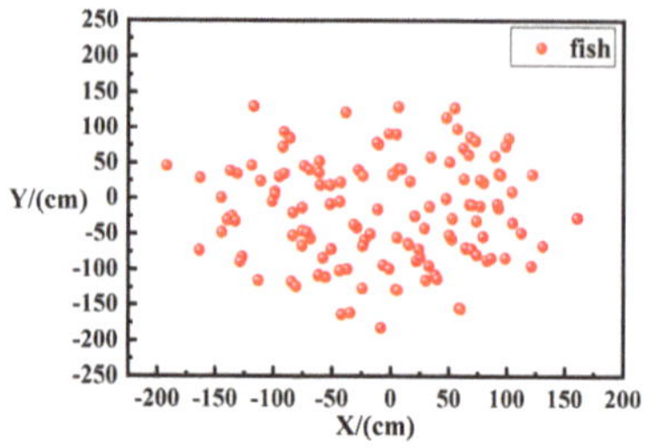

(b)

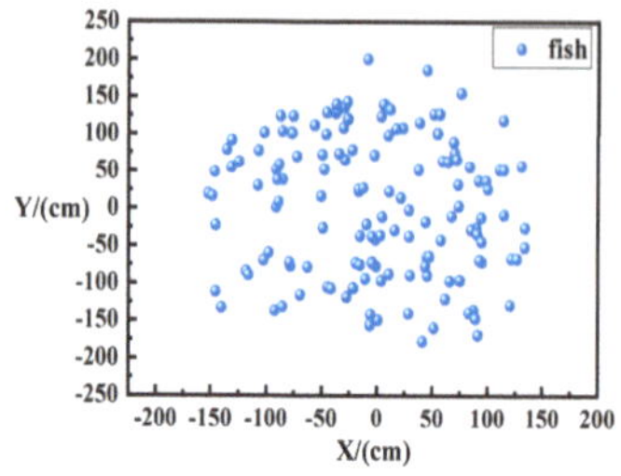

(c)

Figure 19. Horizontal distribution maps of fish. (**a**) Horizontal distribution of 100 fish; (**b**) Horizontal distribution of 150 fish; (**c**) Horizontal distribution of 200 fish.

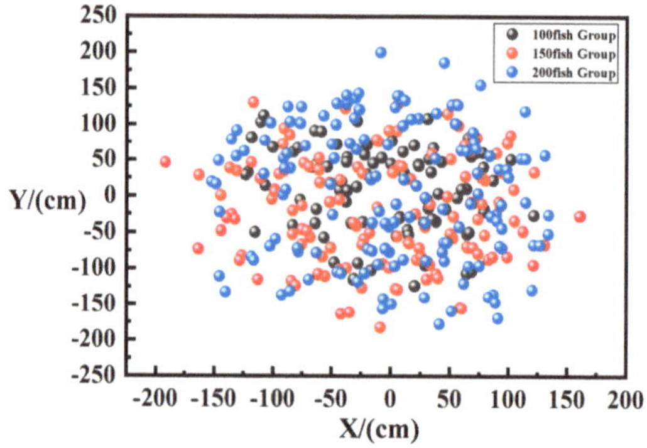

Figure 20. Horizontal distribution comparison chart of fish.

Vertical Spatial Distribution of Fish in Each Layer

To visually display the distribution of fish at different water layers, sets of data from the 100 fish, 150 fish, and 200 fish datasets were selected (numbered 4, 10, and 16) as an example to illustrate the following in this section.

This analysis utilized the 3D coordinates of fish obtained after clustering, resulting in 11 graphs that illustrate the predicted spatial distribution across the 11 water layers. These distributions are displayed in Figures 21–23.

Figure 22a: Two fish were identified in this layer, located near the water surface with a depth of 15 cm and a horizontal distance of approximately 75 cm from the sonar.

Figure 22b: No fish were detected in this layer, indicating that either no fish were present or none were swimming through this water layer at the time.

Figure 22c: Four fish were found in this layer, distributed primarily 75 cm horizontally and 20 cm vertically from the sonar direction.

Figure 22f: Thirty-one fish were observed in this layer, with a concentration of approximately 100 cm horizontally from the sonar, mainly in the middle and lower parts of the water layer, and fewer in the upper part.

The figures demonstrate that the quantity of fish increased with depth up to a certain point and then decreased. The fish were mainly concentrated in the water depth range of 150 to 250 cm, with an even horizontal distribution and no extreme concentrations. The culture density was assessed as reasonable.

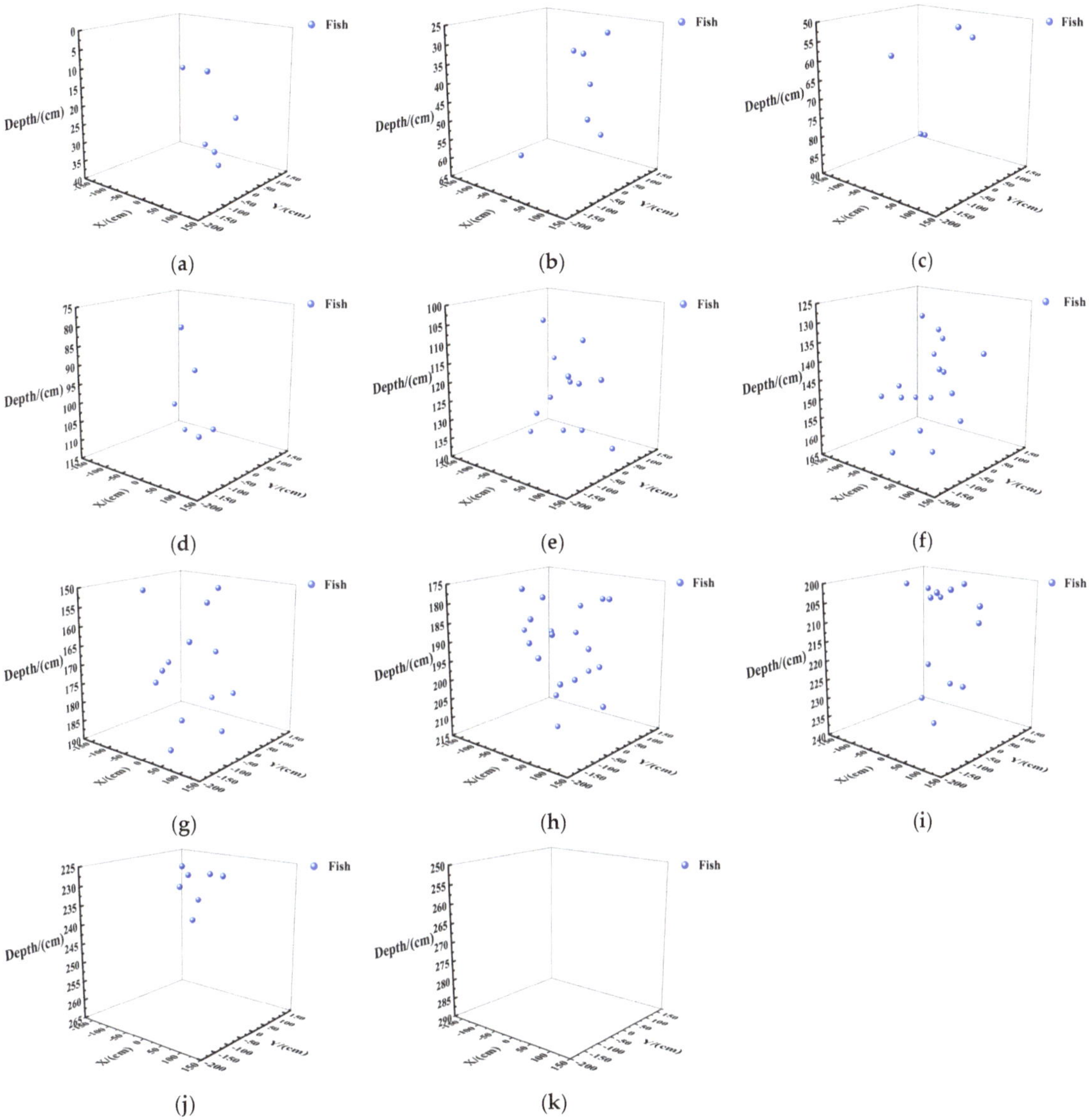

Figure 21. Spatial distribution maps of 100 Fish in each water layer. (**a**) Water layer: 0–40 cm; (**b**) Water layer: 25–65 cm; (**c**) Water layer: 50–90 cm; (**d**) Water layer: 75–115 cm; (**e**) Water layer: 100–140 cm; (**f**) Water layer: 125–165 cm; (**g**) Water layer: 150–190 cm; (**h**) Water layer: 175–215 cm; (**i**) Water layer: 200–240 cm; (**j**) Water layer: 225–265 cm; (**k**) Water layer: 250–290 cm.

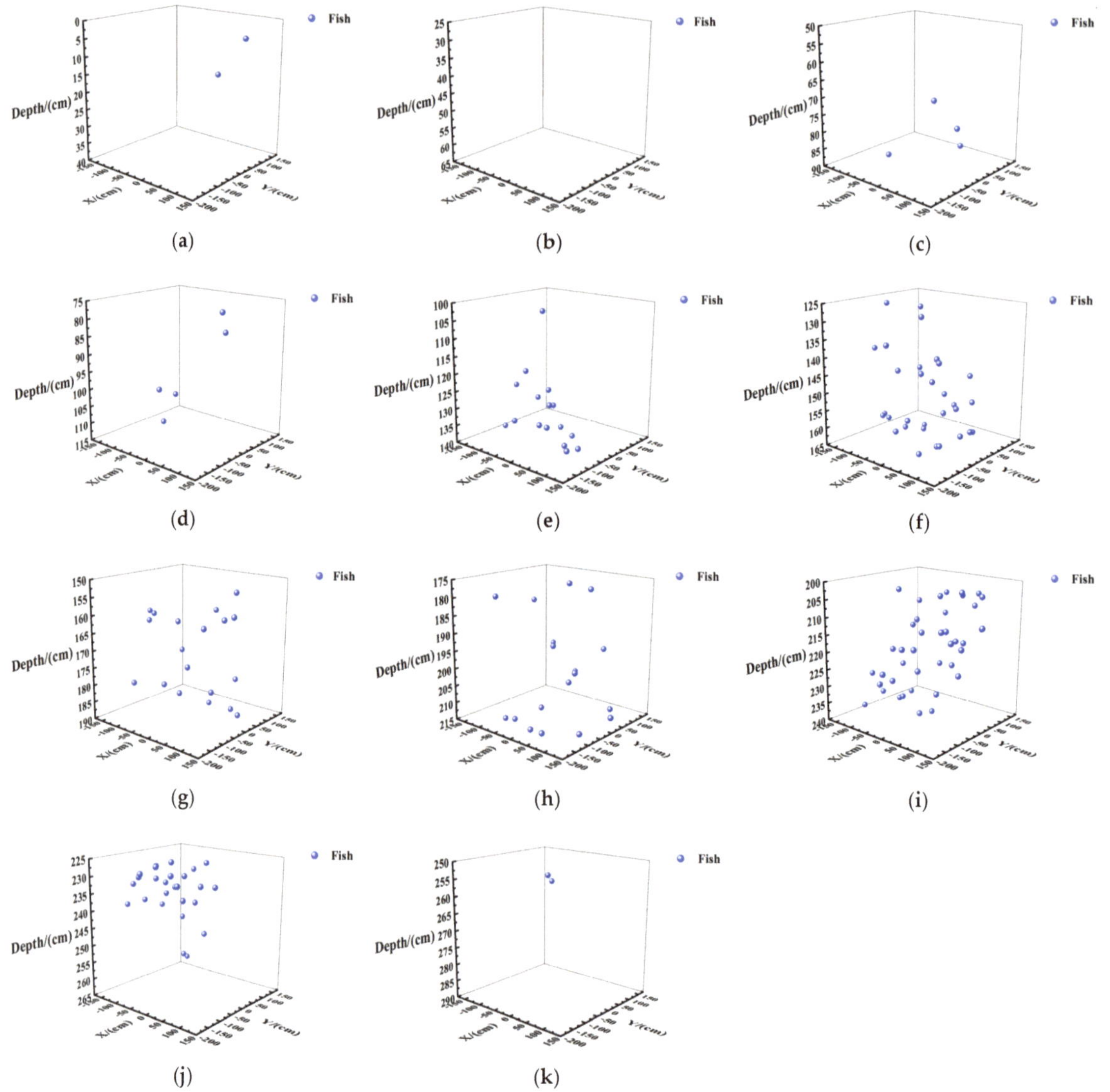

Figure 22. Spatial distribution maps of 150 Fish in each water layer. (**a**) Water layer: 0–40 cm; (**b**) Water layer: 25–65 cm; (**c**) Water layer: 50–90 cm; (**d**) Water layer: 75–115 cm; (**e**) Water layer: 100–140 cm; (**f**) Water layer: 125–165 cm; (**g**) Water layer: 150–190 cm; (**h**) Water layer: 175–215 cm; (**i**) Water layer: 200–240 cm; (**j**) Water layer: 225–265 cm; (**k**) Water layer: 250–290 cm.

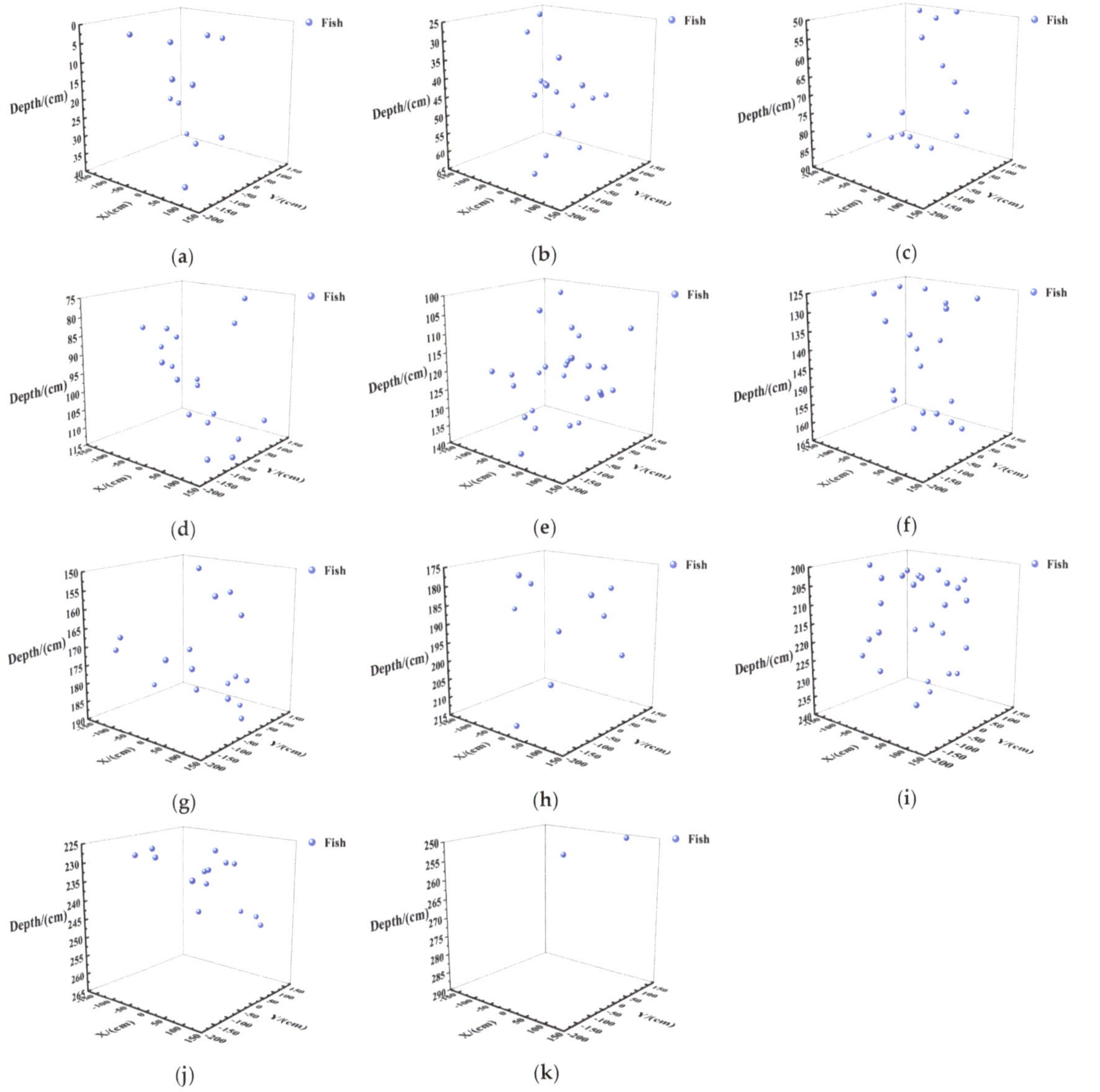

Figure 23. Spatial distribution maps of 200 fish in each water layer. (**a**) Water layer: 0–40 cm; (**b**) Water layer: 25–65 cm; (**c**) Water layer: 50–90 cm; (**d**) Water layer: 75–115 cm; (**e**) Water layer: 100–140 cm; (**f**) Water layer: 125–165 cm; (**g**) Water layer: 150–190 cm; (**h**) Water layer: 175–215 cm; (**i**) Water layer: 200–240 cm; (**j**) Water layer: 225–265 cm; (**k**) Water layer: 250–290 cm.

Vertical Spatial Distribution of Fish

Based on the 3D coordinates of fish in each layer obtained from clustering, the 3D spatial distribution maps of fish in the entire cage were plotted, as shown in Figure 24. The spatial distribution of fish for experimental groups with varying numbers of fish are described in the three panels of Figure 24. In these maps, the balls represent individual fish, with red, green, and blue circles indicating the projections of fish on the XOY, XOD (Depth), and YOD (Depth) axes, respectively. These projections illustrate the positional status of fish in three dimensions. From the projections on the XOD (Depth) and YOD

(Depth) axes in Figure 24, it is evident that the number of fish increases with greater water depth. Additionally, the projection on the YOD (Depth) axis reveals that there are more fish in the positive direction of the *Y*-axis compared to the negative direction. This difference can be attributed to the effects of seawater flow within the marine cage, which causes fish to move downstream more frequently, resulting in a higher concentration of fish on one side of the cage.

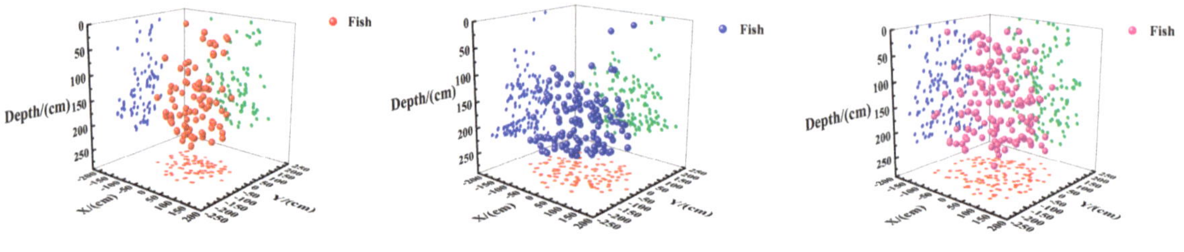

(a) (b) (c)

Figure 24. 3D spatial distribution maps of cage fish. (**a**) Spatial distribution of 100 fish group; (**b**) Spatial distribution of 150 fish group; (**c**) Spatial distribution of 200 fish group.

The comparison of fish spatial distribution in experimental groups with different fish quantities are described in Figure 25. The 3D spatial distribution maps provide insights into the main activity ranges of the fish within the cage. As shown, the primary activity range of the fish spans from 1.5 m to 2.5 m in water depth and within a horizontal radius of 1.5 m, with no significant clustering in corners. As the number of fish increases, their distribution gradually extends from the bottom upwards. In conjunction with the analysis of fish distribution characteristics in culture cages as reported in the literature [32], the primary reasons for these observed distribution patterns can be summarized as follows: Long-term feeding in culture cages typically leads to a domestication effect, causing fish to aggregate around feeding positions. However, in the cages used in this study, which were relatively small, the domestication effect was not pronounced. Consequently, the fish were distributed more evenly throughout the cage rather than clustering around specific areas.

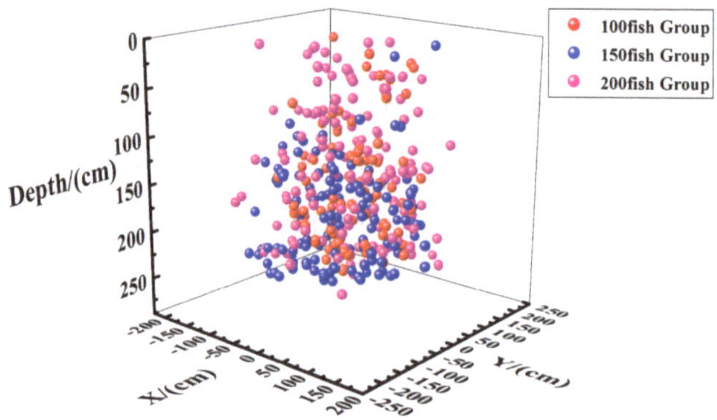

Figure 25. Spatial distribution comparison chart of fish.

3.3.2. Effects of Sea Water Temperature on Fish Distribution

In this section, separate datasets of the same number of fish collected at different times have been selected as examples to illustrate the effects of sea water temperature on fish distribution. Temperature variations in seawater temperatures at three time points, morning, midday and evening, are useful for exploring the effects of temperature on fish distribution. Next, we will compare and analyze the data from the three time points of 8:00, 13:00, and 18:00, respectively. The sea water temperature was approximately 16, 19, and 18 degrees Celsius for the three time periods, respectively.

As shown in Figure 26, with the change of seawater temperature, some changes in fish distribution also occurred. As shown in Figure 26a, at 8:00, the distribution of fish was mainly concentrated in the 2 m water depth range. At 13:00, the distribution of fish gradually shifted to the 1 m water depth range as the temperature increased; finally, at 18:00, the distribution of fish returned to the 1.5 m water depth range. As shown in Table 5, data numbered 1, 7, and 13 are for 8:00; data numbered 3, 9, and 15 are for 13:00.; data numbered 5, 11, and 17 are for 18:00; data numbered 1, 3, and 5 are for 100 fish; data numbered 7, 9, and 11 are for 150 fish; and data numbered 13, 15, and 17 are for 200 fish. From 1, 7, and 13, it can be seen that at 8:00 the fish are mainly distributed at 2 m and the distribution expands to both ends as the number increases, with the peak of the number expanding to 1.25 m and 2.25 m when the number reaches 200 fish. From 3, 9, and 15, it can be seen that the distribution of the number of fish gradually expands towards the surface as the temperature of seawater increases, and the number of fish in the shallow water above 1 m gradually increases. From 5, 11, and 17, it can be seen that the number of fish in the shallow water areas decreased, but still remained higher than the number at 8:00 Among the three sets of data from 13, 15, and 17, the three changes in the distribution of 200 fish were closer to each other, with peaks in numbers mainly in the 1.25 m and 2.25 m areas.

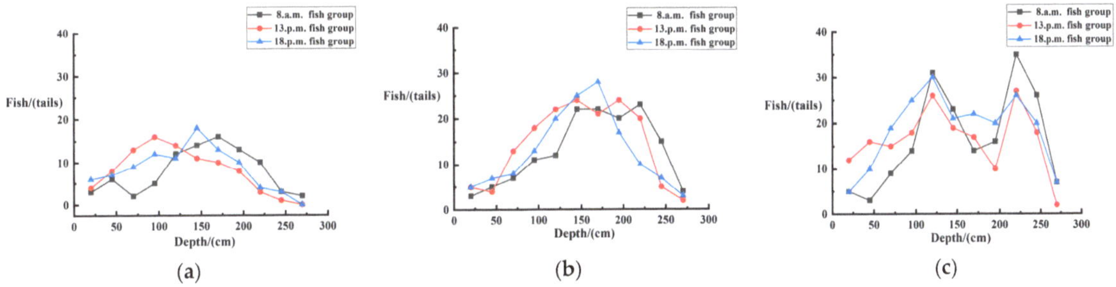

Figure 26. Distribution of fish at different temperatures. (**a**) Distribution of 100 fish; (**b**) Distribution of 150 fish; (**c**) Distribution of 200 fish.

Table 5. Distribution of fish in different datasets.

Water Layer/cm \ Data Number	1	3	5	7	9	11	13	15	17
0~40	3	4	6	3	5	5	5	12	5
25~65	6	8	7	5	4	7	3	16	10
50~90	2	13	9	7	13	8	9	15	19
75~115	5	16	12	11	18	13	14	18	25
100~140	12	14	11	12	22	20	31	26	30
125~165	14	11	18	22	24	25	23	19	21
150~190	16	10	13	22	21	28	14	17	22
175~215	13	8	10	20	24	17	16	10	20
200~240	10	3	4	23	20	10	35	27	26
225~265	3	1	3	15	5	7	26	18	20
250~290	2	0	0	4	2	3	7	2	7
Total	86	88	93	144	158	143	183	180	205

In conjunction with the analysis of fish distribution characteristics in culture cages as reported in the literature [33–35], the primary reasons for these observed distribution patterns can be summarized as follows: The primary factors influencing the vertical distribution of fish are water temperature and dissolved oxygen levels. Variations in these factors affect how fish distribute themselves vertically within the water column. Specifically, fish tend to position themselves in layers where water temperature and dissolved oxygen are optimal for their survival and comfort. Considering that, at 8:00, the temperature is low (16 degrees Celsius) and the temperature of seawater will drop after overnight loss, so the fish will be distributed in the middle and lower layers of the cages, and under the influence of the low temperature the fish's appetite and activity are not high; as the sun rises, the temperature of the seawater gradually rises, and at 13:00 (19 degrees Celsius), the fish tends to be in the middle and upper areas of the cages, and under the influence of the long-term rearing, the fish will tend to be in the upper-middle waters of the cages waiting for food; at 18:00, the seawater has some temperature and has not yet been lost after a day of sunshine, and the seawater temperature is higher although the temperature is close to that at 8:00 in the morning, so the distribution of the fish will show a small amount of folding back to the lower-middle waters.

It was observed that *Trachinotus ovatus* predominantly occupied depths above 1 m. This depth provided protection from the sun and offered suitable water temperatures. This observation aligns with the typical distribution characteristics of cultured *Trachinotus ovatus* and can inform future cultivation practices. Based on these findings, it is recommended to develop a more targeted and scientific approach to fish culture, such as setting feeding stations at approximately 1 m deep. This strategy would ensure optimal feeding conditions and enhance the efficiency of feed usage, contributing to better feed conversion rates and overall culture performance. Compared to direct surface feeding, placing feeding stations at a fixed depth is more efficient. This approach minimizes feed waste caused by waves, enhances feed conversion rates, improves culture efficiency, and reduces overall culture costs.

4. Discussion

The methods for monitoring fish include IoT-based fish monitoring systems [36], deep convolutional neural networks [37], deep learning [38], and species checklist data methods [39]. Due to the impossibility of conducting comparative experiments in the same environment, the instruments, equipment, and methods used in various methods are different. The comparison of different estimation and monitoring methods are shown in Table 6.

Table 6. Comparison of different estimation and monitoring methods.

Methods	Equipment Usage	Monitoring Content	Advantage	Disadvantage
IoT-based fish monitoring system	A pH sensor, a temperature sensor, and some other equipment	pH level, the water temperature, dissolved oxygen level, and ammonia level	Monitoring various indicators of water quality can provide a healthy breeding environment for fish.	Lack of estimation of the number and distribution of fish populations in aquaculture water bodies makes it impossible to comprehensively consider water issues.
Deep convolutional neural networks	Electronic monitoring systems (EMSs)	The number of the fish, the types and body lengths of the fish	Automated, high precision	High computational burden
Deep learning	Sonar (oculusM750d)	Fish counting and tracking	Real-time data	Data limitation
Species checklist data method	Bottom trawl nets	Differences in fish assemblages	Large scale, diverse species	High demand for data and difficulty in data analysis
Our method	omnidirectional scanning sonar	Distribution of fish	Convenient, low implementation difficulty	The estimated range is small, lack of applicability to large sample data

As shown in Table 6, fish monitoring systems are often used to monitor the aquaculture environment, monitor the health of aquaculture water bodies, and adjust them in a timely manner to reduce fish losses. However, there is a lack of estimation of the amount of aquaculture in aquaculture water bodies, making it impossible to conduct comprehensive evaluations; two fish monitoring methods using neural networks are mainly used for estimating fish numbers. The trained models have high accuracy in estimating fish numbers, but their applicability is not high. A large amount of sample data is required to train the model to achieve the desired effect. The method of using species inventory data to estimate the spatial distribution of fish is mainly used for the overall distribution of fish diversity in large water areas and requires a large number of datasets for analysis. By comparison, the method proposed in this article is suitable for the spatial estimation of small-scale aquaculture and can accurately estimate the water depth and location of fish, making it easier for breeders to monitor the density and other conditions of aquaculture for management and adjustment of aquaculture plans.

5. Conclusions

In this study, data from *Trachinotus ovatus* aquaculture cages were acquired using omnidirectional scanning sonar. The distribution of cage fish was estimated using the proposed method, which presents innovative advantages over traditional and existing methods. Compared to traditional and existing methods, the approach proposed in this study was innovative in the identification and distribution estimation of fish. This innovation is highlighted in the following aspects:

(1) The fish sonar image object detection model was enhanced and trained by incorporating an attention mechanism into the YOLOv8s backbone network. This modification aimed to improve the model's capability to detect small objects, making it more suitable for detecting fish in sonar images. Subsequently, ablation studies and model comparison experiments were conducted. The results demonstrated that the improved CS-YOLOv8s network achieved a mAP@0.5 score that was 1.4% higher than that of the original YOLOv8s algorithm, as illustrated by the experimental results.

(2) The quantity of fish was estimated layer by layer using the DBSCAN clustering algorithm, which provided the estimated positional coordinates of fish within each layer. This approach effectively addressed the issue of significant statistical errors associated with existing methods for tracking fish swimming, resulting in more accurate estimations.

An estimation method for the distribution of *Trachinotus ovatus* in marine cages was developed using omnidirectional scanning sonar and two advanced models. The approach involved placing the sonar at the center of the cage's surface to conduct spiral descent scanning measurements. The improved YOLO model was employed to detect fish in the sonar images and extract their image coordinates. Subsequently, the cage was divided into water layers based on depth intervals. Clustering algorithms were used to estimate the actual coordinates of fish within each layer. By summarizing the coordinate data across all water layers, the overall distribution of fish in the cage was determined. Comparison experiments were performed on small cages, and sonar detection was conducted in groups according to the number of fish and the sea water temperature. The method provided distribution estimates that closely matched manual exploration results, demonstrating its effectiveness and suitability for practical applications, as evidenced by multiple experimental results.

However, this study is in the preliminary exploration stage and requires extensive experimentation to verify the accuracy and practicality of the method. Additionally, the fish distribution estimation method proposed in this study has certain limitations. Specifically, when fish quantities are large and dense, occlusion between fish can significantly affect the estimation results. Future research will focus on addressing these high-density occlusion issues to improve the method's adaptability in estimating fish distribution under various cage aquaculture conditions.

Author Contributions: Conceptualization, J.H. and Y.H.; methodology, J.S.; software, X.H. and P.S.; validation, G.Z. and P.S.; investigation, Q.T.; resources, Y.H. and T.Y.; data curation, J.H.; writing—original draft preparation, J.H.; writing—review and editing, X.H., G.L., G.P., and Y.H.; supervision, Y.H.; project administration, X.H.; funding acquisition, X.H. and Y.H. All authors have read and agreed to the published version of the manuscript.

Funding: This research received funding from the Major Science and Technology Plan of Hainan Province (Grant No. ZDKJ2021013), Hainan Province Science and Technology Special Fund (Grant No. ZDYF2021XDNY305, ZDYF2023XDNY066), Central Public-interest Scientific Institution Basal Research Fund, CAFS (Grant NO. 2023TD97), Guangzhou Science and Technology Plan Project (Grant NO. 2023E04J0001), and Key R&D Program Projects in Lianyungang City (Grant NO. 22CY080, 21SH038). Project supported by Southern Marine Science and Engineering Guangdong Laboratory (Zhuhai) (Grant No. SML2023SP204).

Institutional Review Board Statement: Not applicable.

Informed Consent Statement: Not applicable.

Data Availability Statement: All data are contained within the article.

Conflicts of Interest: The authors declare no conflicts of interest.

References

1. Cheng, Z.X.; Hong, G.Q.; Li, Q.B.; Liu, S.H.; Wang, S.; Ma, Y. Seasonal dynamics of coastal pollution migration in open waters with intensive marine ranching. *Mar. Environ. Res.* **2023**, *190*, 106101. [CrossRef] [PubMed]
2. Liu, J.Y.; Gui, F.; Zhou, Q.; Cai, H.W.; Xu, K.D.; Zhao, S. Carbon Footprint of a Large Yellow Croaker Mariculture Models Based on Life-Cycle Assessment. *Sustainability* **2023**, *15*, 6658. [CrossRef]
3. Jeeva, J.C.; Ghosh, S.; Raju, S.S.; Megarajan, S.; Vipinkumar, V.P.; Edward, L.; Narayanakumar, R. Success of cage farming of marine finfishes in doubling farmersâ€™ income: A techno-social impact analysis. *Curr. Sci* **2022**, *123*, 1031–1037. [CrossRef]
4. Banno, K.; Gao, S.; Anichini, M.; Stolz, C.; Tuene, S.A.; Gansel, L.C. Expanded vision for the spatial distribution of Atlantic salmon in sea cages. *Aquaculture* **2024**, *588*, 740879. [CrossRef]
5. Gamperl, A.K.; Zrini, Z.A.; Sandrelli, R.M. Atlantic salmon (Salmo salar) cage-site distribution, behavior, and physiology during a Newfoundland heat wave. *Front. Physiol.* **2021**, *12*, 719594. [CrossRef] [PubMed]
6. Johannesen, Á.; Patursson, Ø.; Kristmundsson, J.; Dam, S.P.; Muleid, M.; Klebert, P. Waves and currents decrease the available space in a salmon cage. *PLoS ONE* **2022**, *17*, e0263850. [CrossRef] [PubMed]
7. Ndashe, K.; Hang'ombe, B.M.; Changula, K.; Yabe, J.; Samutela, M.T.; Songe, M.M.; Kefi, A.S.; Chilufya, L.N.; Sukkel, M. An Assessment of the risk factors associated with disease outbreaks across tilapia farms in Central and Southern Zambia. *Fishes* **2023**, *8*, 49. [CrossRef]
8. Muñoz, L.; Aspillaga, E.; Palmer, M.; Saraiva, J.L.; Arechavala-Lopez, P. Acoustic telemetry: A tool to monitor fish swimming behavior in sea-cage aquaculture. *Front. Mar. Sci.* **2020**, *7*, 645. [CrossRef]
9. Proud, R.; Kloser, R.J.; Handegard, N.O.; Cox, M.J.; Brierley, A.S. Mapping the global prey-field: Combining acoustics, optics and net samples to reduce uncertainty in estimates of mesopelagic biomass. *J. Acoust. Soc. Am.* **2019**, *146*, 2898. [CrossRef]
10. Zhang, X.Y.; Gao, Y.H.; Ji, Y.Q.; Feng, A.W.; Zhao, S.J.; Wang, C.H. Compact multi-spectral-resolution Wynne-Offner imaging spectrometer with a long slit. *Appl. Opt.* **2024**, *63*, 1577–1582. [CrossRef]
11. Duan, Z.; Yuan, Y.; Lu, J.C.; Wang, J.L.; Li, Y.; Svanberg, S.; Zhao, G.Y. Underwater spatially, spectrally, and temporally resolved optical monitoring of aquatic fauna. *Opt. Express* **2020**, *28*, 2600–2610. [CrossRef] [PubMed]
12. Pagniello, C.M.L.S.; Butler, J.; Rosen, A.; Sherwood, A.; Roberts, P.L.D.; Parnell, P.E.; Jaffe, J.S.; Širović, A. An Optical Imaging System for Capturing Images in Low-Light Aquatic Habitats Using Only Ambient Light. *Oceanography* **2021**, *34*, 71–77. [CrossRef]
13. Yang, L.; Chen, Y.Y.; Shen, T.; Yu, H.H.; Li, D.L. A BlendMask-VoVNetV2 method for quantifying fish school feeding behavior in industrial aquaculture. *Comput. Electron. Agric.* **2023**, *211*, 108005. [CrossRef]
14. Helminen, J.; Dauphin, G.J.; Linnansaari, T. Length measurement accuracy of adaptive resolution imaging sonar and a predictive model to assess adult Atlantic salmon (*Salmo salar*) into two size categories with long-range data in a river. *J. Fish Biol.* **2020**, *97*, 1009–1026. [CrossRef]
15. Le, Z.Y. Exploration and suggestion on effective protection of offshore fishery resources based on acoustic technology. *ChinaAquatic Prod.* **2022**, *5*, 58–60.
16. Shen, W.; Peng, Z.F.; Zhang, J. Identification and counting of fish targets using adaptive resolution imaging sonar. *J. Fish Biol.* **2024**, *104*, 422–432. [CrossRef]
17. Helminen, J.; O'Sullivan, A.M.; Linnansaari, T. Measuring tailbeat frequencies of three fish species from adaptive resolution imaging sonar data. *Trans. Am. Fish. Soc.* **2021**, *150*, 627–636. [CrossRef]
18. Wei, Y.G.; Duan, Y.H.; An, D. Monitoring fish using imaging sonar: Capacity, challenges and future perspective. *Fish Fish.* **2022**, *23*, 1347–1370. [CrossRef]

19. Hobbs, D.; Bigot, M.; Smith, R.E.W. Rio doce acoustic surveys of fish biomass and aquatic habitat. *Integr. Environ. Assess. Manag.* **2020**, *16*, 615–621. [CrossRef]
20. Feng, Y.H.; Wei, Y.G.; Sun, S.; Liu, C.; An, D.; Wang, J. Fish abundance estimation from multi-beam sonar by improved MCNN. *Aquat. Ecol.* **2023**, *57*, 895–911. [CrossRef]
21. Stacy-Duffy, W.L.; Redman, R.A.; Roswell, C.R.; Peterson, S.D.; Czesny, S.J. Evaluation of fish spawning habitat at offshore reefs in southwest Lake Michigan using side-scan sonar and underwater video. *Aquat. Conserv. Mar. Freshw. Ecosyst.* **2024**, *34*, e4092. [CrossRef]
22. Jing, D.X.; Zhou, H.Y.; Han, J.; Zhang, J. Fish abundance estimation based on an imaging sonar. *Appl. Acoust.* **2019**, *38*, 705–711. [CrossRef]
23. Chevallay, M.; Du Dot, T.J.; Goulet, P.; Fonvieille, N.; Craig, C.; Picard, B.; Guinet, C. Spies of the deep: An animal-borne active sonar and bioluminescence tag to characterise mesopelagic prey size and behaviour in distinct oceanographic domains. *Deep. Sea Res. Part I Oceanogr. Res. Pap.* **2024**, *203*, 104214. [CrossRef]
24. Cui, Z.; Zhu, H.H.; Chen, C.; Liu, W.; Wang, Q.; Chai, Z.G. Study on Fish Resource Assessment Method Based on Imaging Sonar. In Proceedings of the 2022 3rd International Conference on Geology, Map** and Remote Sensing (ICGMRS), Zhoushan, China, 22–24 April 2022; IEEE: New York, NY, USA, 2022; pp. 327–331. [CrossRef]
25. Talaat, F.M.; Hanaa, Z.E. An improved fire detection approach based on YOLO-v8 for smart cities. *Neural Comput. Appl.* **2023**, *35*, 20939–20954. [CrossRef]
26. Li, Y.H.; Yao, T.; Pan, Y.W.; Mei, T. Contextual Transformer Networks for Visual Recognition. *IEEE Trans. Pattern Anal. Mach. Intell.* **2023**, *45*, 1489–1500. [CrossRef] [PubMed]
27. Jin, X.; Jiao, H.W.; Zhang, C.; Li, M.Y.; Zhao, B.; Liu, G.W.; Ji, J.T. Hydroponic lettuce defective leaves identification based on improved YOLOv5s. *Front. Plant Sci.* **2023**, *14*, 1242337. [CrossRef]
28. Yang, W.J.; Zhu, Z.Y.; Zhou, M.; Li, D.; Zhang, J.T.; Qin, M.Y.; Liu, B.L. Extraction of Tooth Cusps based on DBSCAN Density Clustering and Neighborhood Search Algorithm. *Crit. Rev. Biomed. Eng.* **2024**, *52*, 27–37. [CrossRef]
29. Cong, T.T.; Tong, C.F.; Zhao, C.J.; Chen, Z.T.; He, Z.F.; Liu, M.Y. Community composition and distribution characteristics of the fish assemblages in the rivers of Chongming Island in summer. *Acta Ecol. Sin.* **2021**, *41*, 2067–2076. [CrossRef]
30. Mennad, M.; Bachari, N.E.I.; Ferhani, K.; Chabet Dis, C.; Bennoui, A.; Neghli, L.; Ben Smail, S. Spatiotemporal Distribution of Small Pelagic Fish Schools in the Western Algerian Coast. *Thalass. Int. J. Mar. Sci.* **2023**, *39*, 923–930. [CrossRef]
31. Wu, Y.D.; Yu, X.K.; Suo, N.; Bai, H.Q.; Ke, Q.Z.; Chen, J.; Pan, Y.; Zheng, W.Q.; Xu, P. Thermal tolerance, safety margins and acclimation capacity assessments reveal the climate vulnerability of large yellow croaker aquaculture. *Aquaculture* **2022**, *561*, 738665. [CrossRef]
32. Rioja, R.A.; Palomar-Abesamis, N.; Juinio-Meñez, M.A. Correction to: Associated effects of shading on the behavior, growth, and survival of Stichopus cf. horrens juveniles. *Aquac. Int.* **2021**, *30*, 1087–1088. [CrossRef]
33. Li, Y.; Gao, C.X.; Chen, J.H.; Wang, Q.; Zhao, J. Spatial–temporal distribution characteristics of Harpadon nehereus in the Yangtze River Estuary and its relationship with environmental factors. *Front. Mar. Sci.* **2024**, *11*, 1340522. [CrossRef]
34. Kurbanov, Y.K.; Vinogradskaya, A.V. Distribution, Ecology and Size Composition of the White-Blotched Skate Bathyraja maculata (Arhynchobatidae) in the Northeastern Sea of Okhotsk during the Hydrological Summer. *J. Ichthyol.* **2023**, *63*, 1092–1101. [CrossRef]
35. Li, R.; Amenyogbe, E.; Lu, Y.; Jin, J.; Xie, R.; Huang, J. Effects of low-temperature stress on intestinal structure, enzyme activities and metabolomic analysis of juvenile golden pompano (*Trachinotus ovatus*). *Front. Mar. Sci.* **2023**, *10*, 1114120. [CrossRef]
36. Tamim, A.T.; Begum, H.; Shachcho, S.A.; Khan, M.M.; Yeboah-Akowuah, B.; Masud, M.; AI-Amri, J.F. Development of IoT Based Fish Monitoring System for Aquaculture. *Intell. Autom. Soft Comput.* **2022**, *32*, 56–71. [CrossRef]
37. Tseng, C.H.; Kuo, Y.F. Detecting and counting harvested fish and identifying fish types in electronic monitoring system videos using deep convolutional neural networks. *ICES J. Mar. Sci.* **2020**, *77*, 1367–1378. [CrossRef]
38. Xing, B.W.; Sun, M.; Liu, Z.C.; Guan, L.W.; Han, J.T.; Yan, C.X.; Han, C. Sonar Fish School Detection and Counting Method Based on Improved YOLOv8 and BoT-SORT. *J. Mar. Sci. Eng.* **2024**, *12*, 964. [CrossRef]
39. Zhang, Q.T.; Hu, G.K. Utilization of species checklist data in revealing the spatial distribution of fish diversity. *J. Fish Biol.* **2020**, *97*, 817–826. [CrossRef]

Journal of
Marine Science
and Engineering

MDPI

Article

Study on the Dynamic Response of Mooring System of Multiple Fish Cages under the Combined Effects of Waves and Currents

Fuxiang Liu [1,2], Zhentao Jiang [3], Tianhu Cheng [3], Yuwang Xu [3], Haitao Zhu [4,*], Gang Wang [5], Guoqing Sun [2] and Yuqin Zhang [2]

1 College of Shipbuilding Engineering, Harbin Engineering University, Harbin 150001, China; fuxiang.liu@cimc-raffles.com
2 Yantai CIMC Blue Ocean Technology Co., Ltd., Yantai 264000, China
3 State Key Laboratory of Ocean Engineering, Shanghai Jiao Tong University, Shanghai 200240, China; jiangzhentao@sjtu.edu.cn (Z.J.); chengtianhu@sjtu.edu.cn (T.C.)
4 Yantai Research Institute, Harbin Engineering University, Yantai 264000, China
5 Yellow Sea Fisheries Research Institute, Chinese Academy of Fishery Sciences, Qingdao 266071, China
* Correspondence: zhuhaitao@hrbeu.edu.cn

Abstract: Deep-sea aquaculture can alleviate the spatial and environmental pressure of near-shore aquaculture and produce higher quality aquatic products, which is the main development direction of global aquaculture. The coastline of China is relatively flat, with aquaculture operations typically operating in sea areas with water depths of approximately 30–50 m. However, with frequent typhoons and poor sea conditions, the design of mooring system has always been a difficult problem. This paper investigated the multiple cages, considering two layouts of 1×4 and 2×2, and proposed three different mooring system design schemes. The mooring line tension of the mooring systems under the self-storage condition was compared, and it was observed whether the mooring line accumulation and the contact between the mooring line and the steel structure occurred on the leeward side. Additionally, flexible net models were compared with rigid net models to evaluate the impact of net deformation on cage movement and mooring line tension. Finally, based on the optimal mooring design, the dynamic response of the mooring system under irregular wave conditions was analyzed and studied, providing important reference for the safety and economic design of the mooring system of multiple fish cages.

Keywords: multiple fish cages; mooring system; mooring line tension; dynamic response; net model

Citation: Liu, F.; Jiang, Z.; Cheng, T.; Xu, Y.; Zhu, H.; Wang, G.; Sun, G.; Zhang, Y. Study on the Dynamic Response of Mooring System of Multiple Fish Cages under the Combined Effects of Waves and Currents. *J. Mar. Sci. Eng.* **2024**, *12*, 1648. https://doi.org/10.3390/jmse12091648

Academic Editor: Abdellatif Ouahsine

Received: 8 August 2024
Revised: 28 August 2024
Accepted: 7 September 2024
Published: 14 September 2024

1. Introduction

With the saturation of offshore development, the ecological pressure of marine aquaculture is increasing rapidly. Meanwhile, the demand for high-quality fish is also increasing [1–3]. It has gradually become a trend to develop offshore marine aquaculture and build marine ranches [4,5]. Developing deep-sea aquaculture can produce cleaner and healthier high-quality aquatic products while reducing the occupation of nearshore aquaculture space, thereby alleviating pressure on coastal environments. Major countries in the world also attach great importance to aquaculture, and it has become a hot industry competing for development throughout the world.

The current research of fish cages focuses on the structure strength analysis, the hydrodynamic characteristics of net and mooring system design [6,7]. Fredriksson et al. [8] obtained the material properties of the high-density polyethylene floating coil through tensile tests, carried out the local strength check of the floating collar under the tension of mooring lines, and verified the accuracy of the numerical method through model experiments. Then, the stress of the floating collar and the critical load leading to the local failure of the structure were evaluated by a numerical method under typical working conditions. Liu et al. [9] carried out elastoplastic mechanical tests and finite element analysis on the guardrails of gravity cages. They investigated areas prone to structural failure, analyzed the frequency of loads leading to resonance fatigue failure, and optimized the section shape of the guardrail to enhance horizontal bending performance.

In the aspect of net hydrodynamics, the complete net system of offshore fish cages is usually regarded as a combination of limited net units. Based on this idea, Aarsnes and Rudi [10,11] systematically carried out flow load tests on single netting units, and established a mathematical mapping relationship between the drag coefficient and lift coefficient with net solidity ratio and the angle of attack of the incoming flow; this has been widely recognized in industry and academia. Zhan et al. [12]. further investigated the influence of Reynolds number and net shape on the hydrodynamic coefficients of nets through drag testing, and proposed a new mathematical model. Zhao et al. [13,14] carried out a systematic net unit test, revealing not only the effects of solidity ratio, angle of attack, and Reynolds number but also the influence of net material and structure on hydrodynamic coefficients.

In terms of the mooring system of floating fish cage, it mainly includes a single point mooring system and a multi-point mooring system [15]. DeCew et al. [16]. conducted numerical and experimental studies on the performance of small submerged aquaculture cages under single-point mooring subjected to current. Huang and Pan [17] assessed the risk of rope failure in single point mooring systems based on the long-term environmental load and presented the failure probability during the service life. Hou [18] carried out a reliability analysis on the fatigue damage in the mooring system of the deep-sea fish cage. Shainee et al. [19] studied the submerged characteristics of a self-submersible single-point mooring cage system under the action of wave and current. Fredriksson et al. [20] used numerical models and field-measured data to study the tension of a mooring system in a large fishing ground without waves. Huang et al. [21] used numerical models to analyze the influence of uniform wave and current on mooring line tension and net deformation of multi-point mooring cage system. Liang et al. [22] designed a new type of cage floating collar system, compared it with traditional designs, and conducted a comparative analysis of four common multi-point mooring systems. For multi-cage systems, Cifuentes et al. [23] employed the Morison model to calculate the dynamic response of the cages in wave and current fields and compared the results with experimental data. The wake effects due to the fish net were considered in the analysis. Selvan. et al. [24] investigated the surface wave scattering of multi-cage systems, elucidating the impact of the cages on far-field waves. Xu et al. [25] conducted numerical simulations of the dynamic response of multiple cages under combined wave and current conditions and studied the loads on the mooring lines. Zhao et al. [26] conducted both numerical and experimental studies on the dynamic response of various multi-cage arrangement configurations and evaluated the mooring loads and current velocity distributions.

This paper primarily investigates the performance of mooring systems in different layouts of combined cages. It compares and analyzes the dynamic response of mooring systems under self-storage sea conditions for various cage design schemes, considering both flexible and rigid net models. Furthermore, under the optimal design scheme, the study analyzes the motion response of multiple fish cages and the load characteristics of the cables between different cages.

2. Hydrodynamic Response Analysis Method of Floating Cages

2.1. Motion Response Equation of Cage

For floating cages, the motion equation can be written as follows:

$$[m]\{\ddot{x}\} + [c]\{\dot{x}\} + [k]\{x\} = \{G\} + \{B\} + \{F\} \tag{1}$$

where $[m]$, $[c]$, and $[k]$ are, respectively, the mass matrix, damping matrix, and stiffness matrix of the entire system of the cage. The external forces acting on the structure include gravity$\{G\}$, buoyancy force $\{B\}$, and hydrodynamic load $\{F\}$. x, $\dot{x}$, $\ddot{x}$ represent the cage movement displacement, speed, and acceleration, respectively. For the partially submerged floating frame structure, the influence of instantaneous wave surface is considered in calculating the buoyancy force $\{B\}$. In addition, this study focuses on the performance of the mooring system, treating the floating structure as a rigid body to improve calculation efficiency. The numerical approach for hydrodynamic analysis of the fish cages is shown in Figure 1.

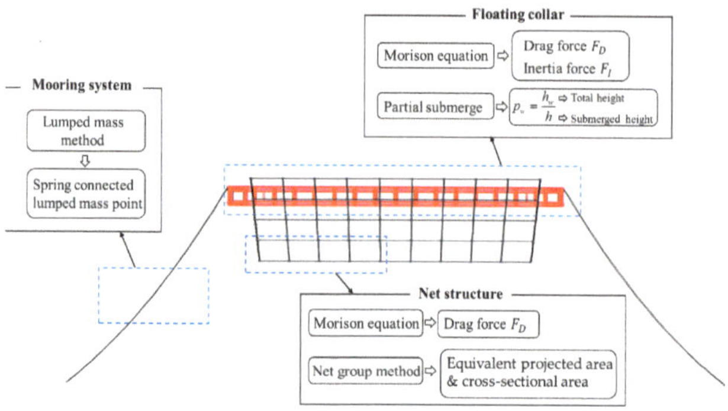

Figure 1. Numerical approach for hydrodynamic analysis of the fish cages.

2.2. Current and Wave Load

The floating structure of the cage is all slender rods. Therefore, the current and wave load subjected to the floating collar, the net system, and the mooring line are calculated by the Morison equation:

$$\begin{cases} \{F\} = \Sigma\{f_{Di}\} + \Sigma\{f_{Ii}\} \\ \{f_{Di}\} = C_{Di}\frac{1}{2}\rho D_i L_i \left| \{u_i^n\} + \{v_i^n\} - \left\{\dot{x}_i^n\right\} \right| (\{u_i^n\} + \{v_i^n\} - \{\dot{x}_i^n\}) \\ \{f_{Ii}\} = \rho C_{Mi}\frac{\pi D_i^2 L_i}{4}\{\dot{u}_i^n\} - \rho(C_{Mi} - 1)\frac{\pi D_i^2 L_i}{4}\{\ddot{x}_i^n\} \end{cases} \tag{2}$$

where, the hydrodynamic load F includes drag force f_{Di} and inertia force f_{Ii}. L_i is the element length and ρ is the fluid density. u_i and $\dot{u}_i$ are the water particle velocity and acceleration caused by the wave, respectively. v_i is the current velocity. D_i is the hydrodynamic diameter of the slender structure. C_{Mi} and C_{Di} are the inertia and drag force coefficient, respectively. Since the net is a slender structure with a very small diameter, the inertial force is negligible. Therefore, only the drag force is considered. For the partially submerged floating body, e.g., the floating collar and the pontoon, its load is adjusted by the immersion ratio and applied to the center of the submerged part. The immersion ratio p_w can be written in the following form:

$$p_w = \frac{h_w}{h} \tag{3}$$

where, h is the total height of the float and h_w represents the height of the float below the instantaneous wave surface.

2.3. Equivalent Group Method of Netting

Due to the large number of real-scale netting units, the equivalent simplification of full-scale fishing nets is needed to improve the computational efficiency. For the simplified model, it is necessary to ensure that the hydrodynamic force, tensile stiffness, and mass of the net remain the same as in the original as the following [27,28]:

$$\begin{aligned} A_{model} &= \sum A_{net} \\ A_{section\_model} &= \sum A_{section\_net} \\ M_{\text{model}} &= \sum M_{net} \end{aligned} \tag{4}$$

where, A_{net} and $A_{section\_net}$ are the projected area and cross-sectional area of the real net, and M_{net} is the mass of the real scale net.

2.4. Mooring Load

The mooring lines of cages studied in this paper adopt catenary form. The dynamic responses of cages and mooring lines are calculated by the fully coupled time-domain method. In the calculation process, the effect of gravity, drag force, and axial elastic tension of the mooring lines are taken into account.

For the mooring line element, the differential equation of motion can be written as the following form under the action of external hydrodynamic load and internal inertial load:

$$\frac{\partial\{T\}}{\partial \mathbf{s}} + \{f_m\} + \{w\} = \rho_m A_m \frac{\partial^2\{R\}}{\partial t^2} \tag{5}$$

where, ρ_m is the density of the mooring line, A_m is the equivalent cross-sectional area of the mooring line, $\{R\}$ is the position vector of the mooring line unit. f_m and w are the hydrodynamic load and gravity of the mooring line unit on the unit length, respectively. $\{T\}$ is the axial tension vector on the mooring line unit, and **s** represents the axial vector of the mooring line unit. The dynamic response of the mooring system is calculated using the lumped mass method in Orcaflex11.2. It is a commercial software developed by Orcina Ltd. (Daltongate, Ulverston, UK) and is widely used in the numerical simulation of mooring line systems and aquacultural structures. As shown in Figure 2, the line elements are converted into concentrated mass points connected by springs for dynamic response calculations. The fluid loads on the line elements are also distributed to the concentrated mass points at both ends.

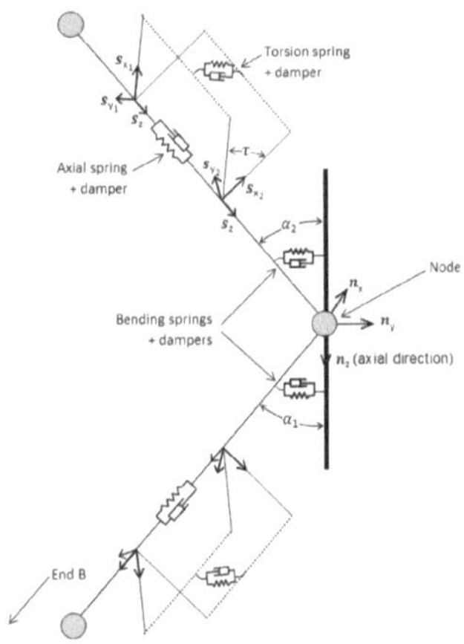

Figure 2. Lumped mass method [29].

2.5. Analysis of Short-Term Sea State Extremum Response of Mooring System

As for the load of the mooring system in multiple cages under irregular waves, there are obvious differences in the load extremums of the mooring system under different random phase conditions due to the phase randomness of the wave components. Therefore, it is necessary to analyze and study multiple different samples. It is generally believed that the load extremum of the mooring system between different samples follows the Gumbel distribution, and its probability function can be written as follows:

$$F(x) = e^{-e^{-\frac{x-\mu}{\beta}}} \tag{6}$$

where, μ and β are parameters of the Gumbel distribution.

3. Numerical Calculation Model

3.1. Single Cage Model

The numerical model of a single cage is shown in Figure 3, which includes the floating collars, the pontoons, the net system, and the mooring system. The pontoons are mainly used to provide sufficient reserve buoyancy. The specific size parameters are shown in Table 1. The drag coefficient C_D of the floating collars is 1.2 and that of the pontoons is 2.0. The inertia coefficient C_M is 2.0. The hydrodynamic coefficient of the net is selected according to the screen model [6], and 1.5 times the marine growth coefficient is taken into account. The main body of the cage is modeled by Orcaflex11.2, and the two situations, rigid net and flexible net, are analyzed.

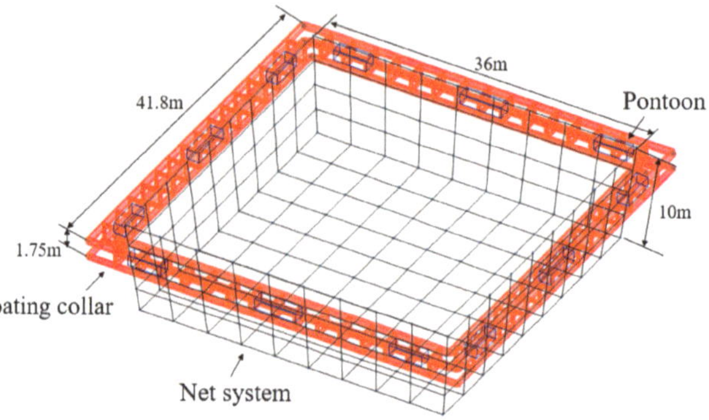

Figure 3. Single cage numerical calculation model.

Table 1. Cage main size parameters.

Floating collar	Length × width × height (m)	41.8 × 41.8 × 1.75
	Pontoon size (m)	3.65/4.6 × 1.3 × 1.1
	Floating collar diameter (m)	0.7/0.6
Netting system	Length × width × depth (m)	36 × 36× 10
	net solidity	0.129

3.2. Layout and Mooring Scheme of Multiple Fish Cages

This study mainly considers two arrangement layouts and three deployment schemes, as shown in Figures 4–6. For the first layout scheme, there are four mooring lines on the outer cage, and there are two chains between each pair of cages through cables, totaling 14 mooring lines. All chains are R3S grade stud-link chains with a diameter of 84 mm. For the second layout scheme, there are similarly four mooring lines on each outer cage, of which two have 45° arrangement. There are two mooring lines between each pair of cages connected through cables, a total of 14 mooring lines. These chains are also R3S grade stud-link chains with a diameter of 84 mm. For the third scheme, each cage has four mooring lines, totaling 16 chains, all using R3S grade stud-link mooring lines with a diameter of 84 mm. In addition, the adjacent cages are connected by cables. In all three schemes, the drag coefficient of the mooring lines are 2.6, the spacing of the cages is 50 m, and the cables are polyester cables with a diameter of 100 mm. Detailed parameters of the mooring lines are shown in Table 2. The safety factor and breaking tension for the mooring lines are determined according to the DNVGL-OS-E302 [30].

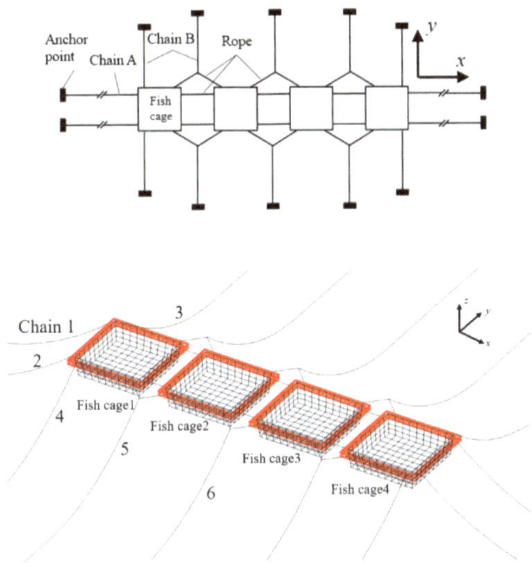

Figure 4. Scheme 1: Layout of 1 × 4 cages and design of mooring system.

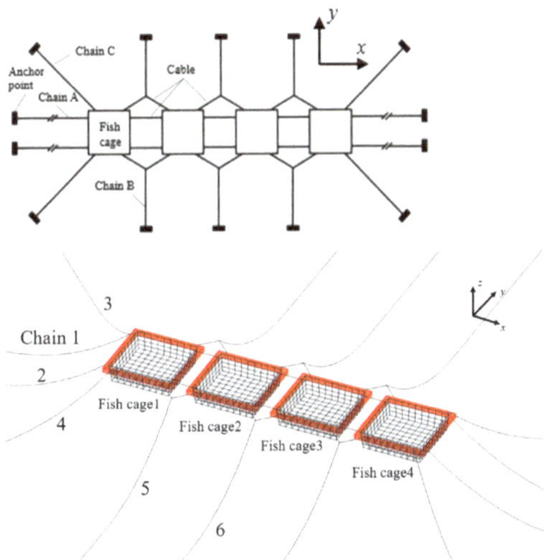

Figure 5. Scheme 2: Layout of 1 × 4 cages and design of inclined mooring system.

Table 2. Cable parameter.

Scheme	Chain Categories	Chain Length (m)	Mooring Radius (m)	Steel Grade
Scheme 1	A	558	550	R3S
	B	508	500	R3S
Scheme 2	A	558	550	R3S
	B	508	500	R3S
	C	508	500	R3S
Scheme 3	A	448	440	R3S

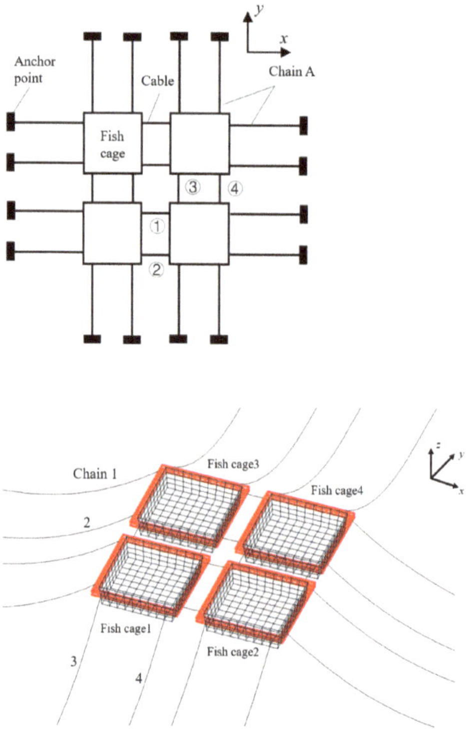

Figure 6. Scheme 3: Layout of 2 × 2 cages and design of mooring system.

3.3. Environmental Parameters

This paper analyzes the performance of multiple cages with different mooring schemes under extreme regular wave conditions and studies the dynamic response of the optimized scheme under irregular waves. The parameters of the extreme regular wave state are shown in Table 3, and those of the irregular wave state are shown in Table 4. The flow field is linear shear flow, and the velocity at the surface and seabed is 1.5 m/s and 1.0 m/s, respectively. Three different wave directions are taken into account, in which 0° are propagated forward along the x-axis and 90° are propagated forward along the y-axis. The operating water depth of the cage is 32.5 m.

Table 3. Mooring system check calculation working condition table.

Condition	Direction (°)	Depth (m)	Wave Period (s)	Wave Height (m)	Flow Velocity (m/s)
Case1	0	32.5	12.5	8.4	Surface 1.5 Seafloor 1.0
Case2	45				
Case3	90				

Table 4. Irregular wave conditions.

Depth (m)	Significant Wave Height (m)	Spectral Peak Period (s)	Flow Velocity (m/s)
32.5	4.7	8.5	Surface 1.5 Seafloor 1.0

4. Result Analysis and Discussion

The maximum tension of each mooring line in the different layout and mooring system design schemes under three wave directions was calculated, and the rigid net model and flexible net model were compared in various working conditions. The motion response characteristics of the cages were analyzed according to the optimal mooring design scheme.

4.1. Analysis of Load Characteristics of Mooring System

4.1.1. Scheme 1: 1 × 4 Layout and Mooring System of Multiple Cages

In this scheme, the direction of the most dangerous wave-current is 0°, and the environmental load is mainly carried by the No. 1 and No. 2 mooring lines. Under the two conditions of rigid and flexible net models, the maximum mooring line tension is 447.90 tons and 438.68 tons, respectively, as shown in Figure 7. The deformation of the net garment is shown in Figure 8. When the net system is deformed, the force of the net becomes smaller, resulting in the tension of the mooring system becoming smaller, but the influence is not significant. In the direction of 0° wave-current incidence, the strain decreases by 2.06%.

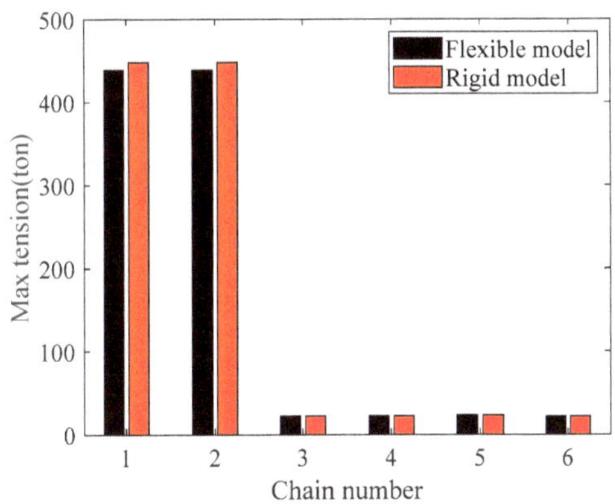

Figure 7. Scheme 1 Maximum mooring line tension under 0° wave-current incidence.

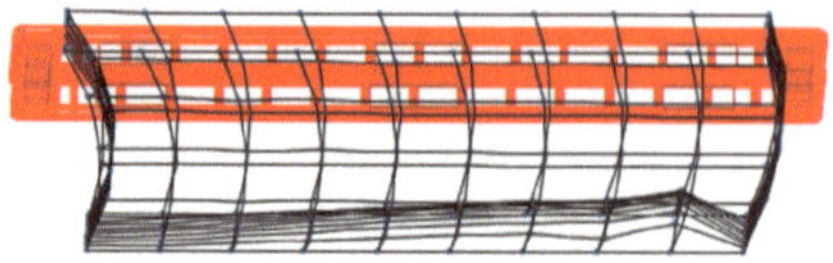

Figure 8. Deformation of nets in Scheme 1.

The distribution of hydrodynamic load is relatively uniform at 45° and 90° wave-current direction, and the maximum load of a single mooring line decreases obviously. The maximum mooring system load of the rigid net model is 314.83 tons and 232.23 tons, respectively. The maximum mooring system load of the flexible net model is 307.50 tons and 224.63 tons, respectively, as shown in Figures 9 and 10.

According to the grade and diameter of the mooring lines, the breaking tension of the mooring line can be determined as 667.8 tons. Under the condition of different inflow direction of wave, the rigid net and flexible net model are considered, and the checking of results of the mooring system are shown in Tables 5 and 6. After calculation, it can be seen that under the arrangement of Scheme 1, in the sea state of 0° incident regular wave, the safety factor of the mooring lines does not meet the requirement that the safety factor of the self-storage condition in the specification be greater than 1.67. Through observation of the motion response process, there is no accumulation of the mooring lines or contact between the mooring lines and cage under the three working conditions. The space attitude of the mooring system under the maximum displacement state is shown in Figure 11.

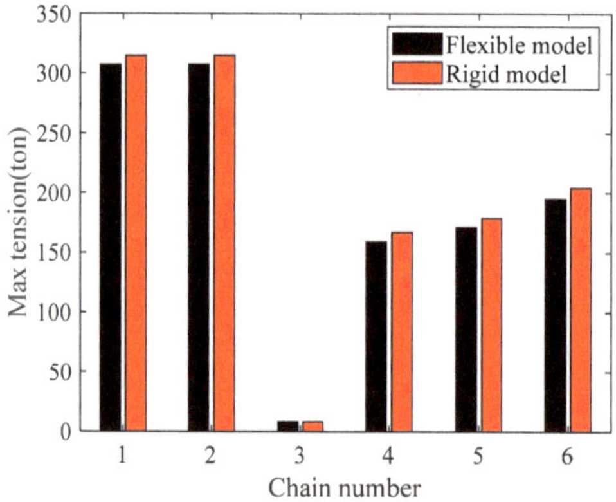

Figure 9. Scheme 1. Maximum mooring line tension under 45° wave-current incidence.

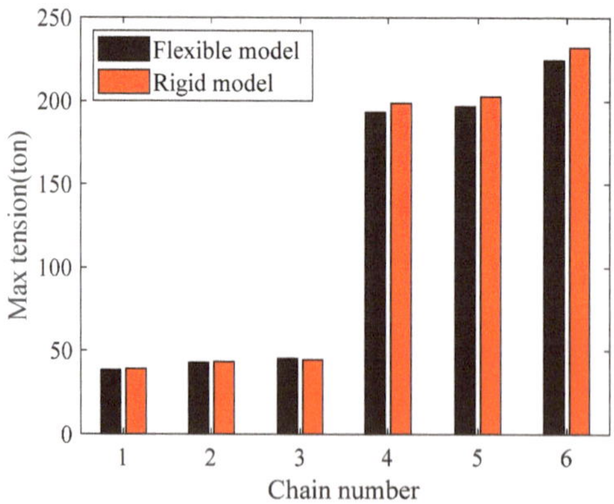

Figure 10. Scheme 1. Maximum mooring line tension under 90° wave-current incidence.

Table 5. Results of mooring line verification in Scheme 1 (rigid net model).

Category \ Degree	0°	45°	90°
Maximum cable tension (tons)	447.90	314.83	232.23
Safety factor	1.49	2.12	2.88
Pile and touch	No	No	No

Table 6. Results of mooring line verification in Scheme 1 (flexible net model).

Category \ Degree	0°	45°	90°
Maximum cable tension (tons)	438.68	307.50	224.63
Safety factor	1.52	2.17	2.97
Pile and touch	No	No	No

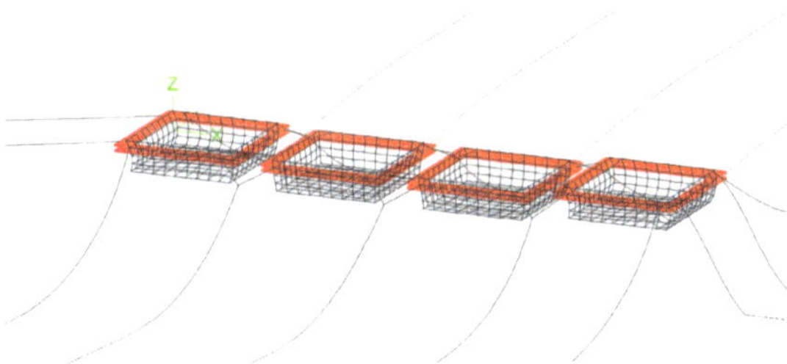

Figure 11. Scheme 1. Side view of cable status when large displacement occurs (0° wave-current incidence).

4.1.2. Scheme 2: 1 × 4 Layout and Inclined Mooring System

In Scheme 2, the diagonal arrangement of No. 3 and No. 4 mooring line shares a large part of the load in the direction of 0° wave-current. In this case, the load of No. 1 and No. 2 mooring lines are significantly lower than that of Scheme 1. Without considering the deformation of the net, the maximum load of No. 1 and No. 2 mooring lines is 349.07 tons. It is 340.81 tons when considering net deformation, as shown in Figure 12.

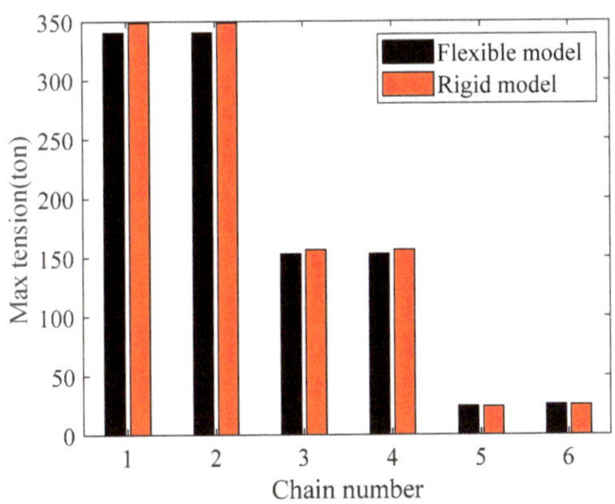

Figure 12. Scheme 2. Maximum mooring line tension under 0° wave-current incidence.

At 45° wave-current direction, No. 4 mooring line is collinear with the wave-current direction and carries the maximum load. Under the conditions of rigid and flexible net models, the maximum loads on the mooring system are 324.81 tons and 313.39 tons, respectively. In the 90° wave-current direction, the No. 5 and No. 6 mooring lines are parallel to the wave direction and carry large loads. The oblique arrangement of part of the mooring lines leads to the low lateral carrying capacity of the mooring system, and the load of the mooring system is greatly increased compared with Scheme 1, as shown in Figures 13 and 14. In Scheme 2, there is also no accumulation of mooring lines or contact between cages. The space posture of the mooring system under the condition of maximum displacement of cages is shown in Figure 15.

The mooring line check results of Scheme 2 are shown in Tables 7 and 8. In Scheme 1 and Scheme 2, four cages are arranged in a straight line, resulting in a large hydrodynamic load of cages in the 0° wave-current direction. Therefore, the maximum tension of No. 1 and No. 2 mooring lines in Scheme 1 are much larger than that of others, which increases the cost. Compared with Scheme 1, Scheme 2 changes the arrangement angle of No. 3 and No. 4 mooring lines. Therefore, No. 3 and No. 4 mooring lines share the load of No. 1 and No. 2 mooring lines in the direction of 0° wave-current, and carry greater load in the direction of 45° wave-current. Due to the reduction of the lateral carrying capacity of the mooring line, the load increases under the condition of 90° incidence, but it still meets the requirement of the safety factor of 1.67. Compared with Scheme 1, the mooring line tension in Scheme 2 is more average under each working condition, which can reduce the construction cost of the mooring system to a certain extent.

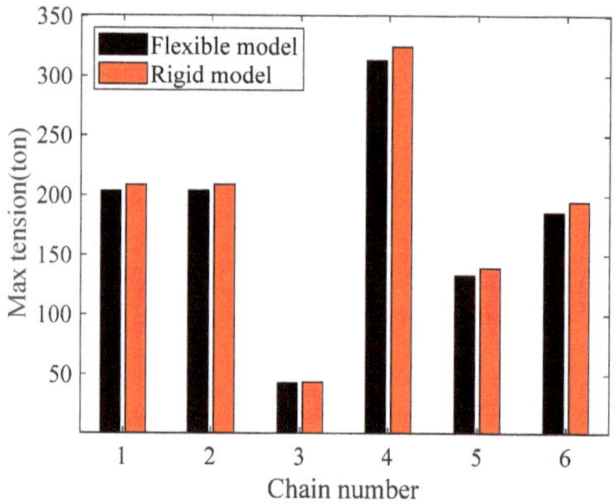

Figure 13. Scheme 2. Maximum mooring line tension under 45° wave-current incidence.

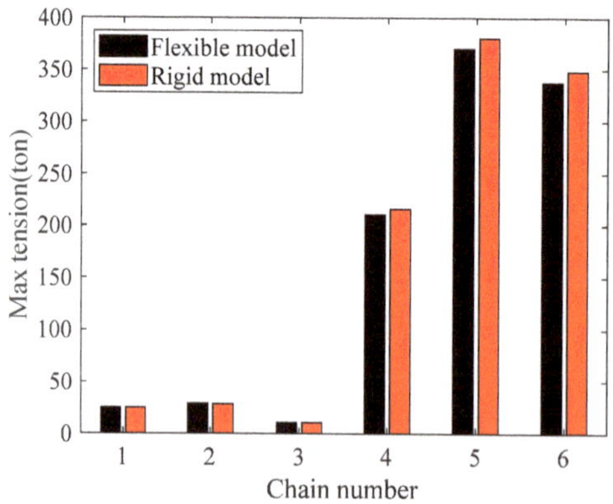

Figure 14. Scheme 2. Maximum mooring line tension under 90° wave-current incidence.

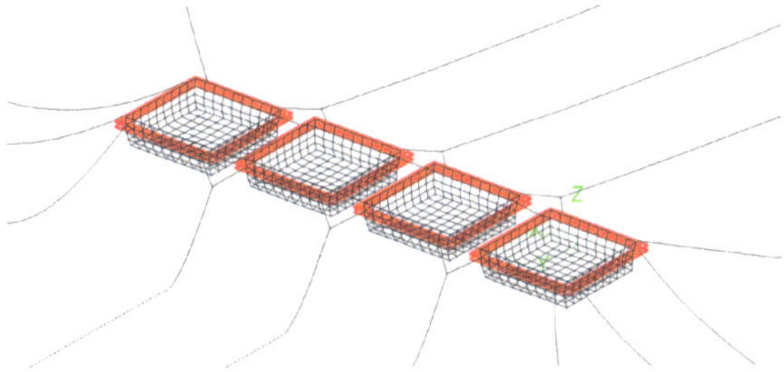

Figure 15. Scheme 2. Side view of cable status when large displacement occurs (90° wave-current incidence).

Table 7. Results of mooring line verification in Scheme 2 (rigid net model).

Category \ Degree	0°	45°	90°
Maximum cable tension (tons)	349.07	324.81	380.06
Safety factor	1.91	2.06	1.75
Pile and touch	No	No	No

Table 8. Results of mooring line verification in Scheme 2 (flexible net model).

Category \ Degree	0°	45°	90°
Maximum cable tension (tons)	340.81	313.39	370.25
Safety factor	1.96	2.13	1.8
Piling and touching	No	No	No

4.1.3. Scheme 3: 2 × 2 Layout and Mooring System of the Tandem Cage

For Scheme 3, the wave directions of 0° and 90° are symmetrical along the diagonal of the whole system, so only the wave directions of 0° and 45° are calculated under this scheme, and the stress of No. 1–4 mooring lines is observed, as shown in Figures 16 and 17. In the 0° wave-current incidence, mooring line 1 and 2 carry the main hydrodynamic loads. The maximum mooring line tension of the rigid and flexible net models is 328.45 tons and 319.86 tons, respectively, so the safety factor exceeds 1.67 and meets the safety design requirements, as shown in Tables 9 and 10. In the working condition of 45° wave-current incidence, since the hydrodynamic load is shared by eight mooring lines, the load is relatively small compared with the working condition of 0° wave injection. In the third scheme, there is also no accumulation of mooring lines or contact between cages.

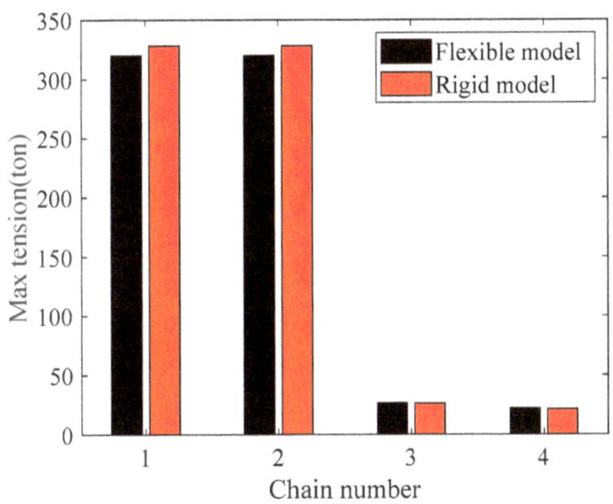

Figure 16. Scheme 3. Maximum mooring line tension under 0° wave-current incidence.

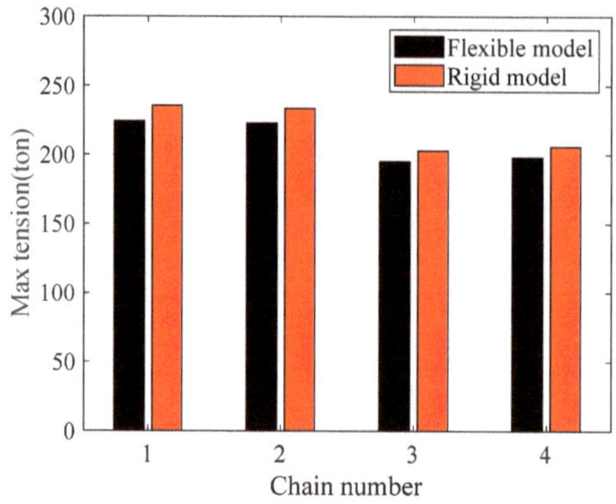

Figure 17. Scheme 3. Maximum mooring line tension under 45° wave-current incidence.

Table 9. Results of mooring line verification in Scheme 3 (rigid net model).

Category \ Degree	0°	45°
Maximum cable tension (tons)	328.45	235.76
Safety factor	2.03	2.83
Pile and touch	No	No

Table 10. Results of mooring line verification in Scheme 3 (flexible net model).

Category \ Degree	0°	45°
Maximum cable tension (tons)	319.86	224.47
Safety factor	2.09	2.98
Pile and touch	No	No

Compared with the first two schemes, the four cages in Scheme 3 are arranged in a square manner, the hydrodynamic load is relatively small, and the mooring lines are arranged in a symmetrical manner, so the maximum tension distribution of the mooring lines is uniform, and the total length of the mooring lines is short, which is conducive to reducing the cost of the mooring system.

4.1.4. Comparative Analysis of Performance of Different Mooring System Design Schemes

In the first scheme, the hydrodynamic load of the cage under the 0° wave-current direction is only carried by the two mooring lines against the wave, resulting in the safety factor of the mooring line being only 1.49, which does not meet the specification requirements. Scheme 2 optimizes the arrangement angle of part of the mooring lines according to Scheme 1, resulting in a more balanced maximum tension of each line under different wave-current directions, thereby increasing the safety factor. However, since the four cages in Scheme 2 are arranged in a linear manner, the hydrodynamic load at 90° is larger, while Scheme 3 improves the cage arrangement to 2 × 2, reducing the hydrodynamic load carried by a single mooring line, making it more economical than Scheme 2. Meanwhile, no accumulation of mooring lines and contact friction between the cage structure are observed. In summary, Scheme 3 is the best in terms of safety and economy.

4.2. Analysis of Cage Motion Response and Cable Bearing Characteristics between Cages under Regular Wave Conditions

For Scheme 3, this paper further studied the motion response characteristics of the cages under the 0° and 45° wave-current directions. The flexible net model was used to analyze the motion response of the cages, as shown in Figures 18 and 19.

In the direction of 0° wave flow, the surge amplitude of the two cages is relatively close, and the displacement difference is far less than the distance between the two cages, thus, there is no collision risk of the cages. Due to the constraints of the mooring line, the heave amplitude of the No. 1 cage is smaller than that of the No. 2 cage.

In the 45° wave-current direction, the horizontal resilience of No. 4 cage on the leeward of the wave is the least, so the surge amplitude is the largest. The horizontal motion amplitudes of No. 1, 2, and 4 cages are not large, and there is no collision between cages. As for heave amplitude, the No. 1 cage on the wave side has four mooring lines to provide the restoring force, so the heave amplitude is the smallest. Only two of the four mooring lines on the No. 2 cage can provide the restoring force, and the four mooring lines on the No. 4 cage are in the leeside, so the heave amplitude of the No. 4 cage is the largest.

Figure 20 shows the maximum tension of the cables between the cages. The cable numbers are shown in Figure 6. At 0° of wave-current incidence, the motion of the cage is mainly surge and heave, so the tension of the cables 1 and 2, which mainly carry the horizontal resilience of the cage, is larger, while the tension of the cables 3 and 4 is smaller. At 45° of wave-current incidence, the mean surge of 1, 2, and 4 cages are close, and the maximum tension of the 1 to 4 cables are also close.

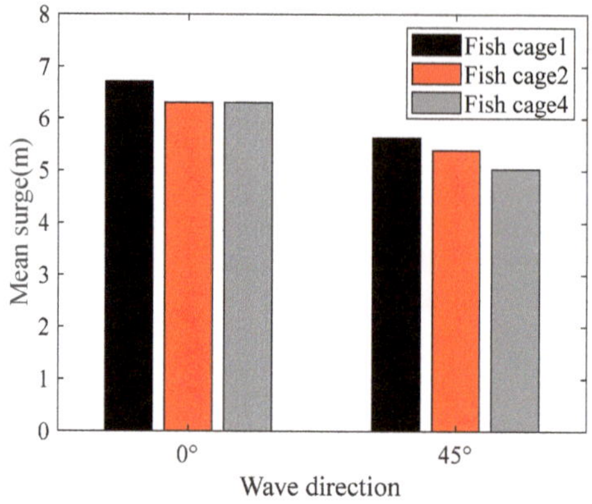

(**a**) The mean values of surge motion

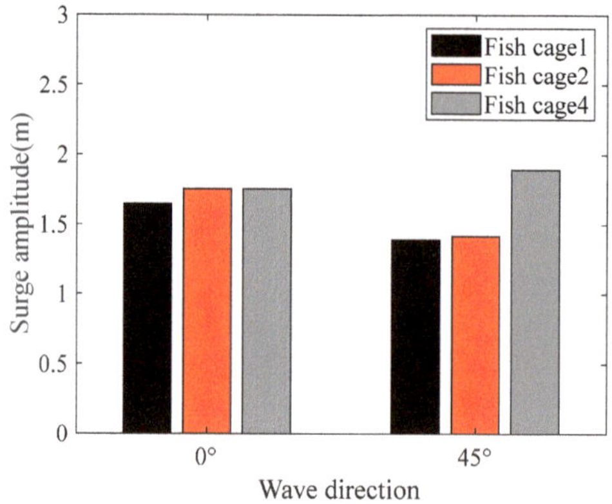

(**b**) The amplitude of the surge motion

Figure 18. Response to the surge motion.

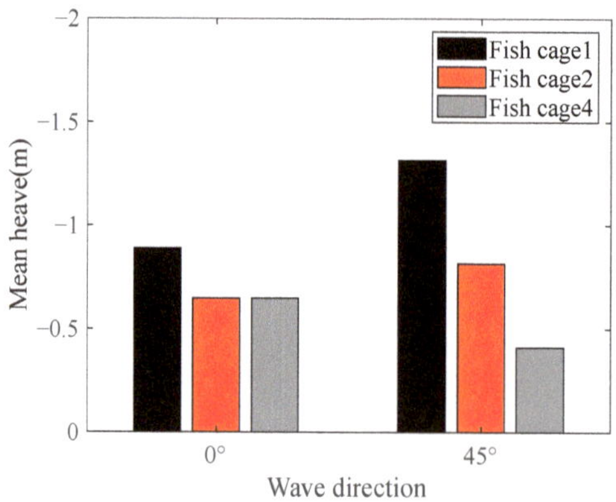

(**a**) The mean value of heave motion

Figure 19. *Cont.*

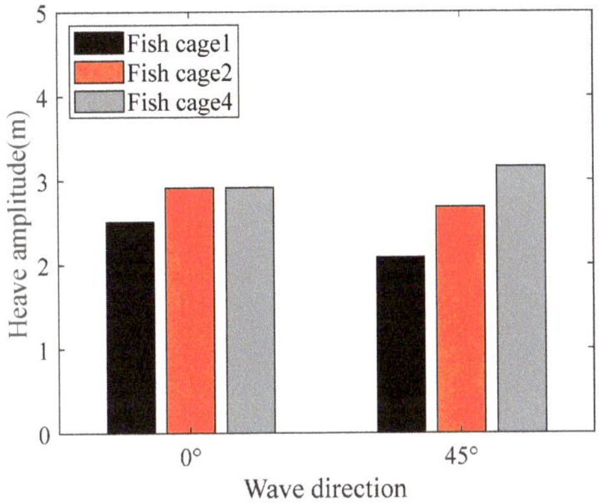

(**b**) Amplitude of heave motion

Figure 19. Heave motion response.

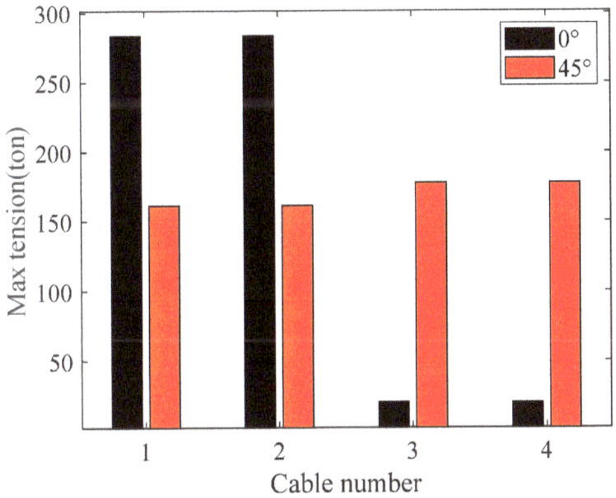

Figure 20. Maximum cable tension.

4.3. Analysis of Cage Motion Response and Cable Loading Characteristics between Cages under Irregular Wave Conditions

In the real marine environment, the wave environment of the aquaculture cages is generally irregular waves. In order to save calculation efficiency, regular wave conditions are adopted to compare the safety and economy of different mooring schemes. According to the third optimal scheme, the dynamic response characteristics of the cage and the load characteristics of the mooring system under the irregular wave condition are studied. For example, under the condition of 0° wave-current incidence, the time history and power spectrum analysis results of surge and heave are shown in Figures 21 and 22. Combined with the time history and its power spectrum, it can be seen that under the action of irregular waves, the surge energy of the cage is mainly concentrated in the low frequency part, while the heave response is more significantly affected by wave frequency. Meanwhile, due to the influence of the Morison force squared term, the power spectrum has the common characteristics of the wave frequency and wave frequency multiplier. The surge amplitude of cages is smaller than the cage spacing, indicating that there is essentially no risk of collision.

The load time history of the cage mooring system and connecting cable with the corresponding power spectrum are shown in Figures 23 and 24. Among them, there is little difference between the peak load of the mooring line and that of the connecting cable under the same sea state. The energy of the mooring line tension is mainly concentrated in the low-frequency part, the energy of the cable tension is mainly concentrated in the vicinity of the spectral peak period, and the low-frequency movement has little influence on the cable tension.

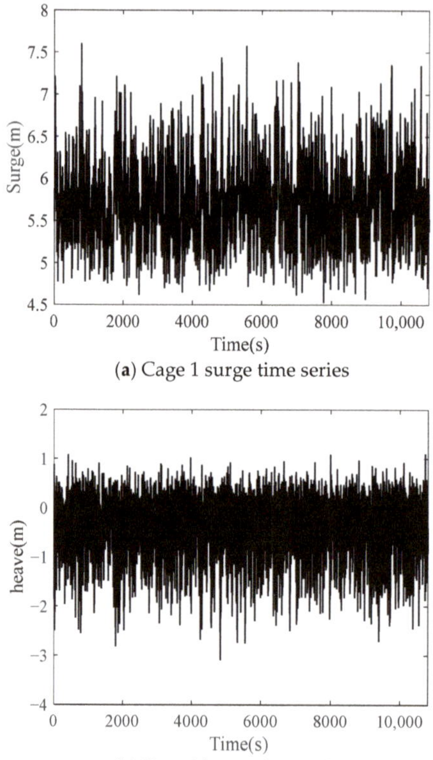

Figure 21. Motion of Fish Cage1.

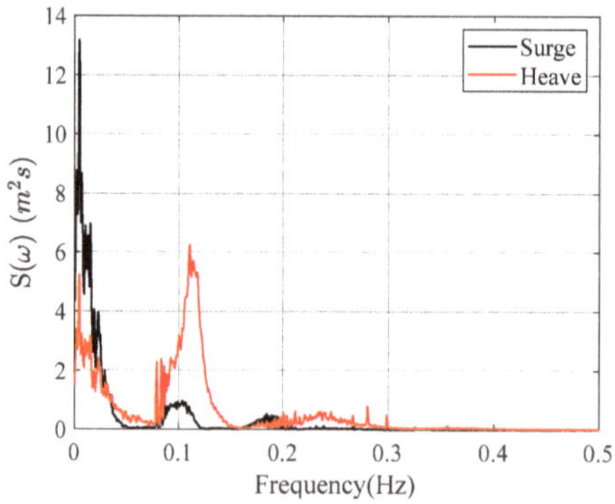

Figure 22. Comparison of motion spectrum.

Aiming at the load of the mooring system under short-term sea conditions, several random seeds were selected, respectively. Gumbel distribution fitting was performed on the extreme value of the mooring line load of each sea state sample under the conditions of 0° wave-current incidence and 45° wave-current incidence, and the extreme value response analysis was performed on the mooring line load corresponding to 0.9 exceedance probability (P90). The results are shown in Figure 25, Figure 26, and Table 11, respectively.

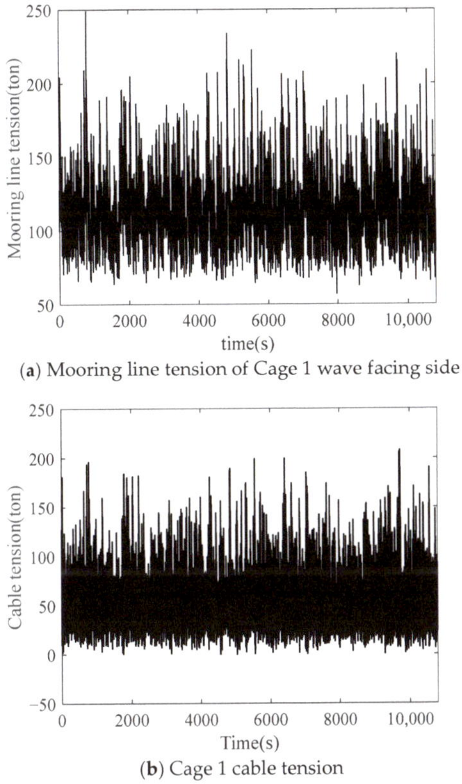

Figure 23. Tension on mooring chain and connect cable.

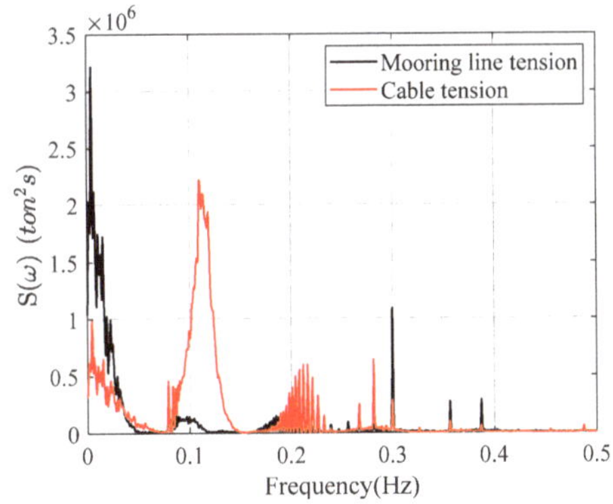

Figure 24. Tension spectrum of mooring chain and connecting cable.

Table 11. Estimation of short-term extreme load of mooring system at different wave directions.

Working Condition	Mooring Line Load (P90)
0° wave-current incidence	288.1 tons
45° wave-current incidence	224.5 tons

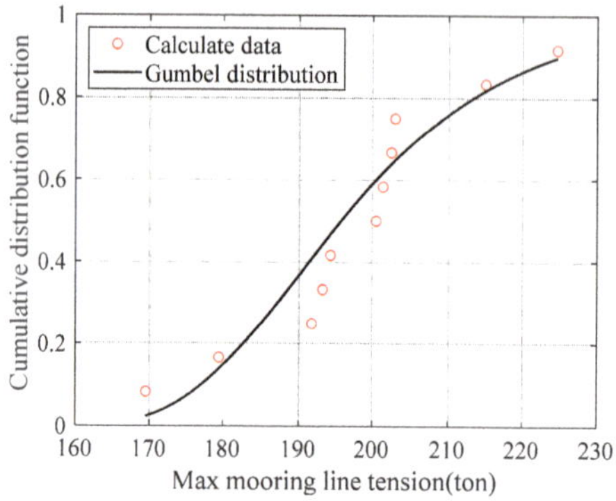

Figure 25. Fitting of Gumbel distribution for mooring line at 0° wave-current incidence direction.

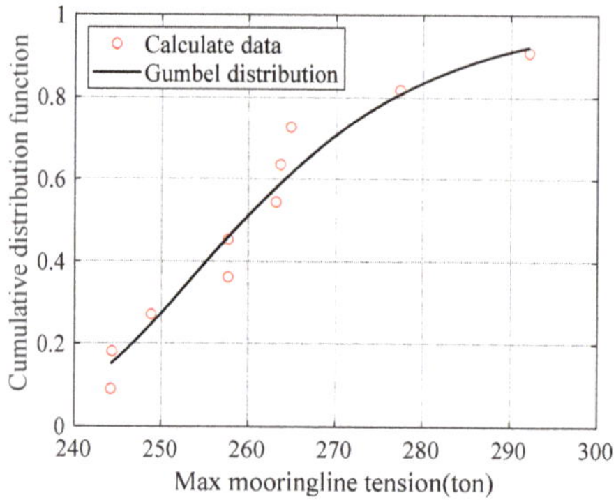

Figure 26. Fitting of Gumbel distribution for mooring line at 45° wave-current incidence direction.

In summary, it is evident that compared to the 0° wave-current incidence condition, the mooring system experiences a lower maximum load and is relatively safer under the 45° wave-current incidence condition in the same sea state.

5. Conclusions

This paper proposes three mooring schemes according to the working environment and hydrodynamic characteristics of floating multiple cages. Their safety under self-storage conditions was numerically simulated and evaluated, and the numerical results of the rigid net model and the flexible net model compared, focusing on the maximum load of the mooring system. The possibility of mooring line accumulation and collisions between mooring lines and the fish cages is also considered. The optimal mooring scheme was selected, and its motion response, mooring line tension, and cable load characteristics between cages studied. The main conclusions are as follows:

1. Based on the three different arrangement configurations mentioned above, the mooring line tension of the fish cages is significantly higher with a 1 × 4 arrangement compared to a 2 × 2 arrangement. Under 0° wave-current incidence, the maximum tension increases by 37.14%, and under 45° wave-current incidence conditions, the maximum tension increases by 46.12%. From a safety perspective for the cages, it is recommended to adopt the 2 × 2 arrangement configuration.
2. The difference between the mooring line tension calculated by the flexible and rigid net model is not more than 5% under different wave-current incidences. Therefore, for this kind of fish cages, considering the modeling workload and calculation efficiency in practical application, the rigid net model can be used to check the safety of the mooring system.

3. Under irregular wave conditions, the load of the mooring system of the cage is mainly controlled by the low-frequency motion, and the load of the connecting cable is affected by the wave-frequency motion. Both surge and heave motions have a low-frequency component, which is due to there being a coupling among surge motion, mooring line tension, and heave motion. In the same irregular wave condition, for the 2×2 arrangement of the combined cage form, the wave-current incident along 0° incidence is more dangerous.

Author Contributions: Methodology, F.L., Z.J. and T.C.; Investigation, F.L., Y.X., H.Z., G.W., G.S. and Y.Z.; Writing—Original Draft Preparation, F.L.; Writing—Review and Editing, Y.X. and H.Z. Resources, F.L. All authors have read and agreed to the published version of the manuscript.

Funding: The research is funded by Taishan Industrial Experts Program, Key R&D Program of Shandong Province (2021SFGC0701, 2023TZXD050) and Young Elite Scientists Sponsorship Program by CAST (2023QNRC001).

Data Availability Statement: The original contributions presented in the study are included in the article, further inquiries can be directed to the corresponding author/s.

Conflicts of Interest: Authors Fuxiang Liu, Guoqing Sun and Yuqin Zhang were employed by the Yantai CIMC Blue Ocean Technology Co., Ltd. The remaining authors declare that the research was conducted in the absence of any commercial or financial relationships that could be construed as a potential conflict of interest.

References

1. Huguenin, J.E. The design, operations and economics of cage culture systems. *Aquac. Eng.* **1997**, *16*, 167–203. [CrossRef]
2. Stickney, R.R. Impacts of Cage and Net-Pen Culture on Water Quality and Benthic Communities. In *Aquaculture and the Environment in the United States*; U.S. Aquaculture Society: St Amant, LA, USA, 2002; pp. 105–118.
3. Colbourne, D.B. Another perspective on challenges in open ocean aquaculture development. *IEEE J. Ocean. Eng* **2005**, *30*, 4–11. [CrossRef]
4. Faltinsen, O.M.; Shen, Y. Wave and current effects on floating fish farms: Keynote contribution for the international workshop on wave loads and motions of ships and offshore structures, Harbin, China, 5–7 November, 2017. *J. Mar. Sci. Appl.* **2018**, *17*, 284–296. [CrossRef]
5. Chu, Y.I.; Wang, C.M.; Park, J.C.; Lader, P.F. Review of cage and containment tank designs for offshore fish farming. *Aquaculture* **2020**, *v519*, 734928. [CrossRef]
6. Løland, G. Current forces on, and water flow through and around, floating fish farms. *Aquac. Int.* **1993**, *1*, 72–89. [CrossRef]
7. Kristiansen, T.; Faltinsen, O.M. Modelling of current loads on aquaculture net cages. *J. Fluids Struct.* **2012**, *34*, 218–235. [CrossRef]
8. Fredriksson, D.W.; Decew, J.C.; Tsukrov, I. Development of structural modeling techniques for evaluating hdpe plastic net pens used in marine aquaculture. *Ocean Eng.* **2007**, *34*, 2124–2137. [CrossRef]
9. Liu, H.; Wang, S.; Huang, X.; Tao, Q.; Hu, Y.; Guo, G.; Song, L. Mechanical property analysis and optimization of deep-water net cage guardrail. *Editor. Off. Trans. Chin. Soc. Agric. Eng.* **2017**, *33*, 248–257.
10. Aarsnes, J.V.; Rudi, H.; Løland, G. Current forces on cage, net deflection. In *Engineering for Offshore Fish Farming*; Thomas Telford Publishing: Thomas Telford, Glasgow, UK, 1990; pp. 137–152.
11. Løland, G. Current Forces on and Flow through Fish Farms. Ph.D. Thesis, University of Trondheim, Trondheim, Norway, 1991.
12. Zhan, J.-M.; Hu, Y.-Z.; Zhao, T.; Sun, M.-G. Hydrodynamic experiment and analysis of fishing net. *Ocean Eng.* **2002**, *20*, 49–53.
13. Zhao, Y.P.; Li, Y.C.; Dong, G.H.; Gui, F.K.; Teng, B. Numerical simulation of the effects of structure size ratio and mesh type on three-dimensional deformation of the fishingnet gravity cage in current. *Aquac. Eng.* **2007**, *36*, 285–301. [CrossRef]
14. Zhao, Y.P.; Li, Y.C.; Dong, G.H.; Gui, F.K.; Wu, H. An experimental and numerical study of hydrodynamic characteristics of submerged flexible plane nets in waves. *Aquac. Eng.* **2008**, *38*, 16–25. [CrossRef]
15. Sui, L.; Huang, X.; Liu, H.; Hu, Y.; Yuan, T.; Wang, S.; Tao, Q. Effects of mooring pattern on dynamic characteristics of a deep-water aquaculture cage. *South China Fish. Sci.* **2021**, *17*, 98–108.
16. DeCew, J.; Tsukrov, I.; Risso, A.; Swift, M.R.; Celikkol, B. Modeling of dynamic behavior of a single-point moored submersible fish cage under currents. *Aquac. Eng.* **2010**, *43*, 38–45. [CrossRef]
17. Huang, C.C.; Pan, J. Mooring line fatigue: A risk analysis for an SPM cage system. *Aquac. Eng.* **2010**, *42*, 8–16. [CrossRef]
18. Hou, H. *Analysis of Fatigue Damage and System Reliability of Mooring System of Offshore Fish Cage*; Dalian University of Technology: Dalian, China, 2020.
19. Shainee, M.; DeCew, J.; Leira, B.J.; Ellingsen, H.; Fredheim, A. Numerical simulation of a self-submersible SPM cage system in regular waves with following currents. *Aquac. Eng.* **2013**, *54*, 29–37. [CrossRef]

20. Fredriksson, D.W.; DeCew, J.C.; Tsukrov, I.; Swift, M.R.; Irish, J.D. Development of large fish farm numerical modeling techniques with in situ mooring tension comparisons. *Aquac. Eng.* **2007**, *36*, 137–148. [CrossRef]
21. Huang, C.C.; Tang, H.J.; Liu, J.Y. Effects of waves and currents on gravity-type cages in the open sea. *Aquac. Eng.* **2008**, *38*, 105–116. [CrossRef]
22. Liang, J.; Pang, L.; Dong, S. Research on Hydrodynamic Analysis and Mooring method of a new net cage floating collar system. *Period. Ocean Univ. China (Nat. Sci. Ed.)* **2021**, *51*, 119–126.
23. Cifuentes, C.; Kim, M.H. Hydrodynamic response of a cage system under waves and currents using a Morison-force model. *Ocean Eng.* **2017**, *141*, 283–294. [CrossRef]
24. Selvan, S.A.; Gayathri, R.; Behera, H.; Meylan, M.H. Surface wave scattering by multiple flexible fishing cage system. *Phys. Fluids* **2021**, *33*, 037119. [CrossRef]
25. Xu, T.J.; Zhao, Y.P.; Dong, G.H.; Li, Y.C.; Gui, F.K. Analysis of hydrodynamic behaviors of multiple net cages in combined wave–current flow. *J. Fluids Struct.* **2013**, *39*, 222–236. [CrossRef]
26. Zhao, Y.P.; Bi, C.W.; Chen, C.P.; Li, Y.C.; Dong, G.H. Experimental study on flow velocity and mooring loads for multiple net cages in steady current. *Aquac. Eng.* **2015**, *67*, 24–31. [CrossRef]
27. Jin, C.; Choi, J.; Kim, M.H. Response prediction and monitoring feasibility of a stow net system using measured environmental data in the southwest coast of korea. *Appl. Sci.* **2018**, *8*, 1517. [CrossRef]
28. Jin, C.; Kim, H.; Kim, M.H.; Kim, K. Monitoring-system development for a bottom-set gillnet through time-domain dynamic simulations. *Appl. Sci.* **2019**, *9*, 1210. [CrossRef]
29. Orcina. Orcaflex Manual Version 11.2. 2021. Available online: https://www.orcina.com/releases/orcaflex-112/ (accessed on 1 August 2024).
30. DNV-GL. Dnvgl-os-e302, Offshore Mooring Chain. Available online: https://rules.dnv.com/docs/pdf/DNVPM/codes/docs/2008-10/OS-E302.pdf (accessed on 1 August 2024).

Journal of
*Marine Science
and Engineering*

MDPI

Article

Applying Neural Networks to Predict Offshore Platform Dynamics

Nikolas Martzikos *, Carlo Ruzzo, Giovanni Malara, Vincenzo Fiamma and Felice Arena

Natural Ocean Engineering Laboratory (NOEL), "Mediterranea" University of Reggio Calabria, Loc. Feo di Vito, 89122 Reggio Calabria, Italy; carlo.ruzzo@unirc.it (C.R.); giovanni.malara@unirc.it (G.M.); vincenzo.fiamma@unirc.it (V.F.); arena@unirc.it (F.A.)
* Correspondence: nikolas.martzikos@unirc.it

Abstract: Integrating renewable energy sources with aquaculture systems on floating multi-use platforms presents an innovative approach to developing sustainable and resilient offshore infrastructure, utilizing the ocean's considerable potential. From March 2021 to January 2022, a 1:15-scale prototype was tested in Reggio Calabria, Italy, which gave crucial insights into how these structures behave under different wave conditions. This study investigates the application of Artificial Neural Networks (ANNs) to predict changes in mooring loads, particularly at key points of the structure. By analyzing metocean data, several ANN models and optimization techniques were evaluated to identify the most accurate predictive model. With a Normalized Root Mean Square Error (NRMSE) of 1.7–4.7%, the results show how ANNs can effectively predict offshore platform dynamics. This research highlights the potential of machine learning in developing and managing sustainable ocean systems, setting the stage for future advancements in data-driven marine resource management.

Keywords: artificial neural networks; offshore platforms; aquaculture platforms; mooring loads; renewable energy

Citation: Martzikos, N.; Ruzzo, C.; Malara, G.; Fiamma, V.; Arena, F. Applying Neural Networks to Predict Offshore Platform Dynamics. *J. Mar. Sci. Eng.* **2024**, *12*, 2001. https://doi.org/10.3390/jmse12112001

Academic Editors: Zhenhua Ma and Jianguang Qin

Received: 27 September 2024
Revised: 5 November 2024
Accepted: 5 November 2024
Published: 7 November 2024

1. Introduction

The transition to renewable energy is crucial for addressing the interconnected challenges of climate change, energy security, and sustainable development. Offshore wind energy, with its higher and more consistent wind speeds compared to land-based alternatives, represents a substantial advancement in reducing carbon emissions [1]. Offshore wind farms take advantage of stronger sea winds, producing electricity for longer periods and avoiding common issues like noise pollution that often arise with onshore wind [2]. As the global population continues to grow, the demand for sustainable energy intensifies, driving innovation in areas like Marine Renewable Energy (MRE) and aquaculture [3]. These industries play a crucial role in addressing energy demands while also contributing to broader socio-economic goals. The concept of "Blue Growth" highlights the sustainable use of ocean resources, with developments like multi-use offshore platforms that combine energy production with other functions, making them more cost-effective and operationally efficient [3,4]. This integrated approach to ocean space addresses immediate energy needs while establishing a foundation for future technological and environmental balance, essential for long-term sustainability.

As the focus on renewable energy grows, there is an increasing interest in the structural dynamics of offshore platforms [5,6]. This research is essential to ensure the reliability and efficiency of these structures, especially in harsh marine environments, as they take on multiple roles, from energy generation to aquaculture. Studies looking into the dynamic responses of these platforms under different environmental conditions are particularly important [7,8]. They help to better understand the limits and capabilities of these structures, leading to improvements in design and technology, especially in mooring systems, which are vital for sustainable development [9]. Moreover, advancements in related areas, such as structural health monitoring and electrical transmission, particularly through sensor applications in offshore cables and multi-terminal high-voltage networks, further highlight the growing significance of offshore infrastructure research [10–13].

The use of Artificial Neural Networks (ANNs) and Machine Learning (ML) techniques in marine engineering has been highlighted by various studies, including applications in the optimization of marine structures [14]. More recently, these techniques have been applied to analyze the dynamic behaviour of multi-use offshore platforms and manage mooring loads. While ANNs are the focus of this study, other methods, such as genetic algorithms, support vector machines, and hybrid artificial intelligence (AI) models, have also shown promise in similar applications [15–19]. These computational tools allow for accurate modelling and the control of complex structures facing varying environmental forces, improving the safety and efficiency of renewable energy platforms. For instance, combining Floating Offshore Wind Turbines (FOWTs) with Oscillating Water Columns (OWCs) helps optimize energy absorption and reduce dynamic responses [20]. Neural network controllers have also been applied to stabilize platforms against wave impacts, showing significant improvements over traditional control methods [21].

There has also been progress in neural simulators for mooring system design, which use historical environmental data and machine learning to improve design accuracy and efficiency [22]. This approach is complemented by using Bayesian networks to enhance the reliability analysis of mooring systems [23], and machine learning models to predict significant wave heights, which is crucial for the operation and maintenance of offshore energy farms [24].

These new approaches enhance the predictive capabilities of dynamic models and support the development of adaptive control systems, improving platform stability and efficiency in variable conditions. Together, these studies underline the key role of machine learning and neural network technologies in advancing the design and functionality of offshore renewable energy systems, promising a more resilient and efficient future for multi-functional marine structures.

The objective of this article is to assess the reliability of an ANN-based algorithm for predicting the mooring load on a multi-purpose floating platform exposed to random sea waves. For achieving this objective, field data collected during an experimental campaign in a natural basin are used. The following sections will provide a description of the experimental activity and of the ANN architecture (Section 2), the results and discussion with the limits of the proposed ANN-based procedure (Section 3), and concluding remarks (Section 4).

2. Materials and Methods

The field experiment took place at the Natural Ocean Engineering Laboratory (NOEL) at the Mediterranean University of Reggio Calabria, Italy. NOEL is known for allowing real-time monitoring of natural sea conditions, which made it an ideal setting for this study. Small, wind-driven sea states were observed with significant wave heights ranging from 0.20 m to 0.80 m and peak wave periods between 2.0 and 3.6 s, with characteristics similar to JONSWAP spectra. The experiment ran from March 2021 to January 2022, with the most focused data collection occurring between May and July 2021.

During the experiment, a 1:15-scaled prototype of a floating multi-use platform (Figure 1) was equipped with an array of sensors. The platform was designed to combine an automated offshore aquaculture plant with wave and wind energy harvesting technologies. It was made up of a rectangular-shaped, steel, semi-submersible hull with a T-shaped cross section. A scaled 10-MW wind turbine [25] and an array of U-shaped OWC [26] wave energy converters were placed in the front side of the hull. The aquaculture cages were instead hosted in the internal moonpool, to be protected from the incoming waves. The scope of the experiment was to investigate the complex dynamic of the platform, considering various dynamic couplings, including those between the floater, the mooring system, the wind and wave energy converters, the internal moonpool, and the cages. Extensive description of the activities can be found in [4,27].

Figure 1. Photo of BGF platform installed at NOEL.

In this paper, a sample of 4458 data records is considered, collected between 26 May and 6 July 2021. Each record is 10 min long and has a sampling frequency of 10 Hz, compatible with the site wave properties [28] and with the dynamic characteristics of the platform model. In more detail, previous experimental identification analyses (see, e.g., [4]) had shown that its natural frequencies lay approximately in the range of 0.12–0.15 rad/s for the horizontal motions (surge, sway, yaw) and 1.35–1.62 rad/s for the vertical motions (heave, roll, pitch). Specifically, the following sensors were used in this analysis:

- Load cells were placed at the fairleads of three out of four catenary mooring lines (i.e., South-East (SE), South-West (SW), and North-East (NE) corners) of the structure to measure forces. AEP CTS1025TC25 load cells were used, with a nominal load of 5 tons. These sensors have a combined error, non-repeatability, and creep at nominal load of over 20–30 min, all under ±0.025%. Sensitivity tolerance is also within ±0.1%.
- Pressure transducers were used to measure the wave head of pressure in an undisturbed field. An STS ATM.1ST model was used, which operates in a range of 0.0–0.7 bar, making it suitable for both normal and extreme wave conditions at NOEL. The sensor's measurement uncertainty is within ±0.25%.
- Ultrasonic probes were used to measure the free-surface displacement. The FAE IRU-2003 model provided direct measurements of the distance between the sensor and the wave surface, which allowed for the determination of the wave profile. This sensor has a default measurement range of 0.30–7.62 m, perfectly suited to the conditions at NOEL.
- A Gill WindSonic M 2-D ultrasonic anemometer (Gill Instruments Limited, Lymington, United Kingdom) was used to measure the wind velocity and direction in the horizontal plane. The wind measurements were acquired independently from the wave ones,

at a reduced sampling frequency of 4 Hz, consistently with sensor capability and local wind dynamic properties.

It is worth noting that the pressure gauges and ultrasonic probes were used to quantify sea state characteristics in undisturbed conditions, i.e., at a sufficient distance from the structure to assume as negligible its diffraction/radiation effects. To achieve this, they were installed far from the BGF structure on two ballasted vertical piles (Figure 2), at water depths of 1.90 m and 3.70 m, respectively, and were spaced 2.75 m apart. This arrangement followed the method developed by Boccotti et al. [29] to estimate the statistical and spectral characteristics of the sea states. Also, the wind was measured in the undisturbed field.

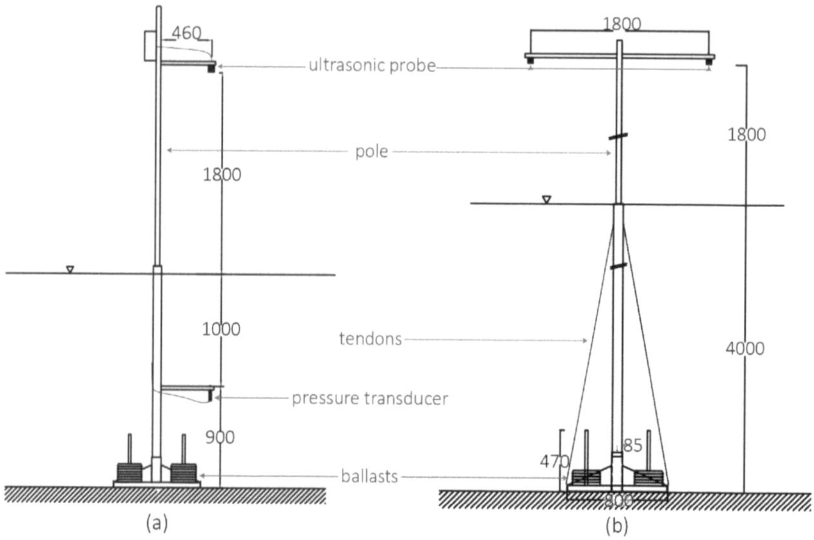

Figure 2. Geometrical characteristics of ballasted vertical piles installed nearshore (**a**) and offshore (**b**). All lengths are provided in millimetres [mm].

These data offered an in-depth view of how the platform interacts with its environment, highlighting the complex relationship between its structural dynamics and the wave climate. This understanding is essential for improving the design and operational protocols of such innovative marine multi-use structures.

In this study, ANNs were trained using a comprehensive set of variables that included wave climate factors, as well as parameters representing the structural loads. The list of the variables used is reported in Table 1.

The choice of the mooring loads (SW, SE, and NE cells) was driven by data availability, as the NW cell had a disconnection period during the experiment. However, the data that were used are deemed sufficient to preliminarily investigate the variability and peaks in the mooring loads. The choice of the wave parameters was aimed to completely characterize the wave properties, including magnitude, wave energy distribution in the frequency domain, and direction. In particular, since bimodal spectra cannot be fully characterized by synthetic variables such as mean and peak wave period, the ratio between wind-generated (periods from $T_1 = 2\pi/\omega_1 = 3.0$ s to $T_2 = 2\pi/\omega_2 = 1.0$ s) and swell (periods from $T_3 = 2\pi/\omega_3 = 11.9$ s to T_1) spectral components (areas) has been introduced and is defined as

$$Ratio_{w,swell} = \frac{\int_{\omega_1}^{\omega_2} S(\omega)d\omega}{\int_{\omega_3}^{\omega_1} S(\omega)d\omega} \tag{1}$$

where $S(\omega)$ is the wave frequency spectrum and ω the frequency.

Table 1. Variables used for training the ANNs, with corresponding symbols, sensors, and units; d represents water depth.

	Variable	Symbol	Sensor	Units
Mooring loads	Load at the south-west fairlead	T_{SW}	Load cells	[N]
	Load at the south-east fairlead	T_{SE}		
	Load at the north-east fairlead	T_{NE}		
Surface wave	Significant height	Hs (d = 3.70 m)	Ultrasonic probes	[m]
		Hs_{coast} (d = 1.90 m)		
	Peak period	Tp_1 (d = 1.90 m—left)		[s]
		Tp_2 (d = 1.90 m—right)		
	Wind-generated/swell spectral area ratio	$Ratio_{w,swell}$		[-]
	Mean zero-up-crossing period	Tz (d = 3.70 m)		[s]
		Tz_{coast} (d = 1.90 m)		
	Mean propagation direction	$\overline{\theta}$ (d = 3.70 m)		[rad]
		$\overline{\theta}_{coast}$ (d = 1.90 m)		
Head of pressure wave	Peak period	Tp_{Ph2} (d = 1.90 m)	Pressure transducers	[s]
	Significant height	$Hs_{Phcoast}$ (d = 1.90 m)		[m]
	Mean zero-up-crossing period	$Tz_{Phcoast}$ (d = 1.90 m)		[s]
Wind	Mean direction	$\overline{Dir_w}$	Anemometer	[rad]
	Mean velocity	$\overline{u_w}$		[m/s]

The methodology that was employed focused on developing and refining ANNs to predict the dynamic response of the floating multi-use offshore platform. Using the dataset collected from the field experiment, this approach involved optimizing the ANN model structure and investigating its predictive accuracy under various environmental conditions.

To build the final model, ANNs were developed and validated following a systematic process:

- Data preparation: a cleaning process of the data was applied to remove any inconsistencies and gaps, ensuring the dataset was reliable for further analysis.
- Dataset division: The cleaned dataset was then split into training and test sets, using an 80–20% ratio. This split is commonly used in the field [30] to allow the model to be evaluated on unseen data, ensuring its performance holds up in real-world scenarios. From the training set, an additional 10% was further split off to serve as a validation set.
- ANN architecture selection: different feed-forward Deep Neural Networks (DNNs) were explored to find the best balance between complexity and learning efficiency.
- Robustness enhancement: to make the models more robust and reduce uncertainty, 10-fold cross-validation was employed, with multiple realizations run for each configuration.
- Training data length optimization: the length of the training data was adjusted to ensure there was enough information for the model to learn from without overloading it.
- Hyperparameter tuning: the model's hyperparameters were carefully adjusted to improve its learning efficiency and overall predictive accuracy.
- Input variable assessment: the input variables were thoroughly evaluated to determine their significance, ensuring that only the most relevant factors were used in the ANN models.

The data cleaning process involved discarding missing values to ensure only complete data were used, detecting and removing outliers above the 99th percentile, and normalizing the data. Most variables were scaled to the range $[-1, 1]$ through min–max normalization, while directional variables (i.e., $\overline{\theta}$, $\overline{\theta}_{coast}$) were transformed using the trigonometric functions (i.e., $sin\overline{\theta}$, $cos\overline{\theta}$, $sin\overline{\theta}_{coast}, cos\overline{\theta}_{coast}$) to account for their cyclical nature.

In this step, TensorFlow [31] in Python was used to build and train the neural networks. The architectures varied in the following key parameters:

- Input layer: 17 features based on the training data.

- Hidden layers: the number of hidden layers was dynamically adjusted between 1 and 5, there being a DNN when more than one layer was used.
- Units per layer: each hidden layer could have between 32 and 512 units, with the specific number of units per layer being determined through hyperparameter tuning.
- Activation functions: both 'ReLU' and 'tanh' were explored as activation functions, with 'ReLU' introducing non-linearity and 'tanh' providing bounded activation.
- Learning rate: the Adam optimizer was used with a learning rate dynamically tuned between 1×10^{-5} and 1×10^{-2}.
- Output: Each approach predicted three load cell-related features for the directions SE, SW, and NE. The first approach focused on predicting the maximum, while the second approach predicted the standard deviation of the load cell ($max(T_{SE})$, $max(T_{SW})$, $max(T_{NE})$ and $\sigma(T_{SE})$, $\sigma(T_{SW})$, $\sigma(T_{NE})$).

The hyperparameter optimization was conducted using the RandomSearch tuner from TensorFlow's Keras Tuner library. RandomSearch performed an initial broad search, minimizing the validation mean squared error across 10 trials, with each trial executed twice. The best hyperparameters identified during this process were then used to build and train the final model on the training dataset for 50 epochs. The model's performance was subsequently validated on a separate dataset and evaluated on the test dataset using metrics such as test loss, Root Mean Square Error (RMSE), and Normalized Root Mean Square Error (NRMSE). In addition to RandomSearch, a Bayesian optimization and Hyperband methods [32,33] were utilized. Bayesian optimization fine-tuned high-performing parameters by constructing a probabilistic model based on previous results, while Hyperband expedited the process by adaptively allocating resources and using early stopping to focus on the best-performing models. The optimization process was conducted in three stages across three different ANNs: starting with a grid search, followed by Bayesian optimization, and finally, the introduction of Hyperband. This sequential approach allowed us to compare RMSEs and ultimately select the best-performing model.

The RMSE and NRMSE metrics were calculated between the predicted $\left(y_{predicted}\right)$ and actual (y_{actual}) values using the following formulas, where n is the number of observations, and y_{max} and y_{min} represent the maximum and minimum values in the dataset, respectively:

$$RMSE = \sqrt{\frac{1}{n}\sum_{t=1}^{n}\left(y_{t,predicted} - y_{t,actual}\right)^2}, \tag{2}$$

$$NRMSE = \frac{RMSE}{y_{max} - y_{min}} \cdot 100 \tag{3}$$

To rigorously determine the impact of each input variable on the predictions made by the ANNs, the permutation importance method [34,35] was used. This approach is crucial for identifying the most influential features in the model, ensuring that only the most relevant predictors are included. The method works by shuffling the values of each feature while keeping the target values fixed, then measuring the decrease in model performance. This decrease indicates the feature's importance. The permutation process was repeated 10 times to account for random variations and ensure robustness. The importance of each feature was quantified by the mean decrease in performance, resulting in unitless scores that reflect the relative importance of each predictor. In this way, the methodology enhances both the transparency of the predictive models and the optimization of the network architecture and hyperparameter settings.

The standard deviation ($\sigma(T_{SE})$, $\sigma(T_{SW})$, $\sigma(T_{NE})$) and maximum ($max(T_{SE})$, $max(T_{SW})$, $max(T_{NE})$) of load cell values for each record were selected as the primary output variables for the ANNs. The standard deviation provides insights into the variability and stability of the structure, while the maximum values highlight the peak load that the structure can withstand. Two separate iterations of the ANNs were conducted, one focused on predicting maximum values and the other on standard deviations. This dual approach captures different aspects of the mooring load dynamics under varying metocean conditions, deepening the understanding of the relationship between environmental forces and structural responses. In Figure 3, both output variables, as recorded before any processing, are represented across the sampling period used in this work, illustrating the variations in maximum and standard deviation load values over time.

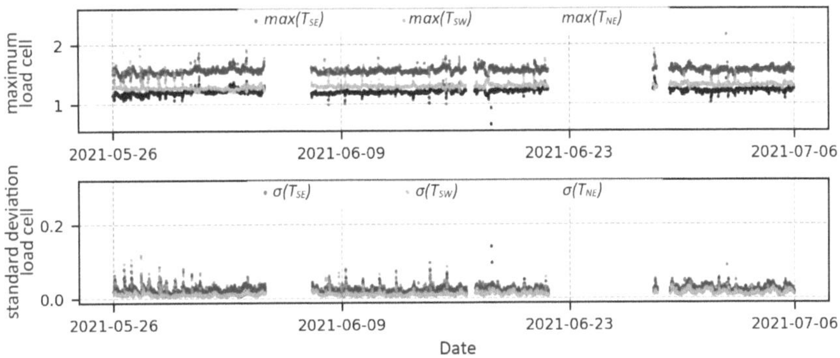

Figure 3. Maximum and standard deviation load cell values over sampling period.

3. Results and Discussion

After applying ANNs to predict both the maximum and standard deviation of load cell values in the SW, SE, and NW directions, the NRMSE for standard deviation was observed as higher compared to that of the maximum values (Table 2). This higher NRMSE indicates a greater difficulty in accurately predicting variability, which reflects the inherently stochastic nature of marine environments more than peak loads. Given the importance of standard deviation in capturing the variability and uncertainties in these conditions, this work focused on its prediction.

Table 2. The errors (RMSE, NRMSE) of the predicted maximum and standard deviation of load cell values through the ANNs.

	$max(T_{SE})$	$max(T_{SW})$	$max(T_{NE})$	$\sigma(T_{SE})$	$\sigma(T_{SW})$	$\sigma(T_{NE})$
RMSE	0.031	0.043	0.051	0.008	0.008	0.010
NRMSE%	3.776	5.307	3.865	6.016	7.098	3.467

Improving the accuracy of $\sigma(T_{SE})$, $\sigma(T_{SW})$, $\sigma(T_{NE})$ predictions is a more technically demanding but highly valuable task. This approach expands modelling capabilities and enhances the structural resilience, enabling better adaptation to variable loads. Prioritizing this focus is expected to significantly improve the reliability and predictive power of the ANNs, which is especially important for dynamic and complex environments where understanding variability is key to ensuring safety and optimizing performance.

To improve the predictive accuracy of the ANNs, a comprehensive analysis of input variable importance was conducted, which revealed several variables with minimal impact on model predictability (Figure 4). The feature importance was evaluated using the permutation importance method [36] from Python's Scikit-learn library [37]. This method measures the impact of each feature by calculating the increase in the model's error when the values of the feature are randomly shuffled. The importance scores reflect the corresponding changes in the performance of the model, with higher scores indicating features that have a greater influence on the predictions. As a result, variables related to the mean wave direction at both the available water depths ($cos_{\overline{\theta}_{coast}}, sin_{\overline{\theta}_{coast}}, cos_{\overline{\theta}}, sin_{\overline{\theta}}$) and to the mean wind direction ($\sigma(Dir_{\overline{W}})$) were excluded to streamline the model, reducing its complexity and enhancing computational efficiency. This exclusion was based on current findings, though additional variables may be revised or excluded in future iterations as part of ongoing optimizations.

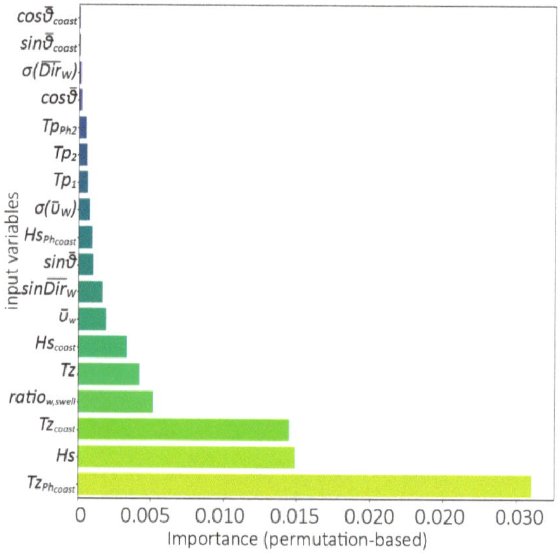

Figure 4. Permutation-based feature importance analysis of input variables.

To evaluate the impact of this refinement, the RMSE and NRMSE metrics before and after the variable exclusions were compared. The results, shown in Figure 5 along with normalized values, indicate minimal differences in performance metrics post-exclusion. This confirms that the streamlined model maintains its predictive accuracy while operating with reduced input complexity. These findings suggest that excluding these variables effectively simplifies the model without compromising its predictive capability, thereby validating this approach to improve model performance by focusing on more impactful variables.

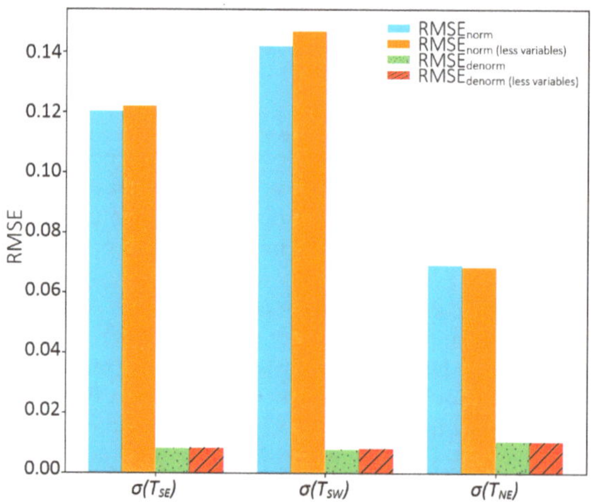

Figure 5. Comparative analysis of RMSE for normalized and denormalized predictions before and after skipping less important input variables.

The model was further optimized using both Bayesian optimization and Hyperband techniques in TensorFlow, with a focus on fine-tuning the hyperparameters to improve predictive accuracy. As shown in Table 3, the effectiveness of both optimization strategies was assessed using RMSE and NRMSE values across the three directions (SE, SW, NE). Bayesian optimization generally achieved slightly lower RMSE and NRMSE values compared to the Hyperband approach. In particular, the SE and SW directions showed consistently lower errors with Bayesian optimization, indicating a more accurate model fit. The stability of the Bayesian approach was further validated through 10-fold cross-validation, confirming its robustness across different data subsets. As a result, the ANN model refined using Bayesian optimization was selected for final implementation due to its better performance and reliability in predicting the standard deviation of load values.

Table 3. The errors (RMSE, NRMSE) of the predicted standard deviation loads through the ANNs.

		$\sigma(T_{SE})$	$\sigma(T_{SW})$	$\sigma(T_{NE})$
RMSE	Bayesian	0.007	0.007	0.010
	Hyperband	0.008	0.007	0.009
NRMSE (%)	Bayesian	5.416	6.518	3.210
	Hyperband	5.795	6.570	2.898

The optimal ANN was used to make the final predictions of $\sigma(T_{SE})$, $\sigma(T_{SW})$, and $\sigma(T_{NE})$. Figure 6 shows a comparison between the actual and predicted standard deviations for a randomly selected test dataset, demonstrating a close alignment between the two.

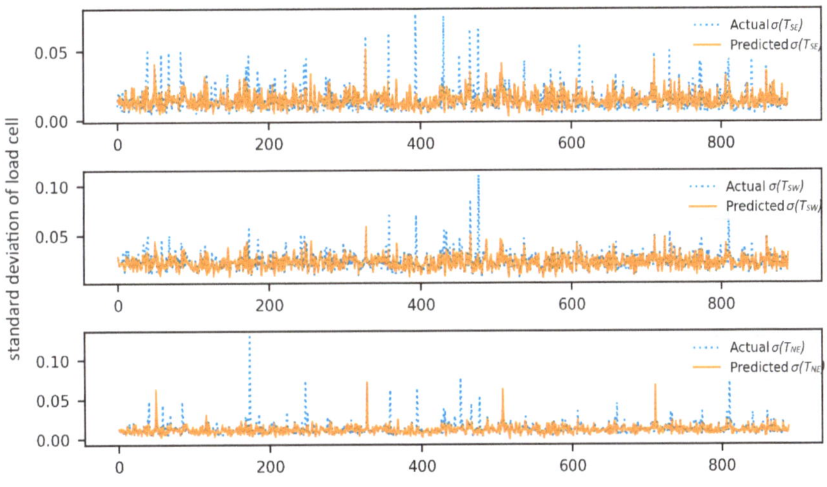

Figure 6. Comparison of actual vs. predicted values through final ANN using Bayesian optimization.

However, spikes detected in Figure 6 revealed significant differences between the predicted and actual values, leading to the consideration that discarding outliers above the 99th percentile might be beneficial. Further investigation identified that these outliers were caused by current-related phenomena occurring at the NOEL site, which are sporadic and not representative of significant platform dynamics. These phenomena were not consistently measured and do not reflect the platform's typical behaviour.

By applying this threshold, only 79 values were discarded, representing less than 1% of the dataset, without compromising the variability in the standard deviation. The removal of these extreme values improved the model's performance by reducing the influence of unrepresentative spikes. Following this adjustment, the final ANN was retrained using Bayesian optimization, and its accuracy was evaluated based on the new predictions

(Figure 7). The RMSE and NRMSE values, shown in Table 4, indicate improved model performance.

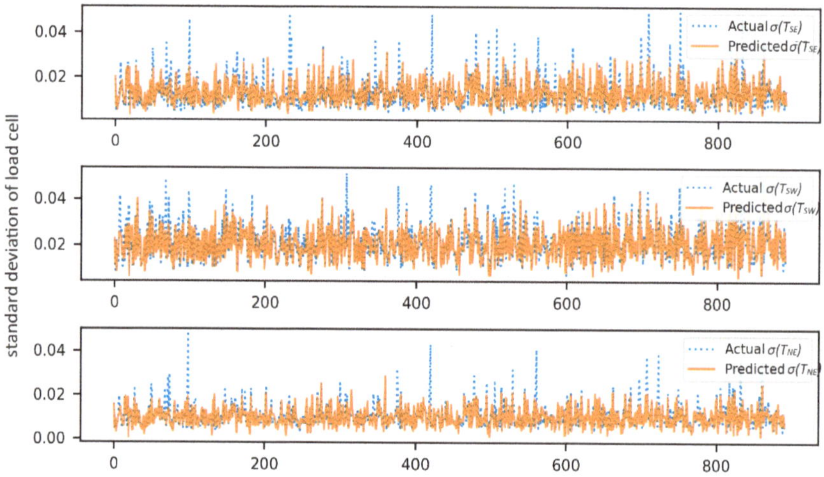

Figure 7. Comparison of actual vs. predicted values through final ANN using Bayesian optimization after discarding outliers above 99th percentile.

Table 4. The errors (RMSE, NRMSE) of the predicted standard deviation loads through the final ANN using Bayesian optimization after discarding the outliers.

	$\sigma(T_{SE})$	$\sigma(T_{SW})$	$\sigma(T_{NE})$
RMSE	0.006	0.005	0.005
NRMSE (%)	4.441	4.720	1.767

Further validation is presented in Figure 8, where a scatterplot between the predicted and actual standard deviations of load cells demonstrates a strong correlation, particularly in the SE and SW directions. This is evidenced by the close alignment of data points along the diagonal line, which indicates that the model's predictions are highly accurate. Although the NE direction exhibits slightly more variability, the overall alignment of points highlights the ANN's high predictive accuracy and efficiency across different directions.

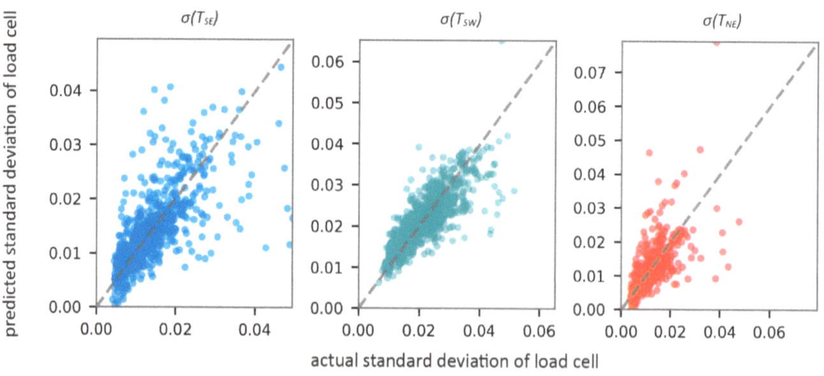

Figure 8. Comparison of predicted vs. actual standard deviation of load cells for SE, SW, and NE directions.

The residual plots in Figure 9 illustrate the difference between the predicted and actual standard deviations of load cells across the SE, SW, and NE directions. For SE, a clear negative monotonic trend is observed, where the error tends to increase with higher actual standard deviation values, indicating the model under-predicts for larger load values in this direction. SW shows a similar but less pronounced trend, while NE exhibits more variability, with residuals spread more widely compared to other directions, and without a clear pattern. This variability could suggest areas for further model refinement. Fine-tuning the hyperparameters or incorporating additional training data for higher load cases could enhance future model performance.

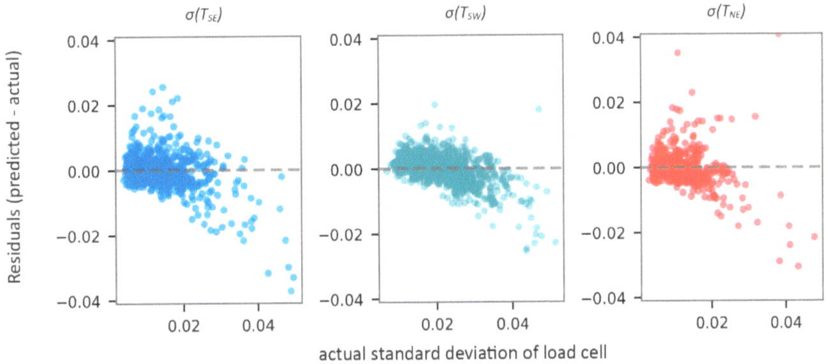

Figure 9. Residual plot showing difference between predicted and actual standard deviation of load cells for SE, SW, and NE directions.

To provide an additional layer of validation for the model's accuracy and consistency, the error distribution plot was examined. Figure 10 illustrates the spread of prediction errors across the SE, SW, and NE directions. In all three directions, the distributions are tightly centred around zero, indicating that the model does not exhibit significant bias in over- or under-predicting the values. The distributions appear approximately symmetric and normally distributed, though the slight negative trend in SE residuals observed in Figure 8 is less obvious here. The error spread, typically ranging between −0.04 and 0.04, indicates that the model's predictions are generally close to the actual values. However, minor variations in the spread may suggest that the model performs slightly differently depending on the direction, with NE exhibiting a narrower error distribution and a slightly higher peak frequency.

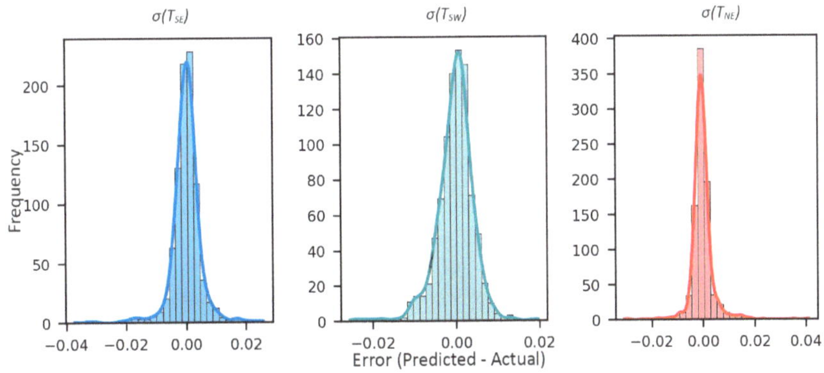

Figure 10. Error distribution of predicted standard deviation of load cells in SE, SW, and NE directions.

To further assess the model's robustness across varying environmental conditions, the relationship between prediction errors and significant wave heights was analyzed. As shown in Figure 11, the data do not suggest a strong correlation between higher prediction errors and larger wave heights, indicating consistent model performance across the range of conditions tested. The distribution of errors across the SE, SW, and NE directions appears relatively symmetrical, suggesting that the model does not exhibit a systematic bias toward overestimation or underestimation. This uniformity is encouraging, as it suggests the model's predictions are unbiased overall. However, although the model maintains consistent performance, some outliers are present across a range of wave heights. These outliers suggest that further refinements may be needed to improve predictive precision in certain cases, although they do not seem to be linked to extreme wave conditions.

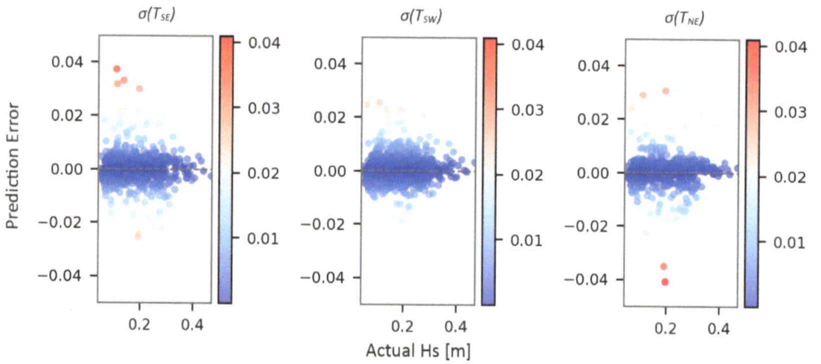

Figure 11. Analysis of prediction errors relative to significant wave height.

4. Conclusions

In conclusion, this study proposed the application of ANNs for the prediction of the mooring loads of a multi-purpose floating platform and demonstrated its effectiveness based on the field data collected at the Natural Ocean Engineering Laboratory (NOEL) of Mediterranea University of Reggio Calabria (Italy), on a 1:15-scaled prototype of the structure. By using advanced optimization techniques, particularly Bayesian optimization, the models were refined, resulting in reliable predictions under diverse environmental conditions. The key outcomes of this study are outlined below:

- ANNs successfully captured the complex dynamics of offshore structures, proving their potential as a valuable tool for predicting dynamic responses.
- Incorporating techniques like Bayesian optimization contributed to improved model reliability, ensuring accurate dynamic response predictions for offshore platforms.
- Optimization techniques enhanced model performance, leading to reliable predictions under a range of operational scenarios and highlighting the potential of ANNs in managing offshore dynamics.
- Further investigation into data extremes, particularly under severe environmental conditions, may be useful. Such analyses could identify additional variables that may enhance model performance and offer valuable insights beyond average responses, potentially enabling predictions of other response variables.
- Ongoing fine-tuning of ANNs is essential for optimizing model architecture and hyperparameters, which will improve the model's predictive power and ensure robust performance in complex environments.
- Given that ANNs have shown effectiveness in this problem, investigating additional techniques, such as genetic algorithms, support vector machines, or hybrid AI models, which have already demonstrated value in similar applications, could further improve model performance in future studies.

- This research represents an important advancement in integrating AI with marine engineering, particularly in the deployment of dynamic response predictions for offshore platforms.

In summary, this study confirms the value of ANNs in addressing the challenges of offshore platform dynamics. However, ongoing refinements—such as focused dataset adjustments and the analysis of extremes—will be essential for improving predictive accuracy. With continued improvements, ANNs have the potential to play a critical role in the sustainable operation of marine infrastructure, representing a significant step forward for renewable energy platforms.

Author Contributions: Conceptualization, N.M., C.R., G.M., V.F. and F.A.; methodology, N.M. and G.M.; software, N.M.; validation, N.M.; formal analysis, N.M.; investigation, N.M.; resources, N.M., C.R., G.M., V.F. and F.A.; data curation, N.M., C.R. and V.F.; writing—original draft preparation, N.M.; writing—review and editing, N.M., C.R., V.F. and G.M.; visualization, N.M. and V.F.; supervision, G.M. and F.A. All authors have read and agreed to the published version of the manuscript.

Funding: This work was funded by the Next Generation EU—Italian NRRP, Mission 4, Component 2, Investment 1.5, call for the creation and strengthening of 'Innovation Ecosystems', building 'Territorial R&D Leaders' (Directorial Decree n. 2021/3277)—project Tech4You—Technologies for Climate Change Adaptation and Quality of Life Improvement, n. ECS0000009. This work reflects only the authors' views and opinions; neither the Ministry for University and Research nor the European Commission can be considered responsible for them.

Institutional Review Board Statement: Not applicable.

Informed Consent Statement: Not applicable.

Data Availability Statement: Data are not publicly available due to privacy restrictions. Data can be made available upon request to the authors.

Acknowledgments: The data used in this work have been obtained during "The Blue Growth Farm" project funded by the European Union's Horizon2020 research and innovation programme (Grant Agreement number 774426). The content of the work does not report the opinion of the European Commission and reflects only the views of the author(s), including errors or omissions. The European Commission is also not liable for any use that may be made of the information contained herein.

Conflicts of Interest: The authors declare no conflicts of interest. The funders had no role in the design of the study and in the writing of the manuscript.

References

1. Li, L.; Lin, J.; Wu, N.; Xie, S.; Meng, C.; Zheng, Y.; Wang, X.; Zhao, Y. Review and Outlook on the International Renewable Energy Development. *Energy Built Environ.* **2022**, *3*, 139–157. [CrossRef]
2. Akhtar, N.; Geyer, B.; Schrum, C. Larger Wind Turbines as a Solution to Reduce Environmental Impacts. *Sci. Rep.* **2024**, *14*, 6608. [CrossRef]
3. Billing, S.L.; Charalambides, G.; Tett, P.; Giordano, M.; Ruzzo, C.; Arena, F.; Santoro, A.; Lagasco, F.; Brizzi, G.; Collu, M. Combining Wind Power and Farmed Fish: Coastal Community Perceptions of Multi-Use Offshore Renewable Energy Installations in Europe. *Energy Res. Soc. Sci.* **2022**, *85*, 102421. [CrossRef]
4. Ruzzo, C.; Malara, G.; Collu, M.; Santoro, A.; Fiamma, V.; Scialò, A.; Lagasco, F.; Arena, F. Field Experiment on a Scaled Prototype of a Floating Multi-Purpose Offshore Platform: Dynamic Response Determination with Uncertainty Quantification. *Appl. Ocean Res.* **2022**, *129*, 103402. [CrossRef]
5. Gutiérrez-Romero, J.E.; García-Espinosa, J.; Serván-Camas, B.; Zamora-Parra, B. Non-Linear Dynamic Analysis of the Response of Moored Floating Structures. *Mar. Struct.* **2016**, *49*, 116–137. [CrossRef]
6. Ransley, E.J.; Greaves, D.; Raby, A.; Simmonds, D.; Hann, M. Survivability of Wave Energy Converters Using CFD. *Renew. Energy* **2017**, *109*, 235–247. [CrossRef]
7. Subbulakshmi, A.; Verma, M. Dynamic Response Analysis of a Semisubmersible Floating Offshore Wind Turbine Subjected to Mooring Line Failure Under Normal and Extreme Environmental Conditions. *Ocean. Eng.* **2024**, *304*, 117907. [CrossRef]
8. Michailides, C.; Gao, Z.; Moan, T. Experimental and Numerical Study of the Response of the Offshore Combined Wind/Wave Energy Concept SFC in Extreme Environmental Conditions. *Mar. Struct.* **2016**, *50*, 35–54. [CrossRef]

9. Castro-Santos, L.; Filgueira-Vizoso, A.; Carral-Couce, L.; Fraguela Formoso, J.Á. Economic Feasibility of Floating Offshore Wind Farms. *Energy* **2016**, *112*, 868–882. [CrossRef]
10. Drissi-Habti, M.; Neginhal, A.; Manepalli, S.; Carvelli, V. Fiber-Optic Sensors (FOS) for Smart High Voltage Composite Cables—Numerical Simulation of Multi-Parameter Bending Effects Generated by Irregular Seabed Topography. *Sensors* **2022**, *22*, 7899. [CrossRef]
11. Drissi-Habti, M.; El Assami, Y.; Raman, V. Multiscale Toughening of Composites with Carbon Nanotubes—Continuous Multiscale Reinforcement New Concept. *J. Compos. Sci.* **2021**, *5*, 135. [CrossRef]
12. Zhang, Q.; Drissi-Habti, M. Electric Cable Insulator Damage Monitoring by Lasso Regression. *Machines* **2024**, *12*, 50. [CrossRef]
13. Normandia Lourenço, L.F.; Louni, A.; Damm, G.; Netto, M.; Drissi-Habti, M.; Grillo, S.; Sguarezi Filho, A.J.; Meegahapola, L. A Review on Multi-Terminal High Voltage Direct Current Networks for Wind Power Integration. *Energies* **2022**, *15*, 9016. [CrossRef]
14. Panda, J.P. Machine Learning for Naval Architecture, Ocean and Marine Engineering. *J. Mar. Sci. Technol.* **2023**, *28*, 1–26. [CrossRef]
15. Pezeshki, H.; Adeli, H.; Pavlou, D.; Siriwardane, S.C. State of the Art in Structural Health Monitoring of Offshore and Marine Structures. *Proc. Inst. Civ. Eng.—Marit. Eng.* **2023**, *176*, 89–108. [CrossRef]
16. Li, D.; Jiang, M.-R.; Li, M.-W.; Hong, W.-C.; Xu, R.-Z. A Floating Offshore Platform Motion Forecasting Approach Based on EEMD Hybrid ConvLSTM and Chaotic Quantum ALO. *Appl. Soft Comput.* **2023**, *144*, 110487. [CrossRef]
17. Chen, Y.; Yuan, L.; Qin, L.; Zhang, N.; Li, L.; Wu, K.; Zhou, Z. A Forecasting Model with Hybrid Bidirectional Long Short-Term Memory for Mooring Line Responses of Semi-Submersible Offshore Platforms. *Appl. Ocean Res.* **2024**, *150*, 104145. [CrossRef]
18. Jiang, H.; Wang, H.; Vaz, M.A.; Bai, X. Research on Dynamic Response Prediction of Semi-Submersible Wind Turbine Platform in Real Sea Test Model Based on Machine Learning. *Appl. Ocean Res.* **2024**, *142*, 103808. [CrossRef]
19. Das, P.; Mashiata, M.; Iglesias, G. Big Data Meets Big Wind: A Scientometric Review of Machine Learning Approaches in Offshore Wind Energy. *Energy AI* **2024**, *18*, 100418. [CrossRef]
20. Ahmad, I.; M'zoughi, F.; Aboutalebi, P.; Garrido, I.; Garrido, A.J. Fuzzy Logic Control of an Artificial Neural Network-Based Floating Offshore Wind Turbine Model Integrated with Four Oscillating Water Columns. *Ocean. Eng.* **2023**, *269*, 113578. [CrossRef]
21. Cai, Z.; Zhang, B.; Yu, X. Neural Network Delayed Control of an Idealized Offshore Steel Jacket Platform. In Proceedings of the Eighth International Conference on Intelligent Control and Information Processing (ICICIP), Hangzhou, China, 3–5 November 2017; pp. 282–286. [CrossRef]
22. Cotrim, L.P.; Barreira, R.A.; Santos, I.H.F.; Gomi, E.S.; Costa, A.H.R.; Tannuri, E.A. Neural Network Meta-Models for FPSO Motion Prediction from Environmental Data with Different Platform Loads. *IEEE Access* **2022**, *10*, 86558–86577. [CrossRef]
23. Abaei, M.M.; Abbassi, R.; Garaniya, V.; Chai, S.; Khan, F. Reliability Assessment of Marine Floating Structures Using Bayesian Network. *Appl. Ocean Res.* **2018**, *76*, 51–60. [CrossRef]
24. Mahdavi-Meymand, A.; Sulisz, W. Development of Pyramid Neural Networks for Prediction of Significant Wave Height for Renewable Energy Farms. *Appl. Energy* **2024**, *362*, 123009. [CrossRef]
25. Bak, C.; Zahle, F.; Bitsche, R.; Kim, T.; Yde, A.; Henriksen, L.C.; Hansen, M.H.; Blasques, J.P.A.A.; Gaunaa, M.; Natarajan, A. The DTU 10-MW Reference Wind Turbine. *Sound/Vis. Prod. (Digit.)* **2013**. Available online: https://orbit.dtu.dk/en/publications/the-dtu-10-mw-reference-wind-turbine (accessed on 4 November 2024).
26. Malara, G.; Arena, F. Response of U-Oscillating Water Column Arrays: Semi-Analytical Approach and Numerical Results. *Renew. Energy* **2019**, *138*, 738–748. [CrossRef]
27. Ruzzo, C.; Fiamma, V.; Scialò, A.; Arena, F.; Santoro, A.; Muggiasca, S.; Taruffi, F.; Di Carlo, S.; Larrea, I.; Corvaglia, P.A.; et al. Field Experimental Campaign on a Multi-Purpose Floating Structure: Set-up Description. In *Trends in Renewable Energies Offshore—Proceedings of the 5th International Conference on Renewable Energies Offshore, Lisbon, Portugal, 8–10 November 2022*; RENEW; CRC Press: Boca Raton, FL, USA, 2022; pp. 817–825. [CrossRef]
28. Boccotti, P. Space–Time Theory of Sea States. In *Wave Mechanics and Wave Loads on Marine Structures*; Elsevier: Amsterdam, The Netherlands, 2015; pp. 115–143. [CrossRef]
29. Boccotti, P.; Arena, F.; Fiamma, V.; Romolo, A.; Barbaro, G. Estimation of Mean Spectral Directions in Random Seas. *Ocean Eng.* **2011**, *38*, 509–518. [CrossRef]
30. Joseph, V.R. Optimal Ratio for Data Splitting. *Stat. Anal. Data Min. ASA Data Sci. J.* **2022**, *15*, 531–538. [CrossRef]
31. TensorFlow Developers. TensorFlow (v2.18.0). *Zenodo*. 2024. Available online: https://zenodo.org/records/13989084 (accessed on 4 November 2024).
32. Li, L.; Jamieson, K.G.; DeSalvo, G.; Rostamizadeh, A.; Talwalkar, A. Hyperband: A Novel Bandit-Based Approach to Hyperparameter Optimization. *J. Mach. Learn. Res.* **2018**, *18*, 1–52. [CrossRef]
33. Feurer, M.; Hutter, F. Hyperparameter Optimization. In *Automated Machine Learning*; Hutter, F., Kotthoff, L., Vanschoren, J., Eds.; Springer: Cham, Switzerland, 2019; pp. 3–33. [CrossRef]
34. Louppe, G.; Wehenkel, L.; Sutera, A.; Geurts, P. Understanding Variable Importances in Forests of Randomized Trees. In Proceedings of the Advances in Neural Information Processing Systems 26 (NIPS 2013), Lake Tahoe, NV, USA, 5–8 December 2013; pp. 431–439.
35. Fisher, A.; Rudin, C.; Dominici, F. All Models Are Wrong, but Many Are Useful: Learning a Variable's Importance by Studying an Entire Class of Prediction Models Simultaneously. *J. Mach. Learn. Res.* **2019**, *20*, 1–81.

36. Breiman, L. Random Forests. *Mach. Learn.* **2001**, *45*, 5–32. [CrossRef]
37. Pedregosa, F.; Varoquaux, G.; Gramfort, A.; Michel, V.; Thirion, B.; Grisel, O.; Duchesnay, E. Scikit-Learn: Machine Learning in Python. *J. Mach. Learn. Res.* **2011**, *12*, 2825–2830. [CrossRef]

MDPI AG
Grosspeteranlage 5
4052 Basel
Switzerland
Tel.: +41 61 683 77 34

Journal of Marine Science and Engineering Editorial Office
E-mail: jmse@mdpi.com
www.mdpi.com/journal/jmse